Technical Writing for Industry

An Operations Manual for the Technical Writer

Larry A. Riney

PRENTICE HALL, Englewood Cliffs, NJ 07632

Library of Congress Cataloging-in-Publication Data

Riney, Larry A.
Technical writing for industry.

Bibliography: p.
Includes index.
1. Technical writing. I. Title.
T11.R57 1989 808′.0666 88-23883
ISBN 0-13-901828-X

Editorial/production supervision and
interior design: Margaret Lepera
Cover design: Joel Mitnick Design Inc.
Manufacturing buyer: Robert Anderson
Page layout: A Good Thing, Inc.

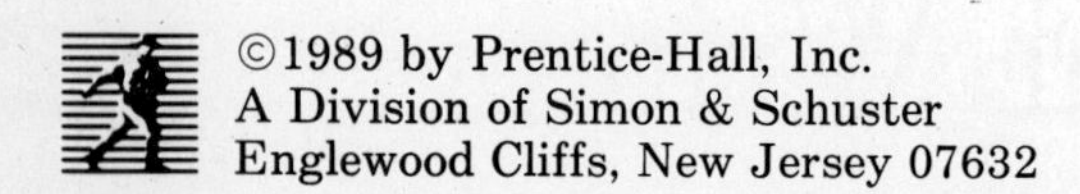

Printed in the United States of America

10 9 8 7 6 5 4 3 2

ISBN 0-13-901828-X

Prentice-Hall International (UK) Limited, *London*
Prentice-Hall of Australia Pty. Limited, *Sydney*
Prentice-Hall Canada Inc., *Toronto*
Prentice-Hall Hispanoamericana, S.A., *Mexico*
Prentice-Hall of India Private Limited, *New Delhi*
Prentice-Hall of Japan, Inc., *Tokyo*
Simon & Schuster Asia Pte. Ltd., *Singapore*
Editora Prentice-Hall do Brasil, Ltda., *Rio de Janeiro*

Dedicated to Julie, Beth,
Family, and Friends

Contents

CHAPTER 3
Develop A Technical Writing Style
41

CHAPTER 4
Develop A Visual Writing Style
61

CHAPTER 6
Know Machine System Concepts
115

CHAPTER 7
Use System Concepts to Outline Ideas
137

Preface

My first day as a technical writer in industry was an unsettling experience. I was hired to help develop a technical writing program for a small midwestern manufacturing company. Although I had taught what I thought was technical writing in a university classroom, I soon learned that I was not prepared for the reality of the job.

Not only were the academic concepts of technical writing for industry inadequate, but also the textbooks that were available only covered letter, memo, proposal, and report writing. The technical writer must be able to create these basic documents. However, they are not the main responsibilities of the professional technical writer. The primary job of the technical writer in industry is to write usable technical operation instructions that explain how to operate a machine safely and efficiently.

This advanced textbook covers the entire process of technical instruction writing, from the development of a highly usable visual technical writing style, to preparing the technical operation instruction for print. It is unfair to the student of technical writing to create a textbook that does not cover both the subjects of technology and writing theory. This book includes the fundamentals of both machine system concepts and writing mechanics, which will help the student survive professionally in any type of industrial technical writing program.

The following pages do not contain an answer to every question that the technical writer will face. The guidelines and information presented offer a personal creative approach to problem solving for the technical writer. This is a book that can be used as teaching material in the classroom and carried confidently into a manufacturing company's technical writing program.

Larry A. Riney

Acknowledgments

The process of assembling and writing this book was a collaboration of many people. I would especially like to acknowledge the companies that have so graciously allowed me to use parts of their operator's manuals and pictures to help illustrate this text. Without their cooperation and help, it would be impossible to present technical writing in a realistic light. Also, I would like to thank the art studios that have worked with me to create artwork for specific topics and ideas that require special illustrations.

A debt of gratitude is owed to all of the people and departments at Deere & Company, Moline, Illinois, who assisted me in collecting and obtaining reference material and artwork for this project, and also to the people at Prentice Hall, who have added their hard work and talents to this book; namely Margaret Lepera, Jim Tully, and Greg Burnell. To all of these individuals and organizations: I credit you with a well-deserved "thank you."

1

Set Your Goals

The information given in this chapter will enable you to:

- Set the objectives that you want to accomplish with this book.
- Better understand the need for technical operation instructions in the worldwide marketplace.
- Explain the responsibilities of the technical writer in industry.
- Explain the job of writing technically as both an art and a skill.
- List the major technical and communication skills that can be used by the technical writer in industry.
- Explain the basic concept of "machine systems."
- List the main parts of a "technical operation instruction."
- Define the terms "direction" and "operation manual."
- Understand the job opportunities in industry as a production technical writer.
- Prepare yourself for a career in technical writing through further study.

INTRODUCTION

Technical operation instructions have become extremely important in our economic system. Every powered machine and toy that is made and sold in our marketplace must be supported by technical information. As our technology advances in complexity, it is the responsibility of the technical writer to explain to consumers "how to operate their machine safely and efficiently." This book is written to help you, the student of technical writing, and the technical writer on the job, to develop and fine-tune the "skills" and "creative methods" that can be used to help you write usable technical operation instructions.

There are many ways to write technically. The following chapters explain *one way* that you can use in order to put together complete information that explains how to use a machine safely and efficiently. It is a book that views technical writing as a step-by-step process that requires the total involvement of the technical writer in a manufacturing company communication program. Although this book is written for you, the technical writer, to help you complete your writing assignments, it puts the needs of the consumer who buys and uses the machine as the foremost reason for writing instructions.

You can use this text to attain four main technical writing objectives. Each of these completed goals can help you meet the demands and responsibilities that will challenge you as a technical writer in industry. First, you can use this book to learn the basic communication and technical skills that you can use to write and put together technical operation instructions. The following practical information can help you survive as a professional writer in industry. Each chapter of this book explains a major step or concept that will help you translate technical data into usable "how to operate" information. It explains the techniques available to make complex ideas as simple and understandable as possible.

Second, this book can help you develop a unique visual technical writing style that will enable you to communicate your instructions clearly and accurately to a mass-consumer audience. The chapters stress the use of visual communication theory to solve the various technical writing problems that you will be faced with in industry. You can use the *guidelines* in this book as a foundation upon which you can build a dynamic *visual technical writing style.* This style can be successfully applied to other types of business communication. The ability to write clearly and accurately is an asset to anyone in industry.

Third, *every* person has a set of skills that he or she has developed throughout one's educational training. The following chapters will show you how to use your particular abilities to their fullest. Technical writing is being done every day in a professional manner by computer programmers, technology, applied science, social science, engineering, and liberal arts graduates of both two-year and four-year college curriculums. No matter what your training or interests are, you can bring useful knowledge with you as you study technical writing. As you proceed through this book, you will discover that the process of writing technical operation instructions is a unique type of communication that demands the skills and creativity found in many disciplines.

Finally, you can use this book to help evaluate the quality of what you write. Technical operation instructions are written for two main reasons: to warn of potential safety hazards and to educate the consumer. Throughout the text, standards and evaluation techniques are presented that can help you decide how well your technical operation instruction is meeting your communication objectives. No matter what you write, you must be sensitive to the needs of the mass-consumer audience in our worldwide marketplace who potentially may use your instructions.

Chapter 1 gives you an overview of what it means to be a technical writer in a worldwide economic system. It introduces you to major themes and ideas that will recur throughout the text. Second, the following information explains some of the primary skills that should be mastered in order to write logically and accurately. It compares the technical writer with such diverse professionals as the computer programmer and the playwright. You must use these comparisons in order to evaluate your own strengths and weaknesses. Third, this chapter gives you a realistic employment picture that explains what you can expect to find in the fast-paced and expanding field of technical writing in today's high-tech industry.

Above all, Chapter 1 defines the term "technical writing" in industrial terminology. Technical writing is not letter, proposal, advertising, report, or article writing. Although these communication skills are necessary, they may not adequately prepare the student for a professional technical writing position. By studying the art and science of instruction writing, you can broaden your practical knowledge and learn a skill that you can apply on the shop floor of a modern manufacturing company.

UNDERSTAND THE INFORMATION NEEDS OF A WORLDWIDE MARKETPLACE

We live in the most technologically advanced time in human history. Manufacturing companies have developed machines that we can use to explore the far reaches of space, the depths of the oceans, and the complexities of the human mind. Our lifestyles depend upon inventions that were the subject of science-fiction stories only ten years ago. Personal computers, cybernetically controlled robots, microwave ovens, satellite communication networks, and lasers have made a great impact on our lives in only a few years (see Figure 1-1).

For the first time in history, state-of-the-art technology is available to everyone. In the past only pharaohs, kings, and heads of state enjoyed the machines and "mechanical apparatus" that make life easier and more enjoyable. Satellite transmission re-

FIGURE 1-1 Artist's concept of space shuttle astronauts building a space structure. (NASA)

ceptors, electronic fuel-injected engines, telephones, and electric can openers are commonplace in our homes, offices, and factories. For less than ten dollars, we can buy a hand-held, solar-powered calculator that is far superior technically than a multimillion dollar computer of only 20 years ago.

These advancements are not localized. They can be found in all parts of the world. Advances in communications and travel technology have added to the shrinking of space and time. Methods that distribute machines to all parts of the world have been upgraded to such an extent that a company can ship its products to virtually every corner of the globe by air, ship, or truck. Important replacement machine parts can actually be delivered overnight from one side of the world to the other. Our own economy has become intertwined with inhabitants of many lands (see Figure 1-2).

FIGURE 1-2 Technology is not isolated to a few countries. (Photo courtesy of Hewlett-Packard Company)

The global village. People of all continents have been brought closer together into what has been called the "global village." Today, most forward-looking manufacturing companies consider people from every land and culture a potential consumer for their developing technology. This attitude has created the first truly worldwide marketplace where machines can be bought and sold by people who speak many languages and who live in a wide variety of cultures.

Without technical operation instructions that explain how to use our advanced technology, these global consumers are unable to appreciate fully the advancements that are at their fingertips. Your job as a technical writer is to write technical operation instructions that help put new technology into the hands of the inhabitants of our global village—your *mass-consumer audience* (see Figure 1-3).

FIGURE 1-3 As technology advances, our "global village" gets smaller. (NASA)

Know the Value of Technical Information

The importance of technical information may be underestimated by the consumer because technical operation instructions have become commonplace. An "operation manual" or "direction" is the first thing that you see when you open a package that contains a new machine such as a kitchen appliance, a lawnmower, a laser printer, an electric drill, a tractor, an electric can opener, or a sophisticated yet inexpensive hand-held calculator (see Figure 1-4).

In fact, you and the millions of people in your mass-consumer audience have made the technical operation instruction a legal necessity. As consumers, you have fought for and won many rights that make the technical operation instruction an integral part of the machine. For example, when we buy a machine we have the right:

To get the safest product that can be manufactured.

To get complete and accurate product information that instructs the consumer on how to use the machine in words that the consumer can understand.

To be warned of possible hazards associated with the operation of the machine.

The Age of Information. In reality, we live in an "Age of Technical Information" where communication and instruction are as important as the product. Information has become so vital that machines are not released to the marketplace for sale by a manufacturing company unless a technical operation instruction is attached. Many countries, for example, will not allow a machine to be imported unless complete technical operation instructions are included in the shipping package. Exporting companies that sell these machines and ignore international laws often are banned from the import market and heavily fined until they produce the necessary technical literature.

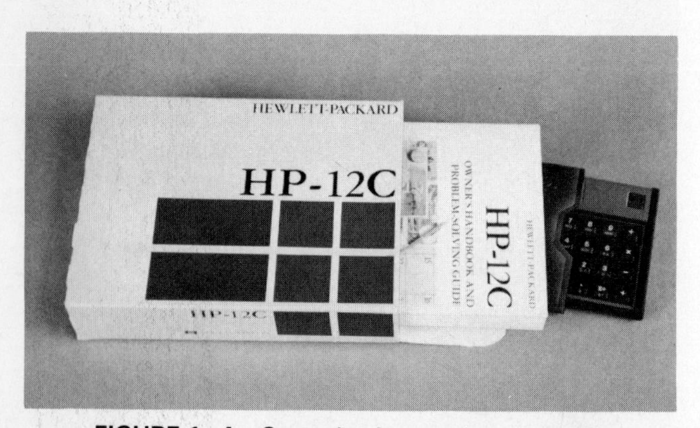

FIGURE 1-4 Operation instructions are packaged with or attached to the machine. (Subject matter courtesy of Hewlett-Packard Company)

You can appreciate the need for usable technical operation instructions if you put yourself in the role of a consumer who has purchased a new personal computer. Without complete, accurate, and understandable instructions, you will not be able to employ your technology. When this happens, you may not have the confidence to buy new and innovative technology in the future. Our profit-oriented supply and demand marketplace must have the support of the consumer in order to work properly. If you and the other members of the mass-consumer audience do not buy the new machines that are developed and manufactured, then our advancing technology and the production machines of our industrial companies will soon come to a grinding halt (see Figure 1-5).

DEFINE TECHNICAL WRITING AS BOTH AN ART AND A CRAFT

The technical writers job in industry was created out of necessity. The writer's role in the manufacturing company is to bridge the communication gap between the machine manufacturer and the consumer. As a technical writer, you are expected to help consumers interface with their machines. Because technology varies so greatly, a short descriptive explanation of how to do this is difficult. You can, however, simplify your thinking by eliminating some common misconceptions that exist concerning the actual job of technical writing.

First, contrary to what many people outside of industry use as a definition, technical writing does not include the creation of annual reports, newsletters, proposals, public announcement news releases, or textbooks that explain technical subjects. Although these types of business communication are important and necessary to the success of a business, they are usually not part of the daily job of the technical writer. Rather, they are the responsibilities of the manufacturing company's business writers, public relation communicators, and technical reporters.

The technical writer does not create data. Second, the technical writer does not create technical documents. This task is the responsibility of the company engineers, technicians, and manufacturing engineers and other *machine experts.* Technical doc-

FIGURE 1–5 Our mass-production-oriented economy depends upon consumer confidence. (Reproduced with permission of AT&T)

uments and information that you must use are the result of the design and manufacturing process that are necessary steps in the creation of a new machine.

The technical writer must be skilled in both the *basics* of language arts and the applied sciences. In industry, people who have been educated in technical curriculums often speak a different language from those formally educated in communication programs. The successful technical writer must be both a general or specific-trained technician and a fundamentally sound writer. The successful technical writer tends to place as much emphasis on writing skills as on technical know-how.

Well-educated engineers, as well as highly trained communicators, do not necessarily write the best technical operation instructions. The art of writing technical operation instructions demands a different approach and attitude from the individual liberal arts and applied science departments that our colleges and universities often teach. The study of technical writing as a subject demands *interdisciplinary* methods of study. The more you know about all aspects of industry, including such diverse topics as interpersonal communication, printing, computer science, robotics, and photography, the more successful you will be in creating highly readable and usable instructions.

Write With Pictures

The visual technical writing style. Unlike the more specialized and well-defined writing jobs such as the public relations communicator, technical writing is both an *art* and a *craft* (or a set of learned skills). Communication, for example, is an art. It is the expression of ideas by an individual through the conscious selection and rearranging of words and pictures. Art is personal and must be developed by the individual. Art is expressed as a style. When we speak of a visual technical writing style, we are defining technical writing as a unique art form. It is the ability to let visual images describe the situation. For example, it is better to show a picture of an astronaut stepping on the moon than to describe it with words (see Figure 1–6). Similarly, why de-

FIGURE 1–6 Pictures are a universal descriptive language. (NASA)

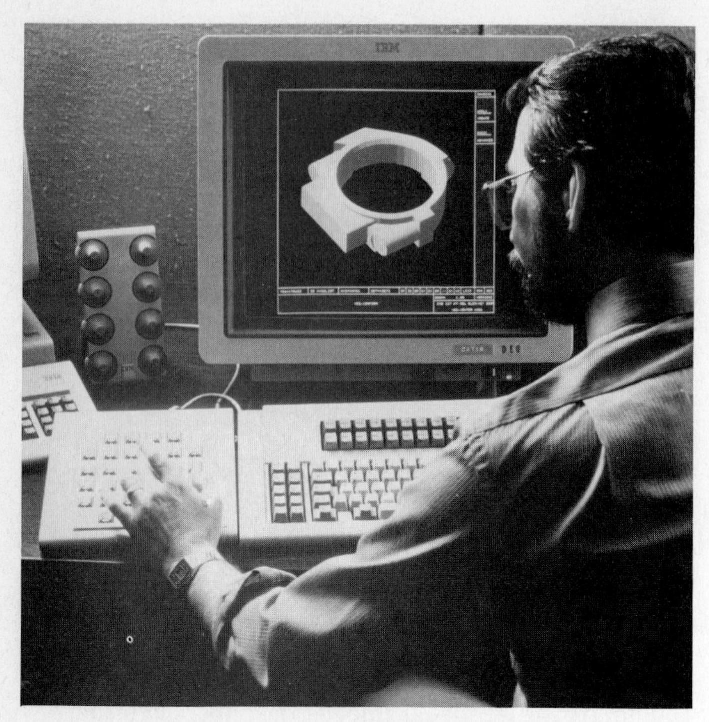

FIGURE 1-7 It may be impossible to describe a complex computer screen by using only words. (Courtesy Deere & Company)

scribe a computer screen when you can show a photograph or drawing of it? (see Figure 1-7).

Learn the "Craft" of Technical Communication

On the other hand, the technical writer must be defined as a craftsman or craftswoman who has many skills. This means that, no matter what your background is, you can use your past education as a starting point in your study of technical writing. But the study of technical writing must include learned expertise, such as:

- The ability to construct mechanically correct sentences and to use verbal logic.
- The ability to use spatial logic.
- The ability to report all of the facts as completely as possible.
- The ability to understand the *basics* of how a machine works.

Look closely at these unique skills one must develop in order to write the best possible technical operation instructions for a mass-consumer audience. To make these concepts more usable, you can draw a direct comparison of these skills with other professional experts who work in other fields of communication and technology.

Become a "Consumer Programmer"

You can begin your study of technical writing by comparing the job of the technical writer with that of the computer programmer. No job outside of technical writing exemplifies the need to develop the mechanical ability to write more than the computer programmer. Although the human brain is infinitely more complex than any present-generation computer, parallels can be made. First of all, the computer programmer must use a specific and well-developed writing *style* (see Figure 1-8).

Second, the programmer's vocabulary comes from such well-structured and controlled languages as BASIC, PASCAL, or COBOL. Whichever vocabulary is used, it must be specific and completely understood by the computer. Sentence structure or task explanations are written as efficiently and clearly as possible. All unnecessary and unclear words are eliminated by the programmer. Similarly, you, as a technical writer, must write with a *basic* English vocabulary that is as accurate and to the point as the words that are used by the programmer.

FIGURE 1-8 Programming the consumer and the computer is similar in some respects. (Courtesy Xerox Corporation)

```
10 ' Demo of various print pitches
20 LPRINT CHR$(15);
30 LPRINT CHR$(27);"M";
40 LPRINT "This line is CONDENSED ELITE pitch."
50 LPRINT CHR$(27);"P";
60 LPRINT "This line is CONDENSED PICA pitch."
70 LPRINT CHR$(18);
80 LPRINT CHR$(27);"M";
90 LPRINT "This line is NORMAL ELITE pitch."
100 LPRINT CHR$(27);"P";
110 LPRINT "This line is NORMAL PICA pitch."
120 LPRINT CHR$(27);"W1";
130 LPRINT CHR$(15);
140 LPRINT CHR$(27);"M";
150 LPRINT "This line is EXPANDED CONDENSED
   ELITE."
160 LPRINT CHR$(27);"P";
170 LPRINT "This line is EXPANDED CONDENSED
   PICA."
180 LPRINT CHR$(18);
190 LPRINT CHR$(27);"M";
200 LPRINT "This is EXPANDED ELITE."
210 LPRINT CHR$(27);"P";
220 LPRINT "This is EXPANDED PICA."
230 LPRINT CHR$(27);"W0"
240 END
```

FIGURE 1–9 A computer program. (Copyright 1986 by Star Micronics Co., Ltd., *NX-10 User's Manual,* PN 80820122. Used with permission)

Like the computer programmer, you must be able to:

Write information in logical tasks.
Write tasks in a numbered sequence.
Write a complete set of tasks in logical order.
Write sentences that are mechanically error free.

Compare the tasks that are written in basic computer language (see Figure 1–9) with the sample instruction that explains how to set an optional clock in a car (see Figure 1–10). The skill used by the writers of these two types of instructions is similar. For both the computer and the human "programmer," it is the ability to write instructions logically, accurately, concisely, and completely. Instructions must be readable and usable so that both the electronic and human "audience" can understand and use them to perform a task.

The human brain can interpret. There is one major difference between your human audience and the "artificial intelligence" of the computer. The human audience is able to *interpret* unclear instructions. Instead of a simple error message from the computer, human readers may, in disgust, throw away their unclear instructions and create their own set of tasks. Unfortunately, vague and poorly written instructions are easily misinterpreted by the human reader. Misinterpreted tasks can put consumers or the machine they are operating in a dangerous position. As a technical writer, you must strive to write instructions that cannot be misinterpreted by the mass consumer audience. Inaccurate and incomplete instructions can create a safety hazard for the operator as well as for the machine.

Become a "Technical Playwright"

As a technical writer you also must be concerned with space and time. Unlike the two-dimensional "on and off" flatland world of the computer, the consumer lives in a three-dimensional environment where one physical action is followed by another ac-

CLOCK (OPTIONAL)

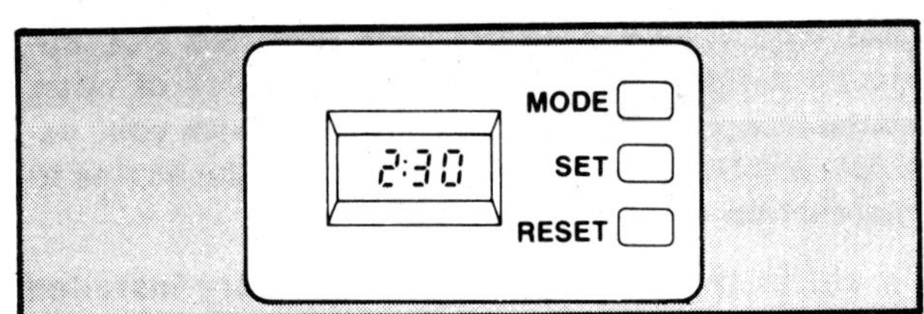

Successive pushing of mode button advances clock through DATE, ELAPSED-TIME, TIME, etc.

To Set: TIME (Colon does not flash) Press

1. MODE button
2. RESET button selects hours
3. SET button to change hours and (AM or PM)
4. RESET button selects minutes
5. SET button to change minutes
6. RESET to return to time

To Set: DATE (No Colon) Press

1. MODE button
2. RESET button selects months
3. SET button to change month
4. RESET to select day
5. SET button to change day
6. RESET to return to date

FIGURE 1–10 The instruction for a "human" audience. (Copyright 1986 by the Ford Motor Company, *Mercury Sable 1987 Owner's Guide,* FPS-12059-87A. Used with permission)

FIGURE 1–11 The film director properly positions the actors for their next lines. D. W. Griffith (in the hat) directs (l to r) Vivia Ogden, Richard Barthelmess, Lillian Gish, Burr McIntosh, and Lowell Sherman in *Way Down East.* (Courtesy the Museum of Modern Art/Film Stills Archive)

tion. If you, like Shakespeare, can view the world as a stage and all of the men and women merely players, then you will be able to appreciate fully the importance of being in the right place at the right time when you operate a machine.

In a play rehearsal or movie "take," there is nothing that is more embarrassing than an actor or actress who enters through the wrong door on the stage, or one who is out of position to take a cue in the middle of a play. The playwright who writes the dialogue and direction for the lively arts must be able to move characters around the stage so that they are in a position to accept dialogue. The playwright must ensure that serious characters do not bump into each other, trip over the props, or sit down where there is no chair. Stage directions that control the movement of the actors are as important as the words they speak. Playwrights must be able to visualize movement in their minds as they write their dialogue. They must safely position their characters through the director for their next line (see Figure 1-11).

You too must develop this skill of positioning. Your consumer should be reading and following your instructions as faithfully as an actor follows a director's commands. You must be able mentally to move the consumer around the machine as you write technical operation instructions. You must master the control of the space around your machine as skillfully as a playwright masters the space of the stage.

Spatial logic. Spatial logic dictates that it is impossible for a consumer to do two or three dissimilar tasks simultaneously. A consumer, for example, cannot do a task on one side of the machine when he or she is spatially positioned by your instruction on the other side of the machine. You cannot expect a consumer to have super abilities such as those of a contortionist, who can stretch the body into uncommon positions. You must keep your consumer actor or actress in the safest and most efficient position possible.

The two examples below are offenders of spatial logic. Instead of telling consumers how to operate their machine safely and efficiently, these tasks create a foundation for a slapstick comedy routine like that used in a Keystone Kops movie. Comedy also often relies on the illogical movements or the out-of-position comedian such as the subtle genius of Charlie Chaplin. But when the consumer is out of position, the results are not humorous. The consumer may not be able to walk away from the situation without a scrape.

EXAMPLE 1

TURN ON YOUR AUDIO COMPONENTS

1. Plug in the electrical cord to a 110-volt supply.
2. Turn on the receiver.
3. Step back 10 feet from the equalizer and adjust the settings for the room.
4. Adjust the speakers both left and right in the room.
5. Move backwards 10 more feet and adjust the speaker balance.
6. Adjust the treble and bass.

EXAMPLE 2

START YOUR AUTOMOBILE

1. Turn the key switch to "start" and start the engine.

2. Release the key switch.
3. Release the hood latch.
4. Open the hood.
5. Check the oil.
6. Close the hood.
7. Put the gear selector in "drive."
8. Check tire pressures.

Caution: Inflate tires only when they are cold. Keep pressure below 35 PSI.

Unlike the obvious examples above, you may not be able to find "subtle" spatial logic errors in your technical operation instruction until the tasks are physically acted out by you or by another actor/consumer who is dress-rehearsing or desk-checking your technical operation instruction. In Example 1, the consumer needs a 10- or 20-foot extension in order to complete the task. In Example 2, the consumer/operator is placed in two unsafe positions. First, the oil should not be checked with the engine running because of the danger of the rotating radiator fan blade. Second, the automobile will move forward if it is in drive. Finally, it is impossible to check the tire pressure while you are sitting behind the steering wheel.

Develop your ability to create spatially logical technical operation instructions. Practice thinking in spatial images. Visually imagine the movement of yourself around an automobile during the following start-up instructions. As a technical playwright, visualize every movement in detail. While doing this exercise, set a stopwatch or other timer. You will notice that it takes the same amount of time to *mentally* start your automobile as you take to physically start it.

EXAMPLE 3

(An instruction that might be written for a start-up procedure for an automobile with a carburetor fuel supply)

START YOUR AUTOMOBILE: COLD ENGINE

Note: Inflate the tires when they are cold. Tires must not be inflated above 35 pounds per square inch (PSI)

1. Check the tire pressure. Inflate if necessary.
2. Pull the hood latch release.
3. Open the hood.
4. Check the oil level.
5. Add oil to the "full" mark on the dipstick.
6. Close the hood.
7. Turn the key switch to the "on" position.
8. Press the accelerator pedal to the floor and release it.
9. Turn the key switch to the "start" position and start the engine.
10. Release the key switch as soon as the engine starts.

Note: Do not crank the starter engine more than 15 seconds at a time.

11. Put the gear selector in gear and drive away.

Work on this skill by first mentally picturing the action and then physically doing the action. The *autogenic* approach to projecting physical action mentally is not an uncommon practice. This exercise has been successfully used by athletes in golf, bowling, track and field events, and other skill sports to develop mental sharpness before their performance. Jack Nicklaus, for example, will stand behind his golf ball and picture his swing and the flight of the golf ball before he actually selects his club. Coaches of many world-class athletes have turned to the study of these mental biofeedback or mental imaging techniques in order to improve the physical skills of their students.

A skilled technician who services or fixes your machine may do the same exercise. By going over a procedure mentally, this person is able to apply logical thoughts to his or her motor skills and tear down an engine or assemble a transmission.

As a technical writer, you can use this mental process to develop your sequence of tasks in the same manner. You can visualize the entire technical operation instruction in your mind before you begin the writing process. You can personally apply your thought pattern to the machine and operate it. This mental and physical spatial skill must be developed and refined in order to help you create a spatially logical sequence of tasks in your technical operation instruction.

Become a "Technical Reporter"

You must remember that technical writers do not *create* information. The documents and test results that the technical writer must research are created during the design and production stages of the manufacturing process. For this reason, the technical writer must interpret available technical data into information that can be used by the consumer.

There are two similarities that you can draw between your job as a technical writer and the reporter for a newspaper or magazine. The first comparison is that both communicators must collect all of the facts, analyze them for accuracy, and select the news that the reader needs or wants to know. Both writing jobs demand that only firsthand sources of information and up-to-the-minute information be used to report the facts.

The reporter must put the facts into a story that is complete and readable. Similarly, you must translate technical facts and data into technical operation instructions that the consumer can easily follow. Both you and the reporter must evaluate your sources of information. You must be able to interview the machine experts who develop and help create the machine in the manufacturing process as well as research the documents of industry. Like the reporter who is about to break a story to the readers, you must ensure that your information is accurate by double-checking all of your sources' information (see Figure 1-12).

Second, as a technical writer, you must work to identify and maintain a communication link between yourself and the machine experts who are your most valuable sources of information. Like the reporter, you must be able to talk intelligently with, sympathize with, and get to know as much as possible about the jobs of those who are your information sources and their role in the company manufacturing process. Like the reporter, you must *study human nature.* Without up-to-date sources, such as the machine expert who can quickly answer your questions, your job will be more difficult and time-consuming.

FIGURE 1-12 The technical writer must report information accurately. (Courtesy Almon Studio)

FIGURE 1-13 Get as much hands-on experience with the machine as possible. (Courtesy Almon Studio)

Become a "Machine Technician"

As a technical writer, you must be able to wear many hats comfortably. First, you must be able to use a basic and easily understood vocabulary with the same skill that a computer programmer uses in employing a specific language to compile a complex program. Second, you must have the skill to control the movement of the consumer just as a playwright controls the actors in a play. Third, you must be able to deal with people, collect information, sort out the facts, and report useful information to the consumer just as a reporter who writes the news for a daily newspaper.

Finally, you also must know your subject. The proper study of the technical writer is the "machine" that they are writing about. You must be able to explain how and why the machine works. You must learn the *basic* skills of a "machine technician." Every writer's first job is to know his or her subject completely, objectively, and accurately. As a technical writer in industry, your primary subject is the technology embodied in the machine. Get as much hands-on experience as you can get with the major parts of the machine. Study how it is put together and how it works. Especially learn as much about the controls as possible (see Figure 1-13).

Machine systems. There is an efficient approach to the study of why and how a machine works. It is the study of *machine systems.* First of all, a machine is made by people in the manufacturing process. Pieces of the machine are fabricated and assembled into parts such as motors, cylinders, computer chips, keyboards, and carburetors. These parts are mounted on the machine frame or skeleton and connected in unbroken *systems.* A machine system transfers a specific type of *power* to a final *point of work.*

Each system does a specific kind of controlled work. For example, hydraulic power in an industrial robot is transferred through pressure reducers, pipes, hoses, and valves that control the flow of the pressurized fluid, to the metal fingers that do the work, such as assemble electronic components. Even the most complex machines, for example, the Apollo II and the machines that it carried to the moon on the first lunar landing mission, can be broken down into basic systems concepts (see Figure 1–14).

FIGURE 1–14 The Apollo 11 Mission was the result of many machine systems working in harmony. (NASA)

Systems concepts are an important part of machine definitions. Machines, for example, are not hand tools such as hammers, saws, and screw drivers. Tools are simple extensions of the human body. A machine, on the other hand, has one or more power systems, and it has one or more points of work like the robot example above. A more complex example of a machine is an automobile.

An automobile is manufactured and sold to do one basic job—to get people from one place to another. Without the electrical, brake, frame, engine, and lubrication systems, the automobile will not move. The parts of a system, such as the power train that includes a clutch, transmission, drive shaft, differential, and final drive, convert power into useful work at the wheels. While each part is important to the system, each system is important to the overall working of the machine. If all systems work except the brake system, then the automobile must not be used to do work.

The technician understands and works with machine systems, and there are many types of technicians in industry. They work in areas such as fine-tuning or refining the operation of the machine, developing production techniques, testing, and other areas that require special machine competence. As machine specialists, technicians must know how and why specific parts of machines work. They measure, evaluate, and help develop the machine design throughout the manufactured life of the machine.

Understand the 'languages' of industry. The technician is a systems expert. Every machine in order to be competitive with similar machines in our worldwide market must be dynamic. It must be constantly redesigned and modified. The technician plays an important role in this process by continually developing, testing, and evaluating parts and systems. Like the technician, you must be familiar with the basic machine systems. You must be able to speak the *system language* of the technician and interpret the technical data generated by these machine experts into usable information for the consumer. General system concepts are covered in detail in Chapters 6 and 7.

In order to define the term *technical writer,* one must say that the technical writer is both a technician and a communicator. As a technical writer, you must be equally at home examining the systems concepts of a machine as well as interviewing the machine experts or reading an engineering drawing. As a technical writer, you must be able to blend the applied science and liberal arts skills together, skills that are usually not thought to be compatible in today's educational and industrial complex.

KNOW YOUR MEDIUM

A common slogan in industry is "the job is not complete until the paperwork is done." Without some form of documentation, the knowledge of the machine expert is unavailable to the other people in the manufacturing process. For example, the technician who tests a machine system must be able to use specific reports and forms in order to communicate actions to those involved in the next level of the manufacturing process. Engineers or machine experts must be able to put their ideas into many types of media, including engineering drawings and reports (see Figure 1–15).

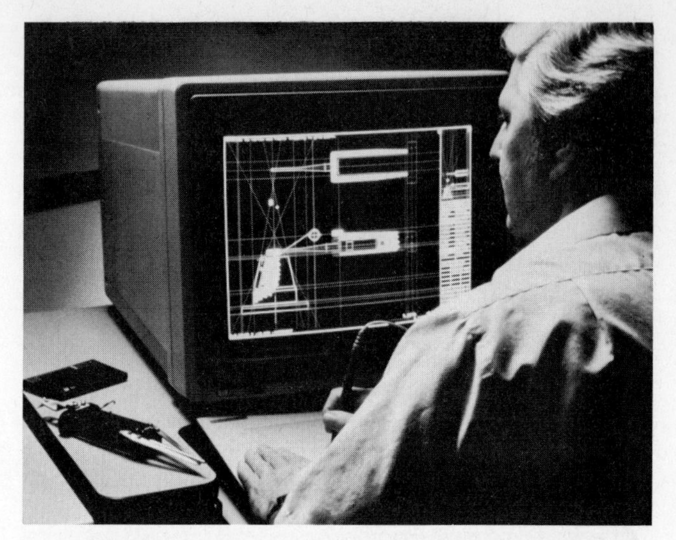

FIGURE 1–15 Engineering drawings are the engineer's and the designer's primary medium. (Photo courtesy of Hewlett-Packard Company)

This also holds true for the technical writer and other communicators. Without the medium of the newspaper, the reporter's work is wasted. Without a keyboard, central processing unit, and CRT, the computer programmer's flowcharts and programs that they conceive are useless. Technical writers, like all communicators, must know how to effectively use their medium—namely the printed technical operation instruction.

Understand the Instruction

The *technical operation instruction* is a set of tasks that are printed, usually on paper. All written instructions have the following basic characteristics:

- A title that explains what machine the instruction supports.
- Subtitles that tell the consumer exactly what job will be completed when all of the tasks under the subtitle are completed.
- Numbered tasks in sequence that explain how to complete the job that is introduced in the subtitle. Tasks include both words and pictures.

The information in an instruction will vary for each machine. Your responsibility when you write an instruction is to include all of the information that the consumer needs in order to operate the machine safely and efficiently. This information may be dramatically different for some machines. A consumer who buys an automobile will need more information than a consumer who buys an electric carving knife. The best way to understand how instructions vary from machine to machine is to look at examples of technical operation instructions. Instructions are easy to find. They are printed on the back of the package of a hand-held calculator, in the carton of a food blender, in the box of a portable cassette tape player, and in the glove compartment of a new automobile.

The major divisions of a manual. In general, you will find some or all of the following basic information divisions in all technical operation instructions. They are:

- A general location picture of the product that points out major parts of the machine.
- Safety warnings that *warn* or caution the consumer about unsafe conditions that can be created with or by the machine.
- Consecutively numbered tasks that tell how to set up, hook up, and assemble the machine.
- Consecutively numbered tasks that tell how to turn on, operate or control, and turn off the machine.
- Some technical operation instructions also may give simple daily maintenance procedures that tell how to adjust, oil, or lubricate the machine so that it runs efficiently for daily operation.
- Some technical operation instructions may contain specifications that describe the quality or quantity of replacement parts, power supply, or machine physical dimensions.

The technical operation instruction is divided into two sizes. They are:

The directions.
The operation manual.

Define the Directions

The *directions* consist of one or two unbound printed sheets of technical operation instructions. All necessary information that the consumer needs to set up, start, and operate the machine are contained in these brief instructions. Directions are used to explain machines that have only a few control points. For example, the shipping carton of a transistor radio may contain a single sheet of instructions that tells the consumer how to insert the batteries, adjust the controls, and use the "machine" safely.

Define the Manual

On the other hand, an *operation manual* is an expanded direction that may include many printed and bound sheets of paper (see Figure 1-16). If the use situation requires that the manual pages be bound or mechanically held together, then binding is usually done with wire staples through the backbone of the manual (wire saddle stitch), staples through the left-hand side or backbone (side stitch), and glued (perfect bound). Operation manuals that must lie flat, such as a computer user's instruction, are often spiral-wire bound. Spiral bound is more expensive, but in some situations necessary. Because the major use of a cover is to protect the manual contents, the use and application of the manual determines the design and strength of the cover.

A bound operation manual can be conveniently carried to the machine by the consumer. If the pages are correctly numbered and correctly titled, consumers are able to refer to parts of the manual as they use the machine. For example, a lawn and garden tractor is a complex machine with many control points. The consumer probably will not be able to memorize all of the tasks, safety cautions, and useful information during the initial machine startup. They can randomly access a manual as the need arises.

There are many pseudonyms that you might encounter for the technical operation instruction;

FIGURE 1-16 The cover of an instruction protects manual contents. (Copyright 1981 by the Xerox Corporation, *Job Preparation and User Guide for the Xerox 8200,* 600P83889. Used with permission)

however, all are similar in their application. Instructions also can be called:

User's manuals. (Bound)
Owner's manuals. (Bound and unbound)
Directions. (Unbound)
Task lists. (Usually unbound)
Technical passports. (Used like a tourist passport in order to get the machine *into* an importing country)
Operator's manuals. (Bound)
Training manuals. (Bound)

With advancements being made in the area of personal communication systems, the media available in the future for the technical writer may include videotapes and software. The prevalence of video equipment today make the VHS or BETA film format a legitimate cost-effective consideration for the technical writer. The video method for outlining and writing a technical operation instruction is actually very similar to the process of visual technical writing.

However, in most situations the printed operation instruction is the most practical medium at this time. The printed direction or operator's manual can be taken to the site of machine operation and can be easily accessed in comparison to the video method. The printed form of the technical operation instruction is thus more convenient and therefore more usable than an electronic format.

Another future development may be "on-board" computer systems on some larger machines. In this case, consumers may be able to take a piece of software to the machine and use it in a built-in computer terminal on the control panel of the machine as they operate the controls. This is definitely a future concept; however, do not overlook a medium such as this if the situation requires a creative approach to getting information to the site where the consumer is using the machine. On-board computers are commonplace in our space program. Whatever technology employed in the space program today is usually commonplace for the average consumer tomorrow (see Figure 1–17).

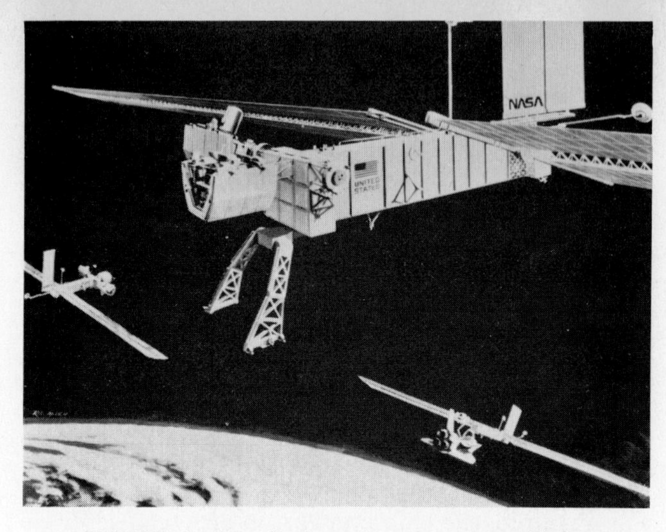

FIGURE 1–17 On-board computers will be common on future machines. This is an artist's representation of the proposed orbiting space station. (NASA)

USE YOUR TECHNICAL WRITING SKILLS IN INDUSTRY

Practical operation instruction training, unlike some formal college or university courses, can be directly applied to a work situation. Many technical instruction writing job possibilities are created each year for both the nontechnical and technically trained person. Four main sources of employment for a trained technical writer include:

Industry.
The federal government.
Technical writing agencies.
Freelance work by the individual.

Industry, especially the "high-tech" area of industry, is by far the best possibility for employment for the newly trained technical writer. Technical writers work for aeronautical, electronic, computer, agriculture and construction equipment, domestic, and other manufacturing companies. The pay and benefits will vary depending upon the financial state of the company, how badly a technical writer is needed, whether the writer is trained to write instructions, and the experience of the writer.

Technical Writing Can Be a Rewarding Career

Not only is technical writing a possible entry-level job but it can also open up new areas for those who are in industry. As a technical writer you can primarily have a rewarding career writing instructions in an ever-challenging job. In your research and study of the machine, you may become a recognized machine expert in the company. Technical writers often know more about the total machine than any one person in the entire manufacturing process. Technical writers also find additional opportunities in such areas of industry as service, sales, training, and marketing. Technical writers also can move into other fields of technical communication including the production and editing of post- and presales literature.

Until training, such as the course work that you are getting with this textbook, a person trained in technical communication might find it more difficult to get an entry-level job in industry as a technical writer. By studying the art and craft of instruction writing, you will be putting yourself in a better position to use technical writing skills as an entry-level job possibility. If you are about to apply for a position in a company, you should stress the fact that your technical writing training included the writing and assembly of technical operation instructions.

This may become more important as our ever-changing manufacturing processes develop. Mark O'Brien, author of *High-Tech Jobs for Non-Tech Grads,* states that developing high-technology companies are rapidly expanding jobs oriented to communication skills and a general knowledge of basic technology. For every technical job created in our future manufacturing companies, there will be approximately 25 nontechnical support jobs. Technical writing leads his list of new jobs that will be created.

Technical writing for the federal government. Another source of employment for the technical writer is the federal government. The U.S. government designs and manufactures weapons and military equipment at many arsenals around the country. The government also contracts weapon and equipment development to outside companies. It is an accepted fact by most historians of industry that the bulk of our technological advancements throughout recorded time has originated from both offensive and defensive weapon and military equipment production.

Besides obtaining work with the federal government, you may find that outside writing agencies that contract for technical operation instruction writing may be a source of employment. Outside writing agencies not only contract services to the government but they often contract to manufacturing companies that are either too busy, understaffed, or otherwise unable to produce the necessary instructions.

A job close to agency work is the "freelance" technical writer. This person works outside of the manufacturing company and the agency. Often the individual technical writer can develop his or her work into an agency as the person's customer list expands. In any situation, the area of technical writing can be very profitable. As an example, look at the information found in the *Occupational Outlook Handbook, 1986-87 Edition,* which is supplied by the U.S. Department of Labor Statistics.

> Demand for technical writers is expected to increase because of the continuing expansion of scientific and technical information. . . . With the increasing complexity of industrial and scientific equipment, more users will depend on the technical writer's ability to prepare precise but simple explanations and instructions. Besides jobs created by increased demand for writers and editors, many job openings will occur as experienced workers in this field transfer to other occupations or leave the labor force. Technical writers [in 1984] had salaries ranging from $19,300 to $31,900. Experienced editors generally earned between $22,000 and $39,000 a year; supervisory editors, $25,300 to $42,500 a year.[1]

Develop a Course of Study

If you are a student or professional technical writer, you may increase your chances for work or advancement in the area of technical writing by studying these areas:

Computer science. Most manufacturing companies are equipped with the latest in microcomputers and electronic word processing hardware. The area of computer publishing is only now being fully explored. Computer-generated drawings are a practical method of creating engineering drawings that tell how a machine fits together. Complete table-top layout and production of instructions and manuals are soon going to be feasible alternatives to the printing process for many companies (see Figure 1-18).

Typography. The study of printing techniques and processes can help you understand the "production" aspects of technical writing. Most instructions are printed and mass-produced after they are written. Many jobs in technical communication in industry require a working knowledge of typography. These are publishing, editing, and management of a technical writing staff. Instructions must be produced at an acceptable quality at the lowest possible cost. A well-trained technical writer who understands cost and quality can become a valuable member of the technical writing group of the manufacturing company.

Photography. Instructions probably do not communicate well unless at least 60 percent of the

[1] *Occupational Outlook Handbook, 1986-87 Edition* (Washington: U.S. Department of Labor Statistics, Bulletin 2250), p. 217

FIGURE 1–18 Computer science is an important area of study for the technical writer. (Courtesy Xerox Corporation)

technical information is artwork, which consists of line drawings and other illustrations. Most artwork used in a modern instruction is in the form of black-and-white photographs. Skillfully taken photographs are good conveyors of information. Technical writers in most companies must either take their own photographs or supervise photographers while they create artwork.

Drafting. The ability to read engineering drawings is a must for the technical writer. These drawings should be a major source of information for the technical writer. Some knowledge of drafting practices will also help you judge the time and cost for line artwork that is drawn by the graphic artist for the instruction.

Graphic arts. Graphic arts, especially layout and "mechanical" creation, can help you understand how to prepare the instruction for the printing process whether it is by the standard printing press or electronic method. The design of any written or visual instruction is as important to the readability of the information as the facts that are contained. Correct type, arrangement of information, and choice of paper and ink are skills that must be learned by anyone in the production area of technical writing.

Technology. A basic study of machine design, especially the hydraulic, electrical, pneumatic, and mechanical systems will help you understand what makes the machine run and how to control it. As you study basic machine systems, concentrate on operation and maintenance *safety* practices.

Psychology. A study of "differential" psychology is a good preparation to the understanding and acceptance of the vast array of human qualities that you will deal with when you research sources for the technical instruction. As a technical writer, you must be able to bring everyone in the manufacturing company together into a tightly knit communication network.

Rhetoric. No matter what you write, you must follow the basics of proper grammar and usage. Words must be spelled correctly according to convention. Sentences must be properly constructed. The technical writer must safely and efficiently use the tools of communication just as a consumer must adhere to the tasks and guidelines contained in a technical operation instruction.

CHAPTER SUMMARY

- The job of the technical writer is to tell consumers how to operate their technology safely and efficiently.
- Comedians who are "out of logical position" are humorous to an audience. Consumers or machine operators that are out of position can be dangerous to themselves or to their machines (see Figure 1-19).
- This book explains *one* way to write and produce a technical operation that is aimed at the needs and abilities of the consumer.
- The chapters in this text explain how to develop a dynamic and visual technical writing style that can be used for more situations than instruction writing.
- No matter what your skills, hobbies, or training, you probably can apply your experience to the job of technical writing.
- For the first time in history, advanced technology is available and is being used by *everyone.*
- We live and compete economically in a worldwide marketplace. Instructions are extremely important in an export-import exchange of goods. Technical instructions are as important as the machine in our "Age of Information."

FIGURE 1–19 Comedy is often an actor in illogical situations. Charlie Chaplin's classic film *Modern Times* is funny because no one really gets hurt. (Courtesy the Museum of Modern Art/Film Stills Archive)

- Technical operation instructions are attached to or packaged with the machine.
- The technical writer must wear many hats and have many communication abilities.
- In order to study how machines can be controlled and operated, study the systems of the machine.
- An instruction is a set of tasks or commands that enable the consumer to interface with the machine.
- The medium most used to communicate technical information is paper and ink.
- *Directions* are a one- or two-page unbound instruction.
- A *manual* is a set of instructions of many pages that are bound together.
- Technical writing is an entry-level job possibility that may help an applicant find work in a manufacturing company.
- Technically and nontechnically trained people can become successful technical writers in industry.
- Besides industry, the federal government, writing agencies, freelance jobs, and educational facilities can offer employment for the technical writer.
- A curriculum for a technical writer might include computer science, printing, photography, drafting, graphic arts, technology, and rhetoric.

EXERCISES FOR CHAPTER 1

Define Your Goal in the Study of Technical Writing

1. Check five common household appliances or machines, such as your automobile or stereo system, that you use every day. Answer these questions:
   a. Where are they manufactured?
   b. Did you get a technical operation instruction when you bought each product or machine?
   c. Are the instruction tasks written bilingually? That is, do they appear in more than one language?
   d. Did you use the instructions when you first operated these appliances or machines?
   e. Are all of the instructions usable? Do they tell you everything that you needed to know in order to set up and safely operate these machines?
   f. Which machines are supported by operator's manuals, and which by directions?
2. Begin a collection of technical operation instructions. Divide them into (1) unbound directions and (2) bound operator's manuals.
3. Write to manufacturing companies and get their list of technical operation instructions that are used to support the machines they make. The service, training, public relations, or communications departments can give you information. Companies usually keep and make available a supply of replacement technical operation instructions in case their customers lose or damage their original copies. If you cannot obtain instructions directly from the manufacturing company, contact the store where the machines are sold.
4. Examine two operator's manuals or direc-

tions. Be critical. How well do they communicate?

   a. Can you rewrite any of the tasks so that you can use them easier?

5. Write five basic instructions that tell how to start, operate, and stop five common machines. Visualize the procedure before you write your instructions. Suggested subjects include CD players, VCRs, motorcycles and microwave ovens.

6. Write an instruction without a diagram that tells your instructor in detail how to get to *your* next class, your car, the nearest water fountain, or the cafeteria.

7. In order to test the "technical" accuracy of communication, form your class into a circle. Begin a simple set of specific instructions at one point in the circle. Write down the instruction and show it to the person next to you. Destroy your instruction. Continue this procedure until the instruction reaches the point where it originates.
   a. Is it still accurate at the point of origin?
   b. If the accuracy is lost, where does the communication process break down?
   c. Would a list of tasks handed from person to person be helpful to keep the accuracy of the information intact?

2

Know Your Audience

The information given in this chapter will enable you to:

- Separate pre- and postsale literature writing techniques.
- Explain the basic developmental process of consumerism in the United States.
- Define "The American System."
- Explain the importance of "interchangeability" in the manufacturing system.
- Explain the important influences that mass production has had on consumerism in the United States.
- Define the term "mass consumer audience" and list the two levels.
- Explain the difference between *measurement* standards and *writing* standards.
- List the important problems that the worldwide audience presents to the technical writer.
- Explain how to write to the "lowest common denominator" of an audience.
- Construct a technical communication model with direct and indirect feedback.

INTRODUCTION

Your success as a technical writer in competition with the students in your class or in a production situation in industry begins with how accurately you can identify and analyze your potential audience. Your writing style, what you include, and how you arrange information in your final technical operation instruction must be compatible with the reading and comprehension abilities of the people who will use your technical literature. A technical writer's professional survival may depend upon how successful he or she is at understanding the information *needs* of the audience.

Your company's success in a competitive marketplace depends upon the dependability and affordability of its manufactured machine. Without complete technical operation instructions, a machine that otherwise may be able to compete successfully can become unreliable and expensive to operate. The competitive edge that is gained with accurate instructions and warnings is lost with poorly written instructions because of the lack of consumer confidence. As a consumer, you probably can remember all of the names of the companies that sold you *what you thought* was an unreliable machine.

On a larger scale, our ability to compete as a manufacturing *nation* depends to some extent upon our dedication to support our manufactured products with technical operation instructions. Technical documents that support our machines must be designed for the needs of the *worldwide* consumer. Support literature that takes the needs of the importing consumer into account can help make our machines more competitive in export markets. In order to compete, we must produce world-class instructions for a worldwide marketplace.

The technical operation instruction is as important as the nuts, bolts, and framework that hold the machine together. If you fail to write effective technical literature designed to interface with your audience, then your information will be as much of a liability to the machine as a defective or broken part. A well-designed machine can never make a poorly written instruction usable; however, a poorly written instruction can help make a machine useless.

The needs of the people who may read your work can change with every type of machine. This chapter will help you develop a method and set of standards that you can use to identify your audience no matter what instruction writing situation you might find yourself in. You must be able to reevaluate your audience and your writing techniques continually in order to adapt to an ever-changing marketplace.

The first step in the process of analyzing the audience is to determine how your ideas, words, and pictures will be used. Technical writing is a unique form of communication in industry. For the most part, the instruction has a highly specialized job to do—it must both instruct and warn the consumer. The instruction must teach the consumer how to "control" the power of the machine in order to do work. It must warn the consumer of any possible danger that can be created if the machine is mishandled or misapplied. For this reason, the instruction is considered to be strictly a *postsales* document. It must not be confused with other forms of business communication such as advertising or public relations literature.

The first topic covered in this chapter is the comparison of the instruction with the presales documents of advertisements and news releases. Although literature of this type is extremely important to the manufacturing process, it must not be confused with postsales literature.

Next, this chapter presents guidelines that can help you determine the people who might make up your potential audience. In order to simplify the job of audience analysis, these guidelines are designed to do two major things. First, they define the role of the consumer in a modern economy, and they then divide the consumer audience into two main categories, the homogeneous level and the heterogeneous level. Although each category of the audience has different needs, the basic technical writing style that you must develop will let you effectively communicate with each type of consumer.

Second, a communication model will be constructed in order to act as a framework for the understanding of the mechanical process of technical communication. This model will help provide a foundation and structure to the general writing guidelines and principles that are given throughout the chapter. By creating a communication model, you can better appreciate the act of technical writing as a total communication *process.* As a technical writer, you must include the input and ideas of many people in your instruction if you are to be successful in a writing assignment.

DEFINE THE USE OF POSTSALE TECHNICAL LITERATURE

You can begin the analysis of an audience by determining *how* your writing is going to be used. First of all, anything that is written and printed on paper and distributed to the consumer is difficult and at times impossible to correct if it is inaccurate. A printed instruction also must be considered a permanent document. You must design an accurate instruction that will be usable in the marketplace for many years.

Technical operation instructions often are used as a reference by the consumer years after the "new" machine design has become well known and accepted. The consumer who purchases the machine may refer to the instruction often. What may seem like appropriate information at the time you are creating the instruction may become extremely out of place and useless to the consumer as time passes.

Do Not Write Presale Information

Presale literature. The problem often arises with the fact that an inexperienced technical writer may not be able to differentiate between pre- and postsale information. A common mistake that writers make is to add presale literature concepts to the technical operation instruction. Presale literature is the words, pictures, and symbols that are used to help *persuade* the consumer or help them decide to buy a particular machine. This literature includes the ads in newspapers, magazines, billboards, and sale brochures. Another type of "business and industry communication" that is often used to introduce a new product is the news release. In some types of news releases, the style or design of the machine is an important part of the information (see Figure 2–1).

A direct sale tool. A definite need exists for presale literature. For example, it is necessary for a manufacturing company to supply information to consumers about the product before they buy. These items are *direct* sales tools that help inform the consumer about what is available in the marketplace (see Figure 2–2). On the other hand, well-written technical operation instructions are primarily used to teach the consumer how to operate the machine. They do not attempt to persuade or inform the consumer.

Technical operation instructions, however, may be an indirect sales tool. The availability and quality of instructions developed by your company may be a deciding factor for consumers when they buy another machine. Poorly written instructions can create an unpredictable machine. If the consumer follows the inaccurate instruction and the machine fails to work properly, then the consumer logically may blame your machine and your company for producing a defective product. The consumer's next purchase may be from a competitor of your company.

Presales literature is a specialized area of communication that demands the training of a professional writer. Presales literature must tell the consumer the truth about what the product can and cannot do. The writer of presale literature also must not exceed the warranty parameters set by the company. A warranty is a contract between the consumer and the machine manufacturer/seller. A warranty contains, among other things, the length of time that the manufacturer will repair the machine in case of defective parts or workmanship. By eliminating presales information from the instruction, the technical writer will reduce the chance of making a statement or *implying* a warranty that the company cannot meet.

Write Postsale Information

As a technical writer who is writing postsale information, you must not attempt to resell the machine. This detracts from the usability of your instruction, adds words, and increases your writing and printing

FIGURE 2–1 The style and other appearance details are important to the potential buyer. (Courtesy Ford Motor Company)

The Tandy 3000 HL: Business Power, Speed and Dependability

In today's business world, if you want to compete, you have to have a reliable business computer system. One that will keep up with your company's expanding needs without compromising power. The Tandy 3000 HL was designed with just that intent.

Designed With Your Requirements in Mind

The Tandy 3000 HL was primarily engineered as an IBM® PC/XT compatible—then we improved upon it. First, we made it faster. With an 80286 processor, it's more than four times faster than the XT. Also, the Tandy 3000 HL comes with 512K memory and can be expanded to 4 megabytes using the built-in expansion slots.

Expansion Capability

For storage, the Tandy 3000 HL is equipped with a thin-line, 5¹/₄″ floppy disk drive that can read 360K formats. You can expand to include two floppy disk drives and a hard disk, or one floppy and two internal hard disk drives. The total capacity of the 3000 HL can exceed 40 megabytes. The keyboard features 84 sculptured keys and includes a numeric entry keypad. The Tandy 3000 HL can expand internally with seven plug-in card slots, as well as four PC/XT compatible slots. An optional 80287 Math Co-Processor can be added for high-speed numerical operations.

TANDY 3000 HL/IBM® PC/XT COMPARISON		
FEATURES	**TANDY 3000 HL**	**IBM XT 267/268**
Microprocessor	80286	8088
Clock Speed	4/8 MHz	4.77 MHz
Standard Memory	512K	256K
Maximum On Board RAM	640K	640K
Expansion Slots Full/Half	7	6/2
Available Expansion Slots*	6	5/2
Power Supply	135W	130W
Number Floppy Drives	1	1
Type of Floppy Drive	Half	Half
Device Slots Available	3	3
Serial Port	Option	Option
Parallel Port	Included	Included
Hard Disk	Option	Option
8087 Socket	Yes (80287)	Yes

*Count shown does not include required video adapter.

The Tandy 3000 HL features a built-in real-time clock with battery backup for automatic posting of jobs, process control and other time-sensitive applications. A standard parallel printer port connection is included as well.

SPECIFICATIONS

Microprocessor: Intel 80286 processor with 16-bit data path. Switchable clock speed, 4/8 MHz. Object code compatible with 8086/8088. Real-time clock with battery backup.

Operating System: Optional Microsoft® MS-DOS 3.2 with BASIC.

Memory: 512K RAM with parity. By using the expansion slots, memory is expandable to 4 megabytes. Includes powerup diagnostics. Sound included.

Keyboard: 84-key sculptured, including numeric entry keypad. Special keys include ESCape, Num Lock, Alt, Ctrl, Caps Lock, Prt Sc, Sys Reg, Scroll Lock, Up, Down, Right and Left arrows. Ten programmable Special-Function Keys. Retractable legs, 6-ft coiled cable.

Video Display: Optional high-resolution, non-glare, non-interlaced 12″ monochrome (green) or 14″ color monitor. 80 or 40 characters per line by 25 lines. Optional high-resolution 80 characters by 50 lines, 640 × 200 graphics and 320 × 200 graphics in 16 colors, or 640 × 200 graphics in 4 colors.

Disk Drives: Built-in thin-line 5¹/₄″ floppy can read 360K formats. Disk storage is expandable to include two floppy disk drives and one hard disk drive, or one floppy disk and two internal hard disk drives. Total internal storage capacity can exceed 40 megabytes.

Internal Expansion: Seven plug-in card slots, including three 16-bit slots and four PC/XT-compatible slots. Optional 80287 Math Co-Processor can be added.

External Connections: Standard parallel printer port.

Dimensions: 6¹/₈ × 17 × 15¹/₂″.

Weight: 32 pounds.

Power Requirements: 120VAC, 60 HZ. UL listed.

FIGURE 2–2 Machine parameters may be an important part of sales literature. (Copyright 1986 by the Radio Shack division of Tandy Corporation, FC-259. Used with permission)

costs. There is nothing to be gained by complimenting the purchaser for buying your company's machine after the fact. After the sale, the wise consumer wants information about how to operate the machine that he or she purchased. This is the time for the consumer to see whether the presale advertisements and news releases have painted an accurate picture of what the machine can and cannot do. You do not need, for example, to patronize consumers by complimenting them for making a wise purchase:

> Congratulations, you have just bought the world's greatest laser-operated machine.

You do not need to continue the comparison of the machine with the marketplace competition that your company's presale literature develops:

> The maintenance schedule points of this machine are the lowest in the industry. You will find that it is easier to maintain than Brand X.

You do not need to repeat or summarize the major technological advancements of the machine:

> The new Teflon-coated metal-on-metal surface contact and the latest in electronic controls make this machine a breeze to operate.

You do not need to put your machine in a historical perspective:

> Since 1920 this machine design has been an economical and efficient commercial machine.

Your job as a technical writer must be to explain, without excess words and ideas, exactly how to control the machine safely in order to get the expected operating results. You can assume that consumers will be offended by material that wastes their time and detracts from their concentration upon the business at hand—to start and operate the machine as safely and efficiently as possible.

UNDERSTAND THE ORIGINS OF MASS CONSUMERISM

To understand the scope and distribution of the modern mass-consumer audience, you can look at the relationship between consumerism and how our machines are made. As an example, look at the development of the mass-consumer market in the United States from the late eighteenth century to the present. This is a revolutionary period in industry, not only for our own country but also the entire Western world.

FIGURE 2–3 Machines were made with hand tools during the first 300 years of United States history. (Courtesy Deere & Company)

Two basic production methods have been used during this time to make machines. First, machines can be handcrafted one at a time. Every machine that is made this way is unique. Very few companies make machines for the modern worldwide market in this manner because labor is costly. A main problem with this method is that replacement parts for those that break or are defective must be remade specifically for that machine. Although each machine that is made this way is an original, the time that the worker spends making the machine adds to the cost. Machines and other products primarily were made this way for the first 300 years of United States economic history (see Figure 2–3).

Outline the Development of Mass Production

Often the consumer during colonial times was the person who designed, made, and marketed the machine. Although there may have been a feeling of accomplishment in making the entire machine, the consumer markets, if they existed, were usually limited to the immediate community of the craftsperson. Volume of sales for handcrafted machines is always limited more by lack of, or supply of, the product than by the demand. Goodwill between the machine maker and the consumer was easier to maintain. If the consumer had trouble operating the machine, the craftsperson was close enough to explain the proper standard operating procedures.

Interchangeability. The second manufacturing process is by the assembly line and automation. Machines built in factories, which move the machine from work station to work station, can be made more rapidly and economically. Mass production is made possible because these machines are made from interchangeable parts.

The roots of mass production began as an idea by French military personnel and arms suppliers in the mid-eighteenth century. Around 1765, Jean Baptiste de Gribeauval proposed to standardize weapon design. He wanted to switch the parts of military weapons as easily as a soldier could be switched in the ranks. His idea was to build single pieces of a musket with close enough tolerances that they could be used interchangeably. This meant that many musket barrels could be made at one time by one person, while other people concentrated on making other parts.

The idea was later brought to the United States by the ambassador to France, Thomas Jefferson. Although Jefferson made these concepts of interchangeability known to the leaders of the country, it took the vocal support of Eli Whitney to create a general interest in the idea. In 1798 the maritime disputes with France forced the young United States government to prepare for war. Desperately needing arms, government leaders granted Whitney a contract to produce 10,000 muskets. Whitney sold his idea to the government on the condition that he would make them with the aid of the new "interchangeability of parts" procedure. Although it is generally believed that Whitney failed to produce an acceptable musket with the procedure, he did create a great deal of publicity for the manufacturing technique.

Others were more successful. For example, Simeon North, contracted to produce 500 muskets in a similar way. He noted that producing muskets with the new methods saved one-fourth the time that it would have taken one person to make one musket. He also proposed to use special-purpose machines to produce the interchangeable parts. The marketplace began to change. The role of the craftsperson who had the responsibility to make a machine was changing. If a machine no longer worked properly, it could now be blamed on the anonymous manufacturing process.

The American System. The process of making many machines from interchangeable parts manufactured by specialized machines became fine-tuned in federal armories in the early nineteenth century. The manufacturing techniques became known as "The American System" to the machine makers of England and Europe. The manufacturing technology that was the foundation of this American System was developed in a number of areas, including the sewing machine, woodworking, agricultural reaper, and the bicycle industries. it culminated and was further refined in the automobile manufacturing and related industries of the early twentieth century. American automobile manufacturers such as Henry Ford created mass-production techniques from the machine technology that was developed in the age of the "The American System" (see Figure 2-4).

FIGURE 2-4 Henry Ford made his 1896 Quadricycle with specialized manufacturing machines. (Courtesy The Henry Ford Museum, Edison Institute)

The success of Henry Ford and his staff, for example, was their ability to organize and create new fabricating machines and technology as the situation demanded, to create a material flow through the factory, and to manufacture what was then highly precision interchangeable parts. The goal of the Ford Motor Company was to make a machine, the Model-T or "Tin Lizzie" car, that everyone could afford. The breakthrough for Henry Ford was the idea of a moving assembly line (see Figure 2–5). Until this development, machine operators were able to make the interchangeable parts faster than they were required for assembly. Robert Lacey documents this breakthrough. In the spring of 1913, Ford and his staff pioneered and developed the first moving assembly line. The concept was first used on the car magneto assembly line at the Highland Park, Michigan, plant.

> Workers lined up side by side, facing the flywheels. . . . Until this date the magneto assemblers had worked at benches with a complete range of magnets, bolts, and clamps, each of them fitting together some thirty-five to forty complete flywheel magneto assemblies in the course of a nine-hour day.[1]

No longer did the skilled workers assemble complete parts of the car. The car "moved" through the factory on an assembly line. Time that was necessary to produce one magneto assembly was reduced from 15 minutes to 13 minutes 10 seconds. With these types of technological advancements in the manufacturing process, Ford lowered the price of his automobile from $825 in 1908 to $345 in August 1916. With an increase and efficiency in the production process, the number of machines made is increased, while the price is usually lowered.

Fordism and mass production. The term "The American System" was replaced by the term "Fordism" after the company and the individual that had revolutionized the production process. In 1925, the American editor of the *Encyclopedia Britannica* asked Henry Ford to define the process that he had developed. In a release issued by the Ford Motor Company, the term "mass production" was explained. Ever since the concept was developed, it has defined the consumer more than the product. Rapidly produced machines in quantity became available to more people at more economic levels than ever before.

In 1924, the ten millionth Model-T was built. The average cost for a Ford automobile was held at $350. The consumer could buy the automobile in one of four basic body styles. Henry Ford also boasted that purchasers could buy the automobile in any color, as long as it was black. Mass production lowered costs and standardized the machine so that there were many clones of the same standard machine design.

There are, however, problems with this type of mass production. David Hounshell notes that even

[1]Robert Lacey, *Ford: The Men and the Machine* (Boston: Little, Brown, 1986), p. 108.

FIGURE 2–5 The first moving assembly line was the magneto line at Ford. (Courtesy The Henry Ford Museum, Edison Institute)

Ford's new concepts were soon outdated. The Model-T design was rigid. There was no regard for options that could be added in order to individualize the automobile. In 1925, General Motors developed the idea of "flexible mass production."[2] The GM concept was to offer a relatively inexpensive automobile, and then make different automobile styles with more options available as the individual "stepped up" into another price level.

Automation attained by robots and computer-aided manufacturing techniques are used to get this result today. With automated material-handling systems and advanced fabrication techniques, a manufacturer can customize the machine as it is being put together on the assembly line.

Assembly line dilemma. The dilemma that faced the Ford Motor Company in 1926 still faces manufacturers today. They must decide whether the consumer wants a machine, such as an automobile, that is built in one style continuously, or will the consumer want options and frequent machine changes. Without change, more automobiles at a lower cost can be made. Changes to an automated mass-production process, or a flexible production system, usually mean fewer machines are produced that sell at a higher price.

Ever since "The American System" manufacturers revolutionized the manufacturing process by making the parts of rifles with machines, the mass-consumer market has benefited from mass production. In the process, the craftsperson has become obsolete and has been replaced by the unskilled or semiskilled factory worker. Because the trend developed rather slowly, our society had time to adapt to the interchangeability and "sameness" of products and of jobs. The craftspeople were for the most part assimilated into the general work force and the factory system.

Mass production is designed to supply machines and enable services to be made that the consumer demands. Imagine how frustrated consumers would be if they were forced to wait for a craftsperson to hand-fit a part on their computer that malfunctioned. Imagine how limited the choice of machines and appliances would be if each one was made by hand with the basic techniques that the colonists used in order to make their crude machines (see Figure 2-6 and Figure 2-7).

The lessons that can be gained by studying the history of consumerism during this time is that:

The machine design must be dynamic and able to change with the demands of the consumer.

The consumer has ultimate control of a free market.

The technical writer must design flexible instructions that can be easily revised, no matter how standard or rigid the design of the machine appears.

The success of mass production depends upon how well the manufacturing company predicts its audience and fulfills its technological needs.

[2]David Hounshell, *From the American System to Mass Production 1800-1932* (Baltimore: Johns Hopkins University Press, 1984), p. 264.

FIGURE 2-6 An assembly shop floor for tractors in the early beginnings of mass production. (Courtesy Deere & Company)

FIGURE 2–7 A modern assembly shop floor is efficient and highly productive. (Photo courtesy of Hewlett-Packard Company)

Define the Mass-Consumer Audience

The mass-consumer audience is potentially anyone who buys the mass-produced machine for an economic or personal need. Often when technical writing is defined, the definition given may attempt to separate the industrial consumer who operates a machine in a factory and the home or domestic consumer. This book treats both types of machine users as the "consumer." The mass-consumer audience and the industrial machine operator both require and deserve the best possible machine support literature that can be developed.

There are a few general observations that we can make about the people who may purchase the products of our mass-production processes and about our consumer-based economy:

> Both men and women may use a product. Traditional roles and strict divisions of labor are no longer followed by many societies.
>
> The mass-consumer audience is a democratic concept. Machines can pass the social boundaries of race, education, religion, and country. Most products will eventually compete in a worldwide market. The mass-consumer audience may contain cross-segments of a worldwide population.
>
> Mass-production techniques like computer-aided design and computer-aided manufacturing can change or modify an existing machine in a very short time in order to keep up with shifting markets and consumer demands. Changes in the machine means that the technical writer must continually reevaluate the mass-consumer audience and update the technical operation instruction (see Figure 2–8).
>
> Technology is not an uncontrollable beast that develops on its own. Technology is the result of an attempt to satisfy a real or perceived need by the consumer. The sale of some mass-produced machines may number in the millions. Often only demand and the ability to supply that demand limit the potential sales and distribution of a product.

The consumer is both the result and cause of our modern mass-production techniques. It is the consumers' real or perceived need to obtain machines for health, happiness, and survival that drives our economy. Modern production relies on the participation of the consumer in the development and distribution of new technology. How well companies can transfer technology to the customer in the form of technical literature can have a great effect on how smoothly our economy runs.

Divide the Mass-Consumer Audience Into Categories

George A. Peters, in his book *Product Liability and Safety,* states:

> Where there are dangers associated with the use of a product, there is an obligation to provide reasonable *product instructions* suitable

FIGURE 2–8 Computer-aided manufacturing has fine-tuned flexible mass-production procedures. (Reproduced with permission of AT&T)

> for the prospective user's age, intelligence, education, literacy and language disabilities, competency, aptitudes, special training in the tasks involved, general knowledge, and understanding of the risks likely to be encountered.[3]

This may seem like an impossible task if you look at the millions of people who may comprise your audience in a worldwide marketplace.

If you think about the makeup of the mass-consumer audience, you may become completely perplexed with the seemingly simple question, "How can anyone communicate with all of the people who can possibly buy the machine?" To find out if there is an answer to this question, divide your audience into categories in order to clarify the question. There are two basic divisions that you can use to help you simplify your instruction audience analysis. They are:

The homogeneous mass-consumer group.
The heterogeneous mass-consumer group.

Understand the Needs of the Homogeneous Audience

There are two absolutes that you can answer about the mass-consumer audience. Either you can measure the ability of your audience to interpret your media, or you cannot completely measure their abilities. The homogeneous consumer group is the ideal audience because these people *can* be measured for their communication abilities. This audience is a known quantity because you can concretely define their:

Comprehension of basic technology or specific machine technology.
Special training and education.
Ability to understand specific technical words associated with the machine.

With this information, you can make knowledgeable assumptions about the information these people must know in order to operate a machine safely and efficiently. You also can use these facts to determine the word choice, physical "layout," and type of media that you will use to present your information to this stratified mass-consumer group.

[3]George A. Peters, *Product Liability and Safety* (Washington, D.C.: Coiner Publications, Ltd., 1971), p. 13.

FIGURE 2–9 A homogeneous audience shares the same educational and intellectual backgrounds. (Reproduced with permission of AT&T)

The specialized audience. This type of audience typically may be created when a product is designed to do a highly specialized job and is produced in limited quantities for a well-defined market. An example of this situation is a laser operating machine that will be sold to surgeons in order to perform a highly technical type of operation in hospitals. The technical writer can confidently assume that certain words and procedures will be known by this closed group of surgeons or "experts" because of their past training and experience. The technical writer can assume that the operating tool will not be sold in a neighborhood hardware store (see Figure 2–9).

Often the technical writer who writes for a homogeneous audience is a professional with the same education, training, and skill as the consumer. Unfortunately, it is easy for technical writers in this example to assume that the audience knows as much as they do about the technology they are explaining. Machine experts who write about the machine must always put themselves in the role of the person who is using the machine for the first time.

No matter who is considered to be in the potential audience, the technical writer must strive to develop the best technical operation that is possible with the resources and information available. When a homogeneous audience situation occurs, you may be able to work with "standards." Standards are attempts by industry or government agencies to help the technical writer create instructions for a limited or highly defined audience.

Symbol standards. There are two types of standards that are important. The first type can be used no matter who your audience is as long as the audience knows what standard you are referring to.

These standards are listings of the measurement and communication symbol standards that can be used to ensure everyone is using the same definition of a word, measurement, or symbol. Matthew Lesko gives a listing of a few sources for standards.[4] Some of the listings are:

- The American National Standards Institute (ANSI)
- The American Society for Testing and Materials (ASTM)
- Association for Computing Machinery (ACM)
- National Bureau of Standards (NBS)
- The National Measurement Laboratory
- Center for Basic Standards
- The Institute for Computer Sciences and Technology (ICST)

Standards include the SI metric equivalent conversion factors that are used to convert standard foot-pound measurements to metric equivalents. Standards also include the USAS Y32.10—1967 "Graphics Symbols for Fluid Power Diagrams," and other listings of schematic symbols. International Standard Organization (ISO) standards are slowly becoming an important factor in worldwide trade. These symbols are a direct parallel to the American National Standards that have been developed in North America.

Manual guidelines. The second type of standard is specialized for the technical writer. This includes *specific* guidelines and rules that must be followed by the technical writer when writing the instruction for a specific audience. An example of a standard that gives the parameters of how an instruction must be written for this type of audience is the U.S. Army MIL-M-63036A (TM) standard entitled *Military Specification Manuals, Technical: Operator's, Preparation Of.* The information is designed to help technical writing agencies and government technical writers write to varied military audiences.

This standard is used to help the technical writer produce operation manuals and technical instructions for all equipment except aircraft in a format of as few words and pictures as necessary. It covers everything from writing style to the size of the completed manuals and instructions, to preparing pictures. It is a complete guideline that the technical writer can use to write and develop instructions. Many industries are too diverse in the equipment and audience they service to create guidelines as specific as the military has prepared.

A similar standard, MIL-M-63038B (TM), covers the specifications or requirements that must be met for someone who writes about aviation unit maintenance and aviation intermediate maintenance. Both standards have been developed by the Department of Defense to explain how to operate the particular machine for two specifically identified groups or audiences. Both groups require different information in order to operate and maintain the machine on a daily basis.

Certain nongovernment industry standards that also may be under development at this time accomplish the same objective as the military standards program. There has been a trend toward standardization of symbols and drawing techniques used on engineering drawings and other types of interindustrial communication. This trend may continue and eventually cover the area of technical writing. The problem confronting any governing industry organization that attempts to create technical writing standards is that the members must first be able to create a specific audience profile before they can develop concrete guidelines.

In conclusion, the homogeneous audience may not need complete *basic* information about the machine. You can assume, for example, that the surgeon knows basic operating techniques, such as preoperation, and other operating procedures that are common knowledge to people in their field of specialty. The surgeon *needs to know* only the procedures that are unique to the new laser operating machine. It is easier to create standard manual guidelines for a homogeneous audience that is well defined.

Know the Needs of a Heterogeneous Audience

The opposite type of audience from the homogeneous, or stratified group of similarly trained or educated people, is the *unidentifiable* group of consumers or heterogeneous audience. They are hard to identify completely because mass-production techniques are used to create many machines. By building quantity, the cost is lowered. As the cost lowers, the machine becomes available to more consumers at different economic and social levels.

Because of the potential size of some audiences, you will not be able to determine if all of these people know the basics of how to operate a certain

[4]Matthew Lesko, *New Tech Sourcebook: A Directory to Finding Answers in Today's Technology-Oriented World* (New York: Harper & Row, Publishers, 1986), p. 554.

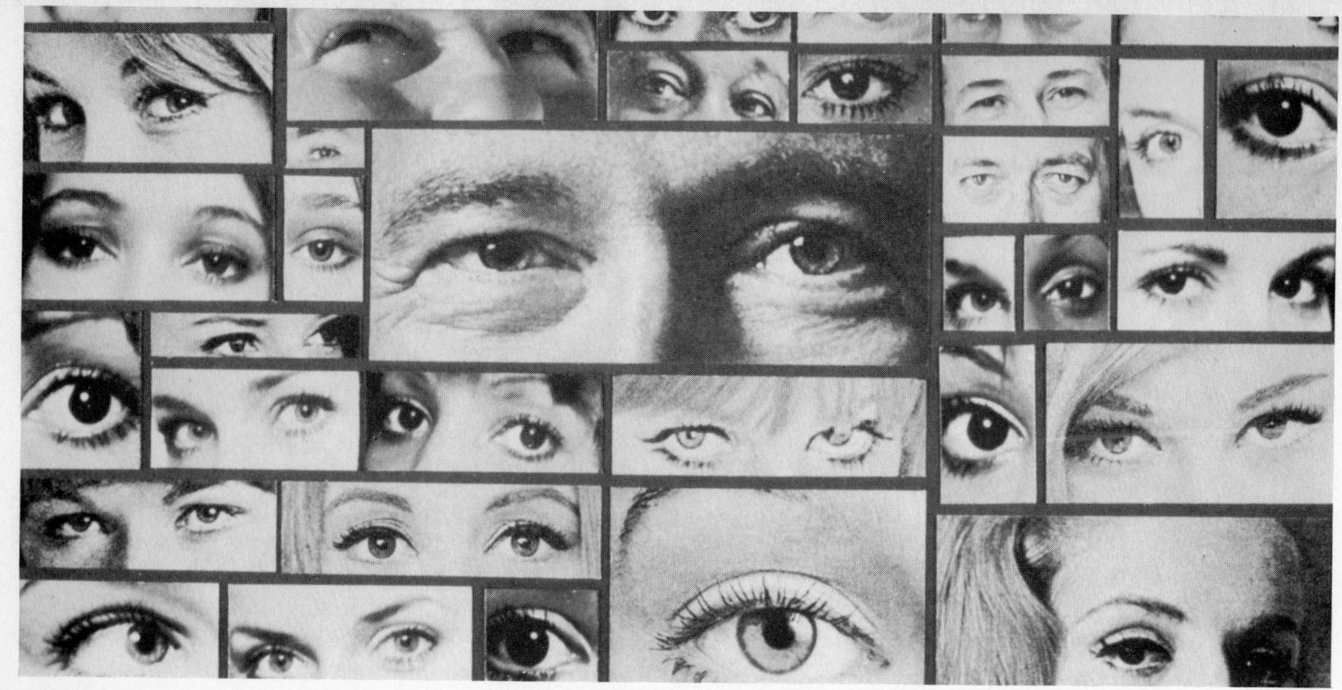

FIGURE 2–10 It is difficult to create a complete mental picture of the typical heterogeneous consumer. (Courtesy Deere & Company)

type of machine. You cannot determine concretely whether they are familiar with certain technical words. You also cannot determine the education or training of the people in this classification, which is so large that measuring each potential consumer's communication abilities is impossible. You will find that it is difficult to create a complete image of the typical person in this audience as you can with the homogeneous audience (see Figure 2–10).

Most manufacturing companies use a domestic market in order to build the sales for their products. They develop the domestic market first. This helps to reduce the risk of overproduction and product failure. Companies tend to have more knowledge of the domestic market than they do of export markets.

The domestic strategy. A good example of this philosophy is the Japanese attack on worldwide markets. If you evaluate the success of Japanese products in international trade, then you will see that there is usually a strong domestic market in Japan for the machines that country exports. Following the strategy of Miyamoto Musashi, a seventeenth-century Japanese swordsman who wrote *A Book of Five Rings*,[5] many Japanese marketing plans call for developing the machine at close hand and fine-tuning it before it is allowed to become a world-class competitor. The Japanese approach the building and support of a machine by creating a militaristic strategy for development and marketing. As a technical writer who may write for a worldwide audience, you can begin to develop your writing strategy by examining the domestic heterogeneous audience. From the conclusions that are drawn from this microcosm of consumers, you can project your analysis findings to the larger worldwide audience.

First, Analyze the Domestic Audience

The reason for the difficulty in measuring the typical mass consumer in the United States is that we are a nation of individuals. It is difficult to create a concrete character sketch from such an immense group of people. Whereas the writer who is able to create a complete character sketch of the small specific audience, the technical writer who writes for the "typical" buyer of a mass-produced machine faces a difficult problem.

The surgeon, for example, who can readily comprehend the new laser operating machine, may not know enough about consumer products to start a lawnmower safely and efficiently. Some technical writers mistakenly rely on their company's marketing studies and projections in order to define a heterogeneous group like the domestic mass-consumer audience.

Untold amounts of money are spent each day by companies to predict what the consumer wants or needs. Target marketing for specific markets, demographics, and strategic marketing practices attempt to uncover the "typical" consumer and his or her buying habits. You can get an appreciation for the immense difficulty encountered by companies to determine where their markets are developing by

[5]Miyamoto Musashi, *A Book of Five Rings* (New York: The Overlook Press, 1974).

reading some of the magazines aimed at the professional marketer.

Consumer labeling. Trends, lifestyles, populations, and consumer buying patterns can be projected with varying degrees of accuracy. Marketing projections can help predict audiences; however, the communication ability of the heterogeneous mass-consumer audience can never be specifically measured by modern marketing procedures. There is, however, no resource that can tell you, the technical writer, exactly how much your potential consumer reader knows or his or her communication abilities. Further, not everyone in the potential audience may need or require the machine that you are writing instructions to support. You cannot rely on isolating certain pockets or groups of potential customers. You cannot write to conveniently labeled groups such as "yuppies," "upwardly mobile," "elderly," or other abstract phrases.

The *Statistical Abstract of the United States*, the *Oxford Economic Atlas of the World*, the *Information Please Almanac*, and other statistical encyclopedias can tell you everything measurable about the habits of the consumer in a given year. The United States Bureau of the Census can tell you the income, the number of cars, the age, the number of televisions, the number of children, and the religious preference of the average U.S. consumer. Statistics once recorded are history. The machine is a dynamic and developing concept.

In order to develop a better understanding of your audience, look at the technological environment that the U.S. consumer has evolved in. First of all, anthropologists have told us that humans have been around for over a million years. The earliest inhabitants were nomads, cave dwellers, and hunters whose technology was limited to stone tools and simple armaments of war and survival. Human existence was not much different from the herd animals that these early people hunted.

About 25,000 years ago, humans entered into what has been called the "Agricultural Revolution." Farming technology began to develop into the tools that were used to cultivate and plant the land. Human, animal, and wind were the basic sources of power for these early people. The agricultural technology passed virtually unchanged through the great Sumerian, Hittite, and ancient Egyptian civilizations to nineteenth-century Europe. Plows drawn by animals and scythes wielded by humans illustrate the farmers' technology during this time. Western civilization was introduced to a revolutionary period that fueled the fire of technology about 150 years ago. The factory system developed and the Industrial Revolution began.

Forty years ago, the Nuclear Age was born near Alamogordo, New Mexico, when the first atomic bomb test was successfully completed. Thirty years ago, the space race began when the first space capsule was launched into orbit outside of Earth's atmosphere.

Our "Personal Computer Age" is only about ten years old. The middle-aged consumer has lived through at least two wars, the placid decade of the latter half of the 1950s, the extraterrestrial 1960s, the economic turmoil of the 1970s, and the social and economic upheaval and migration of the 1980s.

Change is the only constant. The only thing constant in our society is "change." Western technology is indeed picking up speed. High-tech companies are researching the possibilities of a fifth-generation computer that can think and adapt to the environment. Scientists at the National Aeronautical and Space Administration (NASA) are preparing to launch a permanent space station in the near future.

The "average" U.S. consumer is living in a miraculous world that offers many benefits. However, for many people it is an overwhelming time. Technology for these people plays both the part of saviour and destroyer. People who are frustrated with their inability to understand technology may blame our world situation upon uncontrolled technology.

The consumer also is living in an extremely technologically complex time. Thus, the technical operation instruction must be looked upon as a means to soften the great consumer technology "shock" that can influence and completely change the lives of the American consumer literally overnight. It is indeed ironic, then, when the consumer is given a technical operation manual that is unorganized and confusing. Many times, the frustrations of the consumer over technology that is not properly supported with instructions are projected upon the machine the consumer is trying to start or operate.

Because the United States is a nation of individuals, we can measure them only through observed qualities, such as how much money they make and where they live. However, these physical or measurable qualities are not good guidelines. As a technical writer, you must be able to gauge and interface with the mind of the consumer if you are going to write usable instructions. Read what they read; listen to them every chance you get; and ask the consumers who use your machine to evaluate your instructions whenever you get the opportunity.

Understand the Export Audience

A homogeneous stratified audience can go beyond the boundaries of one country. The surgeon in Germany, for example, may find the new laser operating machine used in the previous example as necessary as the American surgeon. If you know that the German surgeon has the same background and ability, then that person can be considered a part of your measurable homogeneous audience.

Countries that may import your company's machine for a heterogeneous consumer market present the same problems to you as does the domestic consumer market. There are, however, special problems that are associated with writing technical operation instructions for a multinational company. First of all, there is the problem of translation, which is an art and not an exact science. Because it is not an exact science, the translation of technical concepts adds more problems to the process of communication. How you physically design your information, what you write, and how much you write become even more important when a translator begins the process of transmitting technology to another culture.

Worldwide consumer facts. You must be aware of the following facts and comparisons when you write.

About 27 million U.S. adults are illiterate according to the Coalition for Literacy. Another 30 million may be functionally illiterate and cannot understand what they read. Although there are a few countries with higher literacy rates among adults, the majority of the world's 160-plus countries have worse literacy rates than the United States (see Figure 2-11).

An estimate that is often made states that during a translation, 10 percent to 15 percent of the meaning of the words can be lost in an instruction if the material is poorly written.

Only an unskilled translator will translate a text word for word. Word-for-word translations cannot capture the meaning of a sentence. There is no one-to-one equivalent set of words for a sentence that is translated.

Not every consumer reads from left to right. The Japanese consumer can also read from right to left or up to down. Translated instructions must be physically designed with this concept in mind.

Translations into languages other than English will expand. In general, a Spanish translation may expand 15 percent. If you have a manual that contains 200 pages of English text, then you may have 230 to 240 translated pages.

Translations are costly. Most translators charge by the page. Extra pages cost your company extra money. Write instructions as clearly and as briefly as possible. If you write a 200-page manual that is hard to translate into Spanish, then you may have 260 translated pages instead of 230 pages.

FIGURE 2-11 About one-third of our population is unable to read and understand a simple instruction. (Courtesy Dover Publications, Inc.)

There are 13 languages with over 50 million *speakers.* They are Chinese, Hindustani, Spanish, Russian, German, Japanese, Indonesian, French, Italian, Portuguese, Arabic, Bengali, and English.

DEVELOP GENERAL WRITING GUIDELINES

In order to write technical operation manuals for both an export and domestic market, you must use short, concrete words that are easy to understand and translate. Terms should be consistent throughout the instruction. Do not call a machine guard a protective shield if you have previously called it a machine guard. By keeping words short, consistent, and concrete, and by constructing short, well-made sentences, you will increase the accuracy of a translation.

You must avoid idioms. Idioms and slang do not translate. For example, there are many dialects of Spanish, because it is a language that is spoken from Baja California to the southern tip of South America. It is a primary language in the Caribbean Islands and Spain. Spanish is divided into many dialects. Idioms and slang differ in the many social cultures where Spanish is spoken.

Within our own society, you can see that idioms and slang are an extremely localized or regionalized phenomenon. As a technical writer, you must approach technical writing as a sublime universal type of communication that transcends the geographic boundaries of dialect and society.

This general principle is true for both the heterogeneous import and export market and the homogeneous audience. *As a technical writer, you must remember that a person's education and training do not necessarily raise the reading and comprehension level of every member in an audience. Machine knowledge and basic technical knowledge are not the same as reading and comprehension ability.*

Write to the Lowest Common Denominator

As a technical writer you must concentrate on writing to the *lowest common denominator* of both the heterogeneous and homogeneous audience. You must strive to make your technical operation instruction usable to the person in your audience who has the least technical machine knowledge as well as communication ability. This person is the last one who can use your instruction to operate the machine safely and efficiently. If you write to the "average" consumer, then you will not be successful at communicating with the bottom half of your audience because you effectively eliminate everyone who has below-average intelligence of the machine technology and communication skill.

As a technical writer you also must prepare yourself for the reality of communication. You may not be able to explain the machine to *100 percent of the audience,* even though that must always be your goal. Some consumers, because of physical and mental handicaps or other barriers, may not be able to read and understand even the simplest instruction. An alternative to the written instruction may be necessary to increase the percent of understanding in some mass-consumer audiences.

This concept applies to both the identified and analyzed audience as well as the unidentifiable audience. Before you begin writing the instruction, you should develop a mental picture of the person who has the least ability and technical knowledge and write to that person.

Avoid Mirror-Image Writing

O. Henry, The American short-story writer at the turn of this century, is famous for his plots that appeal to a universal audience. In one of his stories, a young man becomes intrigued by the phrase "a man about town." The young man, who hates mysteries, is obsessed with the idea of meeting one of these men face to face in order to find out once and for all what type of person they are. The hero begins his quest by questioning everyone he meets.

Among the colorful characters of the time that he talks to are a reporter, a critic, and a waiter. Even though each has his own mental picture, none of these people can point out an average "man about town." After hours of searching, the young man is caught up in the euphoria of the sights and sounds around him, and he absentmindedly steps in front of a horseless carriage as he crosses a street.

As the young man recovers in a hospital the next day, a doctor hands him the morning paper so that he can read how the accident occurred. To his amazement, the young man notices the last lines of the newspaper article that reports his accident:

> . . . his injuries were not serious. He appeared to be a typical Man About Town.[6]

Sometimes the technical writer comes to the same conclusion as the "man about town." As the

[6]O. Henry, *The Complete Works of O. Henry* (New York: Doubleday, 1953), p. 37.

technical writer searches for the average mass-consumer audience, the writer mistakenly *assumes* that the average consumer has the same knowledge about the machine as does the writer. Technical writers see themselves as the embodiment of the average "mass consumer."

Mirror-writing is a barrier to communicating technical information. The technical writer during the course of researching the machine systems will gain a certain amount of technical knowledge about the machine. When the technical writer views the homogeneous or heterogeneous consumer audience as a mirror-image of oneself, the writer can forget that the consumer may be using the particular machine for the first time. The technical writer must become the machine *user* and not the expert technical writer.

This type of writing will create a gap between the technical writer and the consumer that can never be bridged. If a writer is going to communicate ideas clearly and accurately, he or she must assume more of a reader's role than a writer's role. The technical writer must not appear to be a technical writer. The writer must become the audience. The following model of communication demonstrates this concept.

CONSTRUCT A TECHNICAL COMMUNICATION MODEL

To understand both the technical communication process and the difficulty in technology transfer between yourself and the consumer, construct a *technical communication model.* A classical model, for example, includes three dimensions.

First, in order to communicate technical information, there must be a technical writer or *sender.* Second, there must be a *media,* which in industry is primarily a printed technical operation instruction. Finally, there must be a *receiver,* or the mass-consumer audience, who can translate into information the ideas and concepts that are in the media (see Figure 2-12).

No matter what communication model you decide to use, you will find that there is always an unknown space that can add confusion to the most basic attempts to communicate. It is the "gray area" of space between the media and the perception of the consumer. It is an area of symbol (word and picture) interpretation by the consumer. How well you understand the use of this space will determine whether or not your technical communication is successful.

If you cannot use symbols—words and pictures—that are understandable to your consumer audience, then you will fail to communicate. This communication failure, according to scholars such as Ernst Cassirer, is the beginning of mythologies, legends, and political systems in our everyday attempts to communicate. These imagined beliefs have developed from communication problems and misinterpretations by the receiver. For the consumer, however, this misunderstanding can become an immediate matter of life and death or of physical harm caused by misuse of the machine.

Feedback. Like pieces of a jigsaw puzzle, the technical writer, media, and consumer must interface. The consumer must have the ability to translate the media exactly as the technical writer has presented it. The only valid way of determining if your instructions are on the same level as the receiver/consumer is constantly to gather and interpret feedback from the people who are actually using the instruction (see Figure 2-13).

Use Feedback to Evaluate Instructions

A valid communication model also must include a method for the accumulation and interpretation of feedback. *Feedback* is the reaction of your consumer/receiver to your instruction. Feedback tells you if you have properly supported the machine. By using feedback to help you evaluate the effectiveness of your instruction you can find out:

- If the consumer needs more information in order to operate the machine.
- If the instruction is as easy to read and understand as it can be.
- If the instruction covers all necessary safety information.
- What percentage of the potential consumer audience can use the instruction.

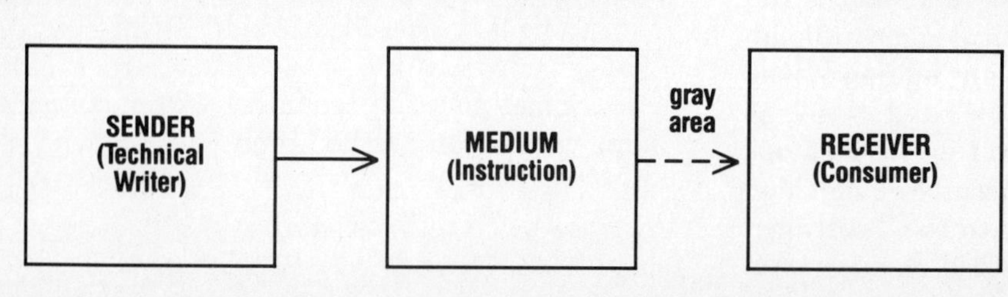

FIGURE 2-12 A communication model must have three basic elements.

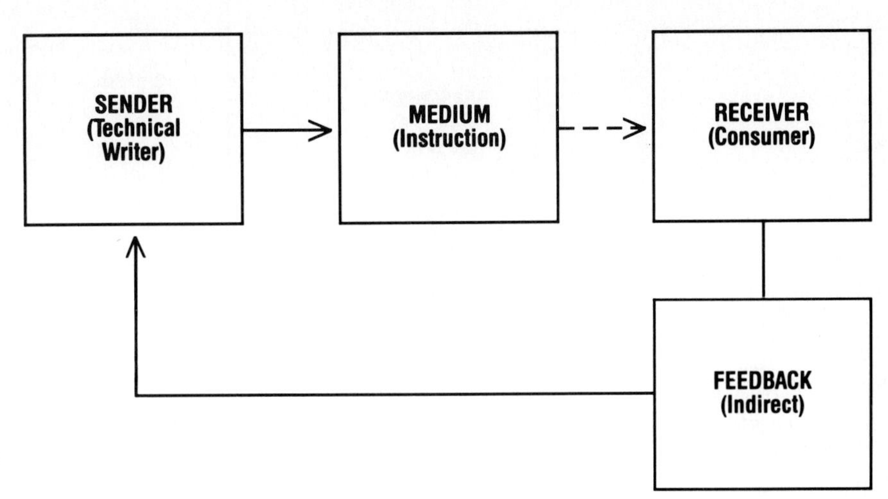

FIGURE 2–13 Feedback helps you evaluate.

There are two basic forms of feedback: direct and indirect. A conversation is an example of a *direct* form of feedback. In an everyday conversation, feedback comes in the form of questions or answers to questions. Feedback helps clarify ideas and helps the sender discover whether or not the receiver knows exactly what the sender is trying to communicate.

> Are you going to the dance tonight?
>
> *(Feedback)* What dance are you talking about?
>
> The freshman and sophomore mixer at the gymnasium.
>
> *(Feedback)* Oh, that dance. I had forgotten about it. Yes I am.
>
> What are you going to wear . . . ?

Direct feedback also can be gathered by watching someone's body language. Physical reactions, such as gestures and other animated movements, can be as important as verbal or written forms of feedback. Feedback is any form of indicator that tells you whether or not your information is understood by the receiver. These types of information are personal forms of communication. The sender is in direct contact with the receiver and can evaluate actions and reactions as they read the instruction.

Direct positive feedback. There are two methods that can be used to help ensure that your instructions are usable and complete. After you have completed the instruction and have "laid it out" or put it in a final form that the consumer will use, you must allow key personnel, or "machine experts," to read and comment on your work. You also can "desk-check" the instruction by monitoring a potential microcosm of users as they attempt to read the instruction and operate the machine.

The review network. The machine experts can help you determine the *accuracy* of your information. From their comments you can determine if the information is complete and up-to-date. Machine experts also can tell you about immediate modification to the machine that you may have missed.

A tendency of some technical writers is to include too many machine experts in the review process. You must limit the review to include such key machine experts as the company attorney, engineer, testing technician, and others who know the *most* about the particular machine you are describing. It is important that these experts become involved in the review process of your communication exercise for their own professional well-being. An unusable instruction can ruin the efficiency and safety of their well-designed machine.

The role-play feedback. Desk-checks are made on computer programs before they are sent through the expensive computer operation or applied to a specific job. Along with obtaining feedback from the machine experts, you can desk-check the instruction by selecting a sample audience that has no particular familiarity with the machine. If the machine is designed to be used every day, you may ask fellow workers or people in your community to operate the machine by following your instructions. As you observe firsthand their physical reactions, or answer their questions, you can evaluate the ability of the instruction to communicate (see Figure 2–14).

From this study of a microcosm of people who are acting out the consumer's role, you can make assumptions about the macrocosm of a homogeneous or heterogeneous audience. After this evaluation, some companies allow their machines to be selectively distributed to people such as those who potentially may buy the product. For example, a limited

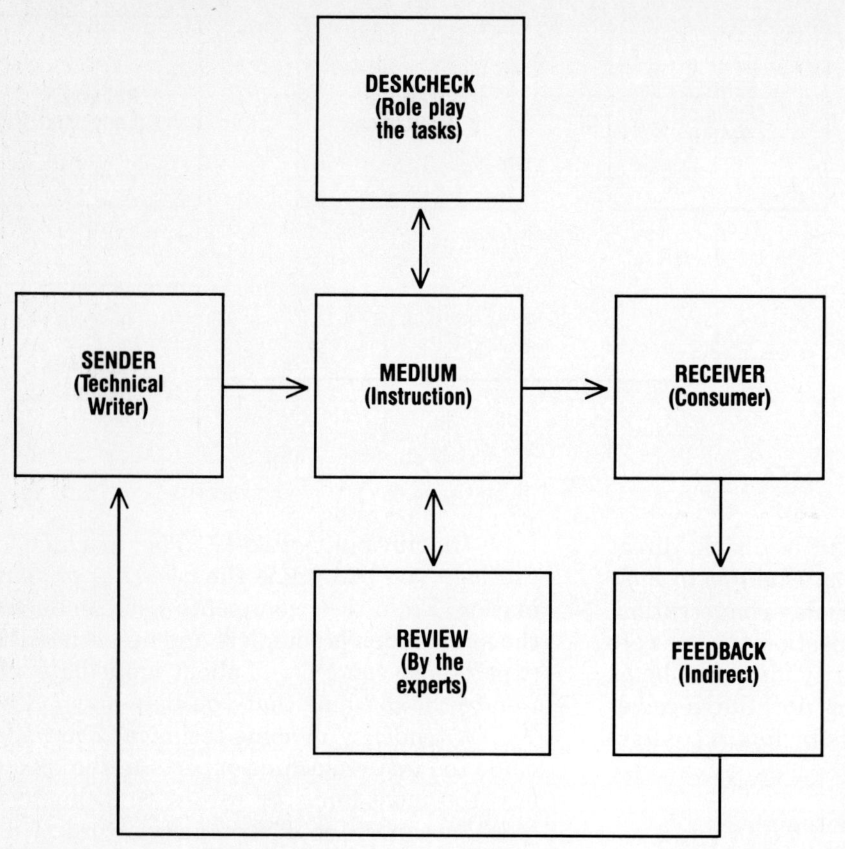

FIGURE 2–14 There are two basic types of feedback—direct and indirect.

number of new lawnmowers may be sent to parks, estates, or golf courses to be used "on the job." In this case, the audience feedback can come in the form of questionnaires or written comments.

The desk-check of the product by a microcosm of people is a hands-on use of the machine by a person who is doing work with the machine that it was designed to do. It is important that the technical writer get as many reactions as possible from a wide variety of people if the intended distribution of the machine is to a heterogeneous consumer audience. For machines that may transcend all social and economic boundaries, a wide test audience will give the most accurate feedback.

In industry there is another form of feedback that must be monitored. You also must be aware of *indirect* types of feedback. This is a negative type of input because it happens *after* the instruction has been distributed with the machine.

Indirect or negative types of feedback. The main types of indirect feedback in industry are the result of *negative* reaction to a product. These indirect reactions to the instruction are discovered *after* the consumer has had a problem with the machine or with interpreting information. Negative types of feedback that are useful to the technical writer are:

Warranty claims.
Liability lawsuits.

Warranty. Warranty is a term that explains the responsibility that the manufacturing company has to supply the consumer with a reliable machine. For example, you receive a warranty statement every time you buy a household appliance, such as a microwave oven. In this warranty statement the manufacturer states how long it is legally obligated to fix the machine if it breaks because of faulty design or poor construction at the factory.

A warranty can be printed on many forms. For example, you may get a warranty statement that is printed directly on the sales slip that the seller uses to record the transaction. This slip will record the item that is sold, the consumer's signature, and the date of the sale.

If the warranty is long, it may be printed on a separate sheet of paper and given to the customer at the time of purchase. Some companies make the warranty a part of the technical operation instruction so that consumers will have all of the information bound together (see Figure 2–15).

Another variation of this is to place a warranty card inside or attached to the instruction. When the consumer buys the machine and records the purchase date or fills in and mails a registration card to the manufacturer if one is provided, the customer activates the machine warranty. By using a registration card, a company can also send out a recall to fix defective parts or to upgrade information to the consumer in case the machine must be modified because of a manufacturing error. In larger companies with a network of distributors or "dealerships" that sell the manufacturer's product, such as an automobile dealer, the dealer often will automatically send the purchase dat^ and warranty information to the manufacturer.

The wording on a warranty statement must be the responsibility of the legal and service departments of the company. The reason for this is that a manufacturing company representative from these groups may be forced to defend or define the warranty in a court of law. A general statement is that every machine type requires a specific warranty.

Usually the main concern of the consumer is: "How long does the warranty last, and what does it cover?" But there are other items that a comprehensive warranty may include as well. An example of this is the outline of the main items covered in a warranty for a popular personal computer:

YOUR ZENITH WARRANTY

Consumer Protection Plan for Zenith Color Television

Welcome into the Zenith family! We believe that you will be pleased with your new Zenith Color TV. Please read this Consumer Protection Plan carefully. It is a **"LIMITED WARRANTY"** as defined in the U.S. Consumer Product Warranty and Federal Trade Commission Improvement Act. This warranty gives you specific legal rights, and you may also have other rights that vary from state to state within the U.S.A.

Service Labor – For a period of 90 days from effective warranty date Zenith will pay for service labor by a U.S., or Canadian, distributor-approved Zenith service center when needed as a result of manufacturing defects.

Parts – New or rebuilt replacements for factory-defective parts will be supplied for one year from effective warranty date (color picture tube – two years). Replacement parts are warranted for the remaining portion of the original warranty period.

Home Service – Warranty service for 19″ diagonal (U.S.A.) or larger screen size models is provided in the home.

Not Covered – Installation, adjustment of customer controls in the home and installation or repair of home antenna systems, cable converters or cable company-supplied equipment are not covered by this warranty; nor is damage due to misuse, abuse or negligence. Any alteration of the product after manufacture voids this warranty in its entirety.

Owners Responsibility

NOTE: Before you ask for Warranty service, check the Operating Guide section entitled, "BEFORE CALLING FOR SERVICE." It may be possible to avoid a service call.

Effective Warranty Date – Warranty begins on the date of original consumer installation. For your convenience, keep the dealers dated bill or sale or delivery ticket as evidence of the purchase date.

Operating Guide – Read your Operating Guide carefully so that you will understand the operation of your set and how to adjust the customer controls.

Carry-In Service – Models smaller than 19″ diagonal (U.S.A.) screen size must be taken to a distributor-approved Zenith service center for warranty service and must be picked up by the owner.

Antenna – Reception problems caused by inadequate home antennas or faulty antenna connections are the owner's responsibility.

Important: Owner Registration – Please fill out and mail your Owner Registration Card. It is imperative that Zenith know how to reach you promptly if we should discover a safety problem that could affect you.

Warranty Service – For warranty service information, contact your Zenith dealer preferably, or any distributor-approved Zenith service center. Parts and service labor that are Zenith's responsibility (see above) will be provided without charge. Other service is at the owner's expense. If you have any problem in obtaining satisfactory warranty service, write or call Customer Service, Zenith Electronics Corporation, 11000 Seymour Avenue, Franklin Park, IL 60131. (312)-671-7550

FIGURE 2–15 A warranty may be printed as part of an instruction. (From Zenith Electronics Corporation, *Operating Guide and Warranty Zenith Systems 3 Color TV,* EP-EDCB 206-01228. Used with permission)

The customer must return the defective machine to an *authorized* repair center. (The store that sold the computer may do this for the consumer.)

The computer manufacturer will replace any defective part and pay for the repairs for 90 days after the sale is made.

The manufacturer will pay for the parts that are replaced for one year after the sale. However, the consumer must pay for the labor that is required to repair the product.

In order to make a warranty claim, the consumer who originally purchased the computer must be the owner.

The warranty will not be honored if the manufacturer feels that consumer neglect and misuse is the reason for the malfunction.

Warranty claims are handled by a claims department in most companies. As a technical writer, you must monitor the results of warranty claims in order to find *trends.* You can question the number of claims on a product. If there are more claims than usual for a certain machine, then you can ask the warranty department if machines are being returned or repaired because consumers are not using them correctly or if the machine is not being properly maintained.

If the technical instruction can help prevent warranty claims in these areas, then the instruction can be revised with new procedures and daily maintenance information. The legal aspects of the language of warranty will be covered in Chapter 8.

The monitor of liability lawsuits. Liability suits against the company are important sources of negative indirect feedback that you, the technical writer, must be aware of. Whereas a warranty deals with machine malfunction or damage, *liability* deals with consumer or machine operator injury. There are many types of liability that a company must deal with during the course of marketing and selling a product. In general, the manufacturing company is somewhat legally responsible for the operation of the machine *after* it is sold to the consumer. In some cases, the manufacturing company must pay restitution to a consumer who is injured by a machine if the courts find that the manufacturer contributed to the situation that caused the accident. Liability cases can develop from:

Unsafe machine design such as improper guarding of dangerous areas, or improperly selected parts that can break and make the machine dangerous.

Poor workmanship or incorrect assembly techniques during the manufacturing process when the machine parts are made and assembled.

A lack of safety warning signs *on* the machine or failure to warn the customer of immediate danger.

Lack of safety information in the technical operation instruction.

Unsafe tasks that either tell the consumer wrong or dangerous ways to operate and maintain the machine.

Incomplete information that fails to explain the operation and care of the machine clearly enough to eliminate foreseeable safety problems.

No matter how safe and reliable a machine is, there may be liability cases brought against it. However, not all liability cases are won by the consumer. No matter how well the technical operation instruction is written, liability suits may occur. A machine that is poorly designed and manufactured can never be made safe by a technical operation instruction. An instruction can only report operation information to the consumer. The technical operation instruction can only be made with the resources and information at hand.

WRITE WITH THE CONSUMER STARING OVER YOUR SHOULDER

One of the most descriptive titles ever given a handbook is *The Reader Over Your Shoulder: A Handbook for Writers of English Prose* by Robert Graves and Alan Hodge (see Bibliography). The advice that is the foundation of the book is that writers must always assume that a reader is standing behind them asking questions as they write. In other words, writers must always put themselves in the position of someone who is reading their material for the first time.

The book has been used by all types of writers ever since it was published in 1943. The 41 principles that are covered in the text provide information on how to build a writing style that can communicate ideas in all sorts of written communication.

Because technical writing is a unique form of written communication, you need to develop a set of principles that you can use to evaluate your technical operation instruction. The following guidelines that conclude this chapter and the ideas and suggestions contained in the chapters that follow can be used by you as a foundation for creating your own standards. When you develop guidelines, always put the consumer's needs first. Write as though that person is scrutinizing every word you write.

ANSWER THESE QUESTIONS EVERYTIME YOU WRITE

Principle No. 1: Is This the Simplest Way That I Can Write?

Your reader will not complain about a technical operation instruction that is too easy to read. Consumers are frustrated when a technical writer tries to impress them with technical knowledge and vocabulary and not express what the consumer must know.

As a consumer, have you read an instruction that was written too simply? Probably the only instructions you remember are the ones that were confusing because they were written in complex words and tasks.

Principle No. 2: Does This Instruction Tell Consumers Everything They Must Know?

Never use the abbreviation *etc.*

Principle No. 3: Am I Assuming That My Audience Is Familiar With the Technology in the Machine?

Always be critical of what you write. Design technical operation instructions that are easy to read and understand by as many potential consumers as possible. The instruction must fit the audience. You must think like the lowest common denominator of your audience and not like a technical writer.

Principle No. 4: Am I Honest With My Reader?

Technical writing for production is a job that requires the technical writer to write to a deadline. Most writing assignments *must* be completed by a certain date in order to coincide with machine production. The technical writer must not take shortcuts in order to meet a deadline. A technical operation instruction that is not complete can be more harmful to the consumer and the manufacturing company than a defective machine.

CHAPTER SUMMARY

- The consumer is both a result and cause of modern production processes.
- Machines can be made one at a time by a craftsperson and sold in a limited market.
- Mass-produced machines are limited in their distribution only by the manufacturer's ability to *supply* the machine and the *demand* of the consumer.
- The mass-consumer audience is a *potential* audience because the technical operation instruction must be written before the machine is sold in the marketplace.
- Homogeneous audiences can be measured for their machine knowledge and vocabulary. Unlike the heterogeneous audience, this audience may not need to know complete basic operation instructions for a machine.
- The heterogeneous mass-consumer audience can cross all social and geographical boundaries and can possibly number in the millions.
- An export market audience presents the problem of translation to the technical writer.
- It is not always possible to write instructions that can be read and understood by 100 percent of a heterogeneous audience. In order to reach the largest percentage of the audience, write to the *lowest common denominator* in the audience that can use the instruction.
- A technical communication module has four distinct parts: the sender, the media, the point where communication occurs, and the receiver.
- *Positive* feedback is a firsthand exchange between the sender and receiver that occurs *before* the technical instruction is printed and the machine is sold in the marketplace.
- *Negative* feedback is the result of warranty and liability claims against the company that manufactures the machine.
- *Warranty* is a contract between the manufacturing company and the consumer. A warranty contract guarantees the consumer that a machine will be reliable and work like it is advertised for a certain length of time.
- *Liability* is a legal term that denotes responsibility by the manufacturer to ensure that a machine works as safely as possible.

- Desk-check the technical operation instruction. This saves time and money by preventing costly revisions.
- Write the technical operation instruction as if the consumer were standing over your shoulder and asking you questions as you write.

EXERCISES FOR CHAPTER 2

Analyze the Audience

1. Collect five technical operation instructions from everyday machines such as an electric can opener, clock, radio, telephone, bicycle, and toy. Critically evaluate these instructions.
   a. Are they complete?
   b. Are they easy to read and understand?
   c. What information is included in each? Do they include extensive assembly information? Do they include maintenance information such as how to change the battery?
2. Choose the instruction from question 1 that is the hardest to use.
   a. How can you improve it?
   b. How many pictures or drawings are in the instruction that you find easiest to use? Compare this instruction with the one that is hardest to read and understand.
3. Survey ten adults. Choose them from all ages and occupations. Ask them the following questions:
   a. What formal education have they obtained? High school? College? Trade school?
   b. What are their hobbies?
   c. What is their job? Have they had specialized on-the-job training in some particular technical area?
   d. Do they consider themselves to be knowledgeable about basic machine operation or technology?
4. If you write a technical operation instruction for the ten people in question 3, who would be the lowest common denominator that you would write to? Why? (Remember, the lowest common denominator may not be the least formally educated person in your audience.)
5. Write two instructions that explain every step of how to do a simple task such as sharpen a pencil, brush teeth, and operate a ball point pen.

3

Develop A Technical Writing Style

The information given in this chapter will enable you to:

- Create a communication model that includes the thinking ability of the consumer.
- Define the "split brain" theory and the impact it has on technical operation instruction writing.
- Define "survival language" and explain why it is important to the technical writer.
- Understand the layered development of our language.
- Explain the basic guidelines that can be used to write visually.
- Separate concrete and abstract nouns.
- Write an active verb.
- Write a simple subject-verb-object sentence in the active voice and in a visual style.
- List the uses for compound and imperative sentences.
- Develop a style sheet.
- Explain the differences between limited and structured vocabularies.
- Determine if "fog indexes" are valid checks for readability and usability of an instruction.

INTRODUCTION

If you are going to write the best possible instruction that you can in order to help support your machine in the marketplace, then you must communicate as much information as possible to the highest percentage of your potential audience. In Chapter 2 you were given guidelines that you can use to write to the lowest common denominator in your consumer audience. In this chapter and in Chapter 4, these guidelines will be explained in greater detail.

The first step will be to construct a communication model that is specific enough to show how the *individual* will react to your instruction. This model will help you develop a *concrete visual writing style* that can be easily adapted to the needs of a mass-consumer audience. It is a writing style that can be used to write instructions for any machine in any type of writing assignment that demands clear and concise instructions. It does not matter whether your audience is heterogeneous or homogeneous, analyzed or not analyzed; you must develop your selection method so that you can choose the most descriptive and understandable word for a specific situation.

In order to define the term "style," look closely at the style of the people whom you meet every day. Each one of them has a unique way that they think, talk, and listen. You can also see another important part of style. Each one of them has a certain visual appearance that includes the clothes they wear and the gestures and emotions that they display. Style is both *verbal* (words) and *nonverbal* (pictures).

The key to success for you as a technical writer is to develop your writing style to interface with the audience. The following information covers two important theories that can help you understand how the individual—the basic unit of the audience—communicates. These theories are by no means the only ones that can help you understand how the individual thinks; however, they are two of the most useful concepts that can assist you in understanding how and why technical operation instructions must be written in an active visual style.

This chapter concentrates on the verbal part of the writing style. This includes the words that you select and use and the way that you pattern them into sentences. In the hands of a skilled technical writer, a few words can be extremely powerful. Your words can create action. They can tell the consumer what to do, how much to do, when to do it, and where. Words are necessary to give the instruction a logical step-by-step continuity that is necessary if the consumer is to use the machine as efficiently and safely as possible.

This chapter also explains why it is necessary for a technical writer to *evaluate* every word before it is put in a technical operation instruction. The difference between safe operation and misunderstanding and possible consumer injury may depend upon one word. Words must always be chosen with care and weighed as to their communication value. Words that the consumer uses in everyday communication are the most understandable. The following guidelines provide you with a practical evaluation technique that you can use to select the right word at the right time. Every word must be the best possible choice that you can make in order to communicate your ideas.

In the final analysis, it is your responsibility as a technical writer to select the most useful word that is available. There are no mechanical rules that you can use to do this. The ability to choose the best possible word will depend upon how well you understand the consumer as an individual. In order to ensure that everyone is communicating with the same symbols and sentence patterns, the chapter ends with a short guideline that explains the basic mechanics that can be used as a framework to develop a style sheet for any writing situation. This will give consistency to the expression of your ideas.

Chapter 4 will continue the guidelines that can be used to write visually. It covers the theory and practice of writing with pictures. In a sense, these two chapters explain how to develop a sophisticated method of picture writing that reaches the basic communication levels of an audience. They will show how you can systematically write tasks with a visual style that can be easily turned into specific physical action or muscle-motor movement by the consumer.

CREATE A COMMUNICATION MODEL FOR THE INDIVIDUAL CONSUMER

In Chapter 2, a communication model was illustrated that showed the process of how information flows to and from the group of individuals labeled the "mass-consumer audience." You might find that you can write more usable technical operation instructions to a heterogeneous or homogeneous audience if you can reduce this mass of people to a single individual. By "personifying" the consumer audience, you can give your audience a face and a form. You can decide what you need to communicate to the lowest common denominator in the audience.

The "enhanced" communication model can help you do this. It can be modified to show how a consumer obtains and decodes information in a technical communication process. With this model, you can create a mental image of the consumer who has the least ability to communicate but who still has the capacity to use your instruction. It is easier to write to an imaginary but fully understood personification of the audience than to write to a group of people who can never be fully analyzed.

Use the Concept of Information Layering to Your Advantage

The "triune brain." Dr. Paul MacLean provides insight into how the consumer's mind has developed over the time that our preconsumer ancestors have been on earth. Dr. MacLean theorizes that the human brain developed in three time periods. He divides the brain into these parts:

The *reptilian* portion developed first. This inner core governs such things as survival instincts, like eating and aggression.

The *limbic* portion developed as a second layer. It gives humans the emotional bonding that is found between parent and child.

The *neocortical* part of the brain developed as the outer portion. With it the preconsumer or early human developed language powers and introspection.

Dr. MacLean's theory can be applied to the craft and art of technical writing. First of all, the human brain can be considered as a model of how knowledge and communication develops. Every idea and every word that the individual knows develops in layers that become increasingly complex. The trend for the development of the body of technical knowledge that eventually put people on the moon began as simple concepts in the survival part of the human brain. Each "layer" of the brain, according to Dr. MacLean, governs specific actions.

During the course of a day, the individual acts and reacts to situations that are both complex and basic according to the situation and the needs of the person. The important concept put forth with this picture of the cerebral development of the human race is that consciousness and thought have developed in layers like an egg. The brain has an inner core, a middle section, and an exterior layer.

The "split brain." How can a consumer read an instruction, translate it into thought patterns, and then use these stored thought patterns immediately or many days later to operate a machine? This is the problem that faces every instructor who attempts to teach a shop skill or similar skill that depends upon motor movement. How well this is answered will depend upon how quickly the instructor can make the point and "teach" the student (see Figure 3–1).

This often is not an easy task. It is commonly accepted that the educational process in the United States is structured around the verbal and mathematical approach to teaching. Unfortunately for some, the words in both books and in instructions do not communicate as well as do pictures or hands-on training and an apprenticeship approach to education.

As a technical writer, you must also answer the question of how instructions can be written so that *every* able consumer can read and interpret them into motor skills and muscle patterns. Instructions must be written so that they have an impact on the consumer. They must etch the safe and efficient operation instructions upon the long-term memory of the consumer. As a group, vocational educators in the United States have begun to develop approaches to teaching that enable them to communicate with all the mental capabilities of the individual.

In an attempt to understand its workings, the brain has often been compared to the technology of the time. For example, it was popular to compare the brain in the early part of this century to the workstations of a manufacturing process. Each stage of the brain was thought to handle information and pass it along to the next station until it became a complete idea.

Later, the technology of the time was the devel-

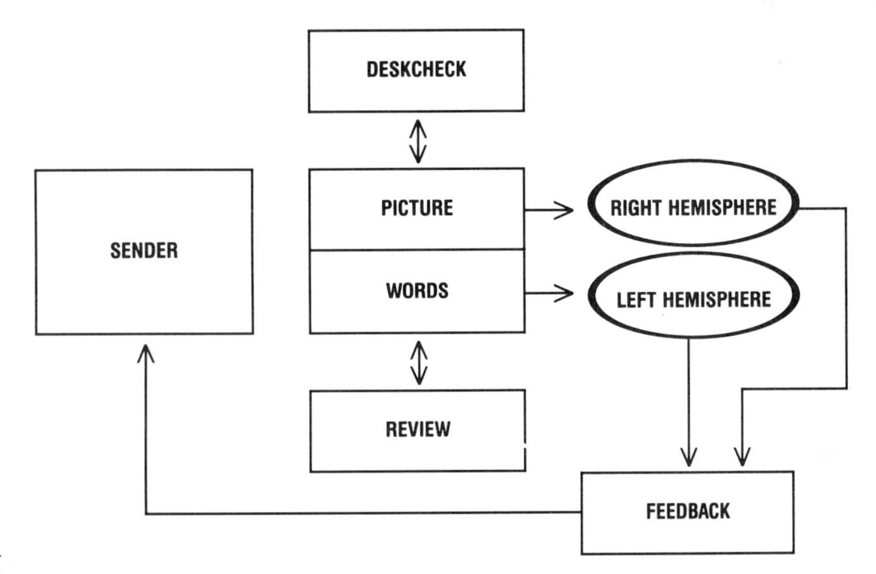

FIGURE 3–1 Both words and pictures are necessary in a technical operation instruction.

oping telephone industry. The brain was viewed as a switchboard of electrical connections that were connected and unplugged to transfer information to the right receiver. Similarly, the computer of today is compared to the inner workings of the brain. While both the brain and the computer can be programmed to a certain degree, the comparison is not wholly accurate. The brain is much more complex than the switches, gates, and silicon chips of the computer.

Brain hemispheres. Further research into the ways the consumer perceives and reacts to a stimulus such as a technical operation manual has provided information that is extremely important to you when you develop the enhanced communication model. A fundamental concept is that every consumer has a primary need for both words and images in order to communicate successfully. This is because the "healthy" brain is divided into two distinct parts, the *left hemisphere* and the *right hemisphere.*

The corpus callosum. Both parts of the brain are connected by a highly specialized group of nerves called the *corpus callosum,* which acts as a pathway for the transmission of information between the two hemispheres. Research begun in the 1950s and 1960s at the California Institute of Technology and other research centers has shown how the two sides of the brain function in decoding individual perception into ideas and actions.

Studies were done on people who had the two lobes of their brain separated at the corpus callosum either by accident or surgical procedures. Through highly sophisticated studies, California Institute of Technology researchers found that not only is the brain divided physically into two halves, but that these halves are areas of specialization that often struggle for dominance (see Figure 3-2).

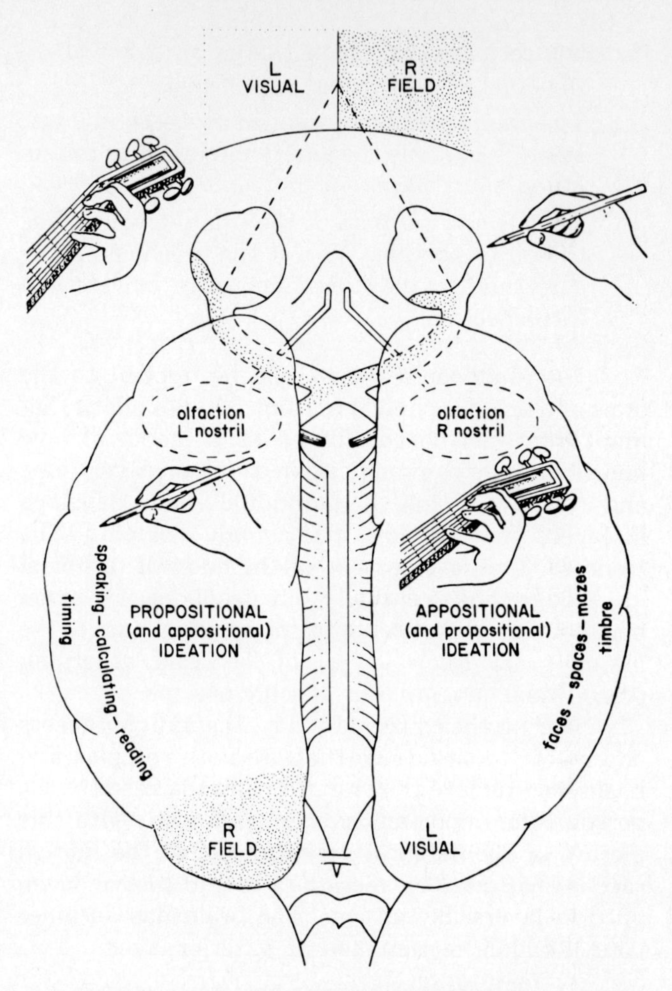

A schematic outline of the brain as seen from above, to suggest the complementary dominance of the cerebral hemispheres for various tasks, summarizing the evidence from cases of lateralized lesions and from testing of patients with cerebral commissurotomy. Based on version updated and redrawn (by J.E.B.) from the original conception of R.W. Sperry (Sperry, Vogel, and Bogen, 1970).

FIGURE 3-2 The human brain is divided into two hemispheres; each one dominates a particular ability.

The left lobe. First, one side of the brain, which in about 90 percent of the mass-consumer audience is the left lobe, perceives and decodes the *logical* information that the individual processes. This logical, noncreative side controls these major functions:

- Language skills.
- Literal interpretation of words.
- One-step-at-a-time thought processes.
- Mathematics capabilities.
- Movement in the right side of the body.

The right lobe. Second, if the left side of the brain is the predominantly logical sphere, then the right side is the nonverbal and *intuitive* side. This side controls such creative and biological functions as:

- Knowledge interpreted from pictures and visual images.
- Developing an overview of the subject.
- The ability to deal with many simultaneous actions at one time.
- Imagination, including intuition, dreams, and fantasy.
- Perceptions of space and time.

Emotions.

Movement on the left side of the body.

Even in a "healthy" brain, either the left lobe or the right lobe tends to dominate the individual. Words can interfere with perceptions of space and time, and intuition and imagination can interfere with one-step-at-a-time logic. This is graphically illustrated in Betty Edwards' book, *Drawing on the Right Side of the Brain: A Course in Enhancing Creativity and Artistic Confidence* (see Bibliography). In her book, she teaches techniques for using the right lobe, or the creative side of the brain, to draw line art.

An exercise that the author uses in her classes on drawing is to turn a picture of someone upside down before she has her students draw the form. For example, if the student is drawing the face of a person, they tend to lose the ability to draw a line if they think of the parts of the face *in words.* If the artist verbalizes that he or she is drawing a nose, then the artist will tend to draw a preconceived nose. The logical verbal side will interfere with the creative side of the brain that must be used to draw the picture accurately.

When a picture of the person that is used as a model is turned upside down, the drawing ability of the art student increases. The verbal concepts are neutralized. A nose simply becomes a series of lines that the right side of the brain can interpret and turn into action on the drawing tablet. Depending upon the information that is being communicated, and the resulting action that is expected, the left and right lobes tend to struggle for dominance.

Satisfy both sides of the brain with information. Both sides of the brain must be used if the technical communication process is going to be as effective as possible. A graphic illustration of this is the method that many people employ to get as much out of a textbook as possible. The first actions are to:

Look at the front and back covers.
Read the author's biography if one is included.
Review the table of contents.
Read the preface.
Look at the pictures and read the captions.
Scan the headings.
Read page one.

Begin the logical verbal process after an overview of the subject is developed. In this method, one gives the right hemisphere a chance to visualize the book as a complete object before beginning the process of reading the logical sequence of words and ideas.

Use pictures. The valuable concepts shown by the "split brain" research is that technical instruction must be written for both sides of the consumer's mind. Pictures must be integrated with logically sequenced words in order to communicate with the two distinct sides of the consumer's thought process. Without words and pictures the consumer may not be able fully to understand and transfer into physical action the tasks that you present (see Figure 3-3).

It may be that a consumer is completely verbally oriented, or, on the other hand, completely oriented to spatial concepts. By writing in a visual style, you will communicate at least a percentage of information to each of these people. At least they

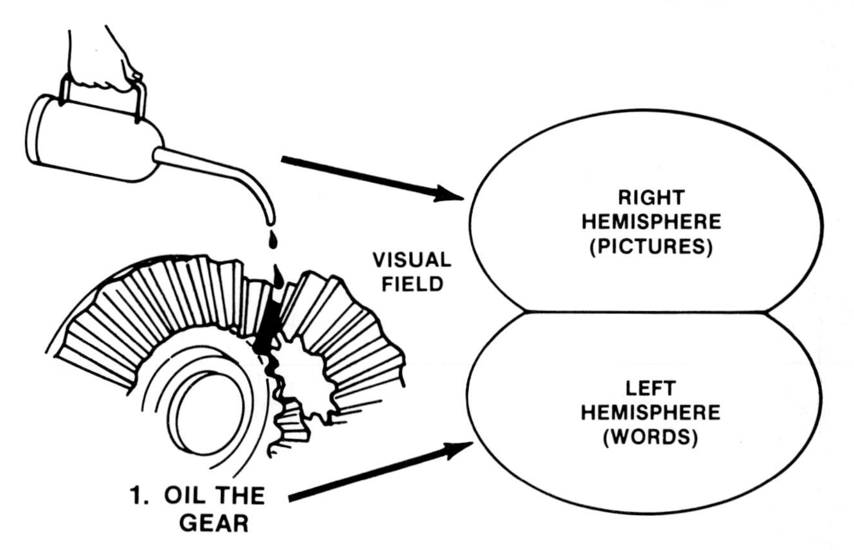

FIGURE 3-3 Pictures are usually interpreted by the right hemisphere of the brain. Words are usually interpreted by the left hemisphere.

will be able to use a percentage of your information. For those who can use their entire thought process, however, a technical operation instruction consisting only of words is as difficult to read and understand as an instruction containing only pictures.

The ultimate goal of the technical writer as communicator is to use words and pictures in a *selected order* to obtain the desired *physical actions* in the consumer. You can use the split brain concept to develop a concrete visual writing style that eliminates misunderstanding and misconception in as many of your potential consumer readers as possible. The first and most logical step is to begin with the selection of words.

SELECT YOUR AUDIENCE'S MOST FAMILIAR WORDS

You may not be able to get direct feedback from every consumer who buys your company's mass-produced machine. You must develop an efficient and accurate method of word choice in order to meet writing assignment deadlines, widen your potential audience, and still communicate complex technical data in a usable format. One way to help ensure that the words you use are translatable to the consumer is to use words that the *consumer* is familiar with. To illustrate this, look closely at the following highlighted history of how our language developed.

First of all, every language consists of a base vocabulary that has developed in a particular culture. Second, every language is exposed through interaction with foreign vocabularies and social advancement. English is no exception. The English language began as a basic language and has through the years become layered with more complex words and phrases.

The beginning of the English language. The roots of our English vocabulary were brought to Britain by the Germanic tribes who invaded the Celts, the earliest known inhabitants, around A.D. 500. It was St. Augustine, a Roman monk who went to Britain a hundred years later to spread Christianity to the Celts and Germanic tribes, who added the first layer of complex vocabulary to our language. Augustine and the 50 monks who accompanied him brought Latin and Greek words and concepts to the mixture of dialects and languages of the time. The importance of this cultural event is documented by Robert McCrum, William Cran, and Robert MacNeil in their book *The Story of English.*

> The importance of this cultural revolution in the story of the English language is not merely that it strengthened and enriched Old English with new words, more than 400 of which survive to this day, but also that it gave English the capacity to express abstract thought. Before the coming of St. Augustine, it was easy to express the common experience of life—sun and moon, hand and heart, sea and land, heat and cold—in Old English, but much harder to express more subtle ideas without resort to rather elaborate, German-style portmanteaux like *frumweorc (fruma,* beginning and *weorc,* work = creation). Now there were Greek and Latin words like *angel, disciple, litany, martyr, mass, relic, shrift, shrine and psalm* ready to perform quite sophisticated functions. The conversion of England changed the language in three obvious ways: it gave us a large church vocabulary; it introduced words and ideas ultimately from as far away as India and China; and it stimulated the Anglo-Saxons to apply existing words to new concepts.[1]

The last great influence that shaped what we now know as the "Old English period" ended with the invasion of the inhabitants of Britain by the Vikings in the ninth century. This Old English period of change upon our language lasted from approximately A.D. 500 to A.D. 1100.

Middle English. Again in A.D. 1066, the Britons were conquered by the Normans, who invaded the island from France. After the Norman victory, the basic "English" language that was spoken at the time became a second-class language. The government, the church, and various social circles spoke French, while the professional, scientific and "educated" people used formal Latin to conduct their business. The basic English language survived in the lower classes of peasants for the next 300 years. This "Middle English" layer lasted from about 1100 to 1500.

Modern English. The "Modern period" of English is marked by an assimilation of the Norman conquerors into the English culture. The original language of the common person found a rebirth. Great writers, such as Chaucer, author of *The Canterbury Tales,* turned to the basic English vocabu-

[1]Robert McCrum, William Cran, and Robert MacNeil, *The Story of English* (New York: Viking, 1986), p. 65.

lary to express their thoughts. A return to the vocabulary of the people that began at least 150 years earlier culminated around 1500.

For the first time, changes in the language were not caused by war and conflict as much as social and economic developments. The greatest influence during this time on the English language was the Industrial Revolution. Advancing technology and the development of the factory system further molded our language into modern-day English.

Some important observations can be made from studying the development of a language. They are:

> Language does not develop in a vacuum. It develops in layers that are often marked by a collision of one culture against another (see Figure 3-4).
>
> The most basic way to communicate is to strip away the layers of complex language to get to the basic vocabulary.
>
> Words can stand either for concrete things or abstract thoughts and ideas.
>
> In comparison, children develop their vocabulary in layers. Language becomes more complex as the child is influenced by social and economic interaction.

A language, such as English, is a living and evolving thing. As long as it is actively spoken and written in everyday communication, it will grow and change. For example, words that were not in your vocabulary five or ten years ago may now form an important part of the words that you rely upon to express your ideas. They might include:

Synthesizer
RAM memory
ROM memory
ASCII language
Acid rain
Basic language
Bar code readers
CAD (Computer aided design)
CAM (Computer aided manufacturing)
Compact disk player
Video cassette recorder
Desktop publishing
Laser printer
Mainframe computer
Byte
Just in time manufacturing

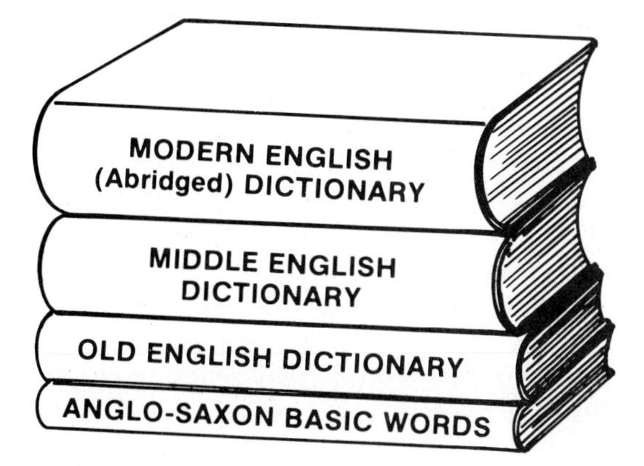

FIGURE 3–4 Language develops in layers.

Avoid "Artificial" Language

The problem with the elasticity of language, however, is that words are easy to create, modify, and redefine. Often language develops for such a specific situation that it is not usable outside of the situation. As Charles E. Witherell explains:

> An increasingly common shortcoming with user instructions, operating and maintenance manuals is their incomprehensibility. The trouble lies not in lack of intelligence of the user or reader. The problem is that the writer of the manual often fails to realize that technical terms and industry jargon familiar to him may be meaningless to the user or reader.[2]

Examples are:

Jargon	Specialized words such as specific industry terms or localized factory terms that are known only by a few people. Jargon sample: Choke the venturi and fire the spark. Let up on the nitro and set the gear.
Legalese	Language so complex that the person outside of the law profession finds impossible to understand. Many contracts and legal forms are

[2]Charles E. Witherell, *How to Avoid Products Liability Lawsuits and Damages: Practical Guidelines for Engineers and Manufacturers* (Park Ridge, N.J.:Noyes Publications, 1985), p. 184.

	written so complexly that they are useless as a form of communication for the "average" person. Legalese sample: Henceforth, the party of the first part, or operator and the licensee, or the party of the second part or dealer . . .
Bureaucratese	The language of government; hinders more than it helps the process of communication between the government and the citizen. Bureaucratese sample: The consumer must be regulated in order to control and regulate the machine in such a way as to make the consumer as consciously safe as possible . . .
Gobbledygook	This is the language famous in the book *Alice in Wonderland.* It is the art of talking without saying anything understandable or usable (see Figure 3-5). Gobbledygook sample: Start the engine, or the motor prime mover, by turning or rotating thc kcy somewhat slowly and deliberately . . .

When you write the English language to a heterogeneous mass-consumer audience, you must choose the shortest and simplest word that your audience can understand. Usually the best word to choose is the basic German root word. The Latin, French, and Greek root words *tend* to be more abstract, longer, difficult to read and understand, and translate. Words derived from Latin and French tend to be the vocabularies of homogeneous audiences. These foreign words must be used only when the situation demands it.

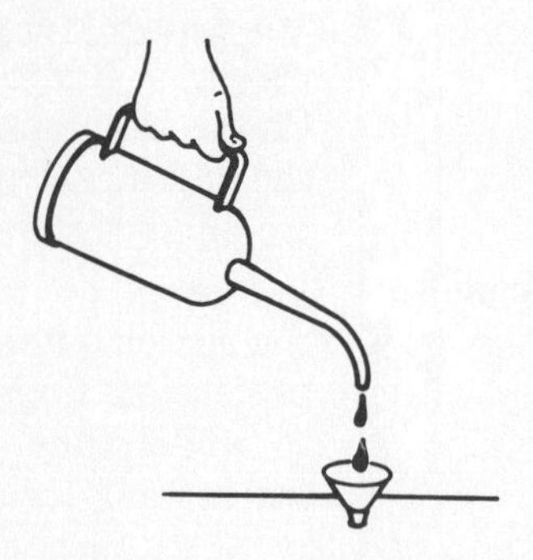

LUBE THE SPANGLE ON THE SYNCHROMESH.

FIGURE 3–5 Avoid nonsense and specialized words.

Avoid "fake" simplicity. A good rule to follow is given by George Orwell, an English political writer who wrote a classic essay on communication entitled "Politics and the English Language." The only important thing is to make your meaning clear. Do not use "fake simplicity" to write. Do not eliminate the more complex word from your selection process. It is not necessary to *always* choose the "Saxon" word over the "Latin" word.

> What is above all needed is to let the meaning choose the word, and not the other way about.[3]

Analyze the Layering of Your Consumer's Vocabulary

Another useful concept for the technical writer to know is that individuals also tend to layer their language as their education, training, and interests develop. In order to make the job of selecting words that are familiar to your audience more efficient, you must analyze the way your personal vocabulary has developed. To avoid mirror-writing and to choose the words that are most familiar to your audience, you must know what makes written communication successful. Fortunately, research in linguistics and language development has shown that everyone who has the ability to communicate verbally has developed verbal skills in similar patterns out of necessity.

The survival language. If you have ever watched a child learn to speak and read, you can make a few practical observations about how the youngster develops its vocabulary. The first stage is devoted to learning words that are necessary for the child's survival and well-being. Simple one-syllable words dominate the language of the child and allow them to communicate exactly what they want.

Arthur Janov, author of *The Primal Scream,* says that between the ages of five and seven, the child moves away from simply reacting to and dealing with the present.

[3]George Orwell, *Shooting an Elephant and Other Essays,* (New York: Harcourt, Brace and Company, 1950) p. 91.

> This is when the child learns to generalize from concrete experience. This is the time when a child can begin to understand the significance of each disparate event that has happened to him before.[4]

For the sake of simplification, you can view this development as the beginning of the middle period of vocabulary building. The child begins to layer its vocabulary with concepts that are moving away from the concrete symbols that the youngster used in the first phase of development.

Albert Einstein, who formulated the general theory of relativity at the beginning of this century, was probably referring to this period of development when he was reminiscing why it was he who conceived of the theory:

> I sometimes ask myself, how did it come that I was the one to develop the theory of relativity. The reason, I think, is that a normal adult never stops to think about problems of space and time. These are things which he has thought of as a child. But my intellectual development was retarded, as a result of which I began to wonder about space and time only when I had already grown up. Naturally, I could go deeper into the problem than a child with normal abilities.[5]

The last period of vocabulary may begin when the child has the opportunity to choose his or her direction of study. For the youth of the United States, this begins in about the 10th grade. From this time on, young adults can elect their formal education in both the secondary and postsecondary classroom. Along with their formal education, they develop special hobbies, interests, and take specialized training that further specializes their talents for the job market.

The common language. In general, the survival vocabulary that is developed in the first period of vocabulary building is a "common" language that everyone who develops a vocabulary shares. After the age of five or six, the child becomes a student and vocabulary gains abstract qualities. From this time on, the child seems to struggle for individuality in both verbal and *written* communication.

[4]Arthur Janov, *The Primal Scream:* Primal Therapy, The Cure for Neurosis. (New York: Putnam Publishing Group, 1981), p. 29.

[5]Ronald W. Clark *Einstein: The Life and Times* (New York: World Publishing Company, 1971), p. 27.

FIGURE 3–6 Assume the role of the machine operator no matter what the subject of the instruction is. (Courtesy Deere & Company)

As a technical communicator, you must remain completely *objective* when you write a technical operation instruction. Your individual drive for *self-expression* must be tempered and dictated by what words your audience can understand and use. In a sense, if you are writing an instruction for domestic households, you must assume the attitude of one who is going to use the machine in that environment. If you are writing a technical operation manual for a space vehicle, then you must assume the role of the astronaut. If you are writing an instruction that explains how to operate a lawn and garden tractor, then you must become the lawn and garden tractor owner and operator (see Figure 3–6).

Carefully Select Words

Students spend money and time to learn the specialized vocabularies of the various professions. Besides learning physical and mental skills, the biochemist, anthropologist, computer scientist, psychotherapist, endocrinologist, lawyer, doctor, and other professionals must master the vocabulary of their occupation. Careful selection of words is extremely important when you write to a homogeneous audience such as the ones above.

As you work with a machine that you must explain to this homogeneous audience, you must be keenly aware of the language of the machine experts. Language that explains exact ideas to a specialized audience should not be much different from the vocabulary that you use to explain a machine to a heterogeneous audience. You must write clearly and accurately. To do this, you must develop a visual writing style that attempts to eliminate the word that is too complex. As with every technical operation instruction, the consumer *must be able to read and understand the information after the first reading.*

For those times when you must write specialized instructions remember to use words that are common to the profession. Do not use the jargon of the profession. A guide that can help understand the origins and use of specialized words in some professions is Mario Pei's book, *Language of the Specialists: A Communications Guide to Twenty Different Fields.* This work, published in 1966 by Funk & Wagnalls, is outdated in some areas, but it offers insight into specialized vocabularies. Twenty-two specialists from the social sciences, business, the arts, and sciences explain the origins of specialty languages.

Case Study

Beth Sommers reviews her first attempt to communicate technical information. She is writing instructions about a fabricating machine that will be sold to another manufacturing company. June Davis, Beth's instructor, has returned the instructions with the comment that they must be more readable and understandable by her audience. June thinks, for example, that the first line of the safety message is not as descriptive or visual as it could be. The message warns the machine operator not to remove a transmission gear case protective cover. The message accompanies a short set of tasks and a photograph of the cover. Beth must rewrite the caution statement and submit it and the photograph to the company legal department for review. The sentence is:

Caution: Keep the Metal Transmission Protective Cover in Place.

To get a good mental picture of her idea, Beth finds the location of the sentence in her instruction. She begins rewriting by analyzing the noun of the sentence, and knows that the moving gears on this part of the machine are a potential hazard to the operator if the cover is removed. She refers to her thesaurus and selects words that are close synonyms to the word "cover." The alternatives include:

safeguard
shield
armor
shroud
protection
cushion
sheath
bonnet
cowl
guard

First, Beth eliminates the words that she thinks are foreign to her audience and do not convey an image of safety. They include "shroud," "sheath," "bonnet," and "cowl." Next, she eliminates "protection" because it does not create a strong visual image. Since the object is made out of steel, it is not a "cushion." The words "armour" and "shield" create an image of protection worn. Beth's word choice is narrowed to "guard," and "safeguard." The origin of safeguard is Middle English and old French, according to Beth's dictionary. "Guard" is an old French root word. These words are not from Germanic origin however; they are highly readable and can be understood by the audience. Beth decides to choose the single-syllable "guard." It is the shortest and most descriptive word that fits the situation.

Next, Beth eliminates "protective" from the sentence because it is a redundant word. A guard is a protective device. Beth also eliminates "metal." The photograph that she selected for the instruction shows the part itself, and this allows Beth to eliminate the descriptive word. She can also eliminate the location word "transmission."

After selecting a noun and reviewing the descriptive words, Beth then analyzes the verb. She knows that "keep" is a basic survival word that cannot be improved upon. Finally, after analyzing the verb, she concentrates upon supporting words and phrases. She feels that "in place" is interchangeable with "on." Beth's revised sentence reads:

Caution: Keep the Guard on.

Beth thinks this shorter sentence is easier to read and be understood by the audience. The photograph that will appear in the instruction, along with the rewritten sentence, creates the image that she wants to communicate. Word selection is a time consuming process for a novice writer like Beth. As she gains experience, she will automatically limit her vocabulary to the best choice for her audience. At this time, Beth must review every sentence by analyzing the noun and descriptive words, the verb, and then the support words and phrases. By following this procedure, she can systematically select the most appropriate words to report her ideas and information.

DEVELOP A VISUAL WRITING STYLE

The visual technical writing style can be defined as a method of technical writing that uses both concrete nouns, active verbs, and artwork to communicate an *action* to a consumer. The writer must carefully *select* symbols that communicate with both sides of the consumer's mind in order to eliminate confusion and misconception. It is a writing style that can be read once and applied to a specific task by a consumer who is using the instruction to operate the machine. The tasks and information are written objectively in order to *express* information and not to *impress* the consumer with the technical writer's machine expertise.

The verbal half of visual technical writing includes the use of the most basic, shortest, and most accurate words that can be used. Concrete nouns that symbolize concrete things must be employed. *Active voice* verbs must be used to show *action.* In a sense, the visual technical writing style is an honest and sincere attempt to communicate complex technical data as simply as possible to the largest percentage of a potential audience as possible.

Use Concrete Nouns

Nouns represent things like *robots, computers, hammers,* and *tools.* They are used as either the subject or the direct object in a sentence. For example:

Robots (subject) need to be maintained daily.
Use the *hammer* (direct object).

Nouns can be divided into two types. They are:

Abstract
Concrete

Abstract nouns are all-inclusive. They represent unspecific things. You describe abstract nouns in general terms as though you are looking at the object through a hazy or cloudy glass. All of the specific details are absent. For this reason, abstract nouns do not create the same picture in the minds of different people. Examples of these are:

Wheel
Tool
Hydraulics
Electronics
Computer
Hardware
Software
Input
Output
Consumer audience
Liability
Gears (see Figure 3–7)

You would get ten different descriptions for each of these words if you asked ten people to describe them to you. Abstract nouns are out of place

FIGURE 3–7 "Gears" is an ambiguous word. (Courtesy Deere & Company)

FIGURE 3–8 "Rotary engine" is a concrete noun. (Courtesy Deere & Company)

in a technical operation instruction. The consumer needs specific information. You force the consumer to interpret your meaning when you use abstract nouns. *Concrete* nouns must be used whenever a specific task is written. Examples of *more* specific concrete nouns are:

Diode
Hammer
Tire
Voltmeter
Carburetor
Floppy disk (a compound noun)
20 weight oil (a compound noun)
Gasoline
Rotary engine (see Figure 3–8)

These nouns create a concrete image in the mind than do those in the abstract list. You use concrete nouns to make your writing more *visual.* You can see the object more clearly in your mind. This concrete image is transmitted more clearly to the consumer. A visual concrete word is easier to read, and understand, than is an abstract noun. Here are a few examples:

Abstract: Make the *adjustment* with the *tool.*
More Concrete: Turn the screw with a *screw driver.*
Abstract: Test the *electrical circuit* with a test apparatus.
More Concrete: Test the *negative wire* with the *voltmeter.*
Abstract: Hit the *fastener* with a blunt *object.*
More Concrete: Hit the *nail* with a *hammer.*

Use Active Verbs

Verbs are tools that you can use to *create action* and build life into the tasks that you write. Verbs are divided into tenses such as:

Present Tense: I *run.*
Past: I *ran.*
Future: I *will run.*
Future Perfect: I *will have run.*

They also are categorized into active and passive voices. Verb choice determines the voice in a sentence. In the passive voice, the subject of the sentence is secondary to the action. The passive voice is constructed with a form of the verb *to be* and a *past participle.*

Passive Voice: The adjustment *was made* by the operator.
The adjustment *must be made* by the consumer.
The button *should be pushed* (by you).

In the active voice, the subject is primary. The action is explained after the subject is identified. Choose the first person, present tense of the verb when you select words for a visually written operation instruction.

Active Voice: The oil *is* 30 weight.
Turning blades *are* dangerous.
Hydraulic pressure *is* 50 pounds per square inch.

Present-tense verbs, such as *is, are, turn, run, hit, bend,* and *press* create *action* (see Figure 3-9). When you write active sentences with present-tense verbs, you do not waste the consumer's time. You explain a task quickly, efficiently, and clearly. *Technical operation instructions are written for the consumer who is using your instruction and the machine together in present time.*

Use adverbs to fine-tune the action in your sentence:

The oil pours *slowly.*
Pressurized hydraulic oil escapes *rapidly.*

Active verbs, like all verbs, must agree in number with the concrete nouns that you select. For example:

One consumer *turns, hits, pushes, replaces,* and *adjusts.*
Two consumers *turn, hit, push, replace,* and *adjust.*

Write Simple Subject-Verb-Object Sentences

Develop your style with well-written sentences. Use *one* simple sentence to tell *one* basic task or *one* bit of useful information to the consumer. As with the

How to remove a printer or cutter recording paper jam

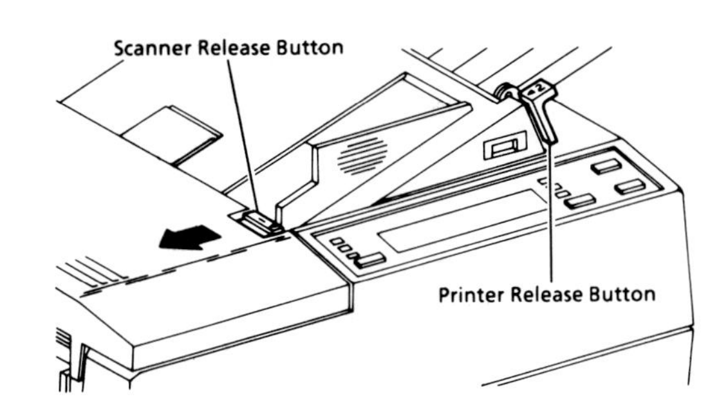

1. **Open the scanner cover**
   a. Press the scanner cover release button number 1.
   b. Lift the scanner cover until it is completely open.

2. **Open the printer cover**
   a. Press the printer cover release button number 2.
   b. Lift the printer cover until it is completely open.

3. **Remove the recording paper**
   a. Turn the green cutter thumbwheel counter-clockwise to cut any remaining paper.
   b. Remove the recording paper.
   c. Remove the remaining jammed paper.
   d. Replace the recording paper (refer to "Operator functions").
   e. Push the jam release lever (number 3) down.
   f. Remove the jammed recording paper.
   g. Close the printer cover.
   h. Close the scanner cover.
   j. Try the operation again.

FIGURE 3-9 Present-tense verbs can show action. (Copyright 1985 by the Xerox Corporation, *Xerox Telecopier 7010 Facsimile Terminal Operator Handbook,* 600P88404 Revision A. Used with permission)

languages used in computer programming, a simple sentence is economical and can be read more easily than lengthy and complex sentences You must remove all excess words and make each sentence structure simple. Use this sentence structure as a foundation upon which to build your instructions (see Figure 3-10). The basic word arrangement in a simple sentence is:

Subject — Verb — Object
Examples:
Knives cut the chaff.
S V O
The oil is *SAE Grade CC.* (Compound object)
S V O
A battery contains acid.
S V O
The alternator charges the battery.
S V O
Pressurized *hydraulic fluid* can be a hazard.
S V O
("Hazard" is the predicate object that is used to explain the subject. "to be" is an active verb that describes the compound noun "hydraulic fluid.")

Articles such as *a, the, an* announce a noun. They also keep your technical operation instructions from reading like a Tarzan or Frankenstein movie script. Use "a" before concrete nouns that begin with a consonant sound such as *a bolt, a nut,* and *a desk.* Use "an" before concrete nouns like *an ignition, an emulsion* that begin with a vowel sound. Use "a" and "an" the same way when you use a modifier before the noun such as *an important note, a wing nut, an eye bolt.*

The task sentence. You will use the simple *task sentence* more than the simple sentence when you write a technical operation instruction. Task or imperative sentences command or demand *action.* In task sentences, the subject "you" is always understood. For example:

(You) Start the machine.

(You) Grease the robot every 5 hours of operation.

(You) Keep the machine pressure off.

(You) Turn on the central processing unit.

REPLACING GUIDE BAR SPROCKET NOSE

1. Remove guide bar cover, chain, and guide bar. (See Cleaning Chain and Bar Guide for removal.)

CAUTION: Wear gloves when handling chain.

2. Remove tip guard from guide bar.

3. Support guide bar.

4. Drill out the three rivet heads.

NOTE: Heads of rivets have a center punch mark.

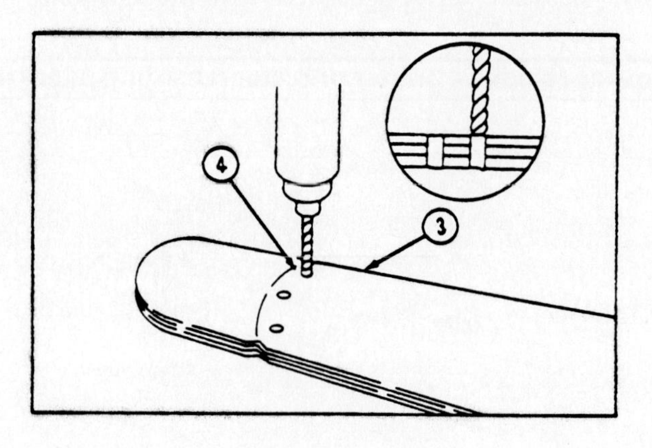

1A3;TY7631 Y05;;69SE A0 020786

FIGURE 3-10 The simple sentence is the foundation for all instruction writing. (Copyright Deere & Company, *450V and 800V Chain Saws,* JDM Division OM-TY20829-F6, 1986. Used with permission)

(You) Stop!

(You) Change the spark plug.

(You) Wear protective clothing.

(You) Slowly turn the adjustment.

(You) Quickly change the diode.

(You) Cautiously connect the negative ground wire.

The compound sentence. Use compound sentences when you are telling the consumer to do *two or more tasks at the same time.* These actions are held together by a conjunction such as "and." You are telling the consumer that one task cannot be done if the consumer does not do a second or third task at the same time. You must remember the rules of spatial logic. Do not expect the consumer to be in more than one place at a time. *Do* use a compound sentence to tell the consumer to do a job that normally requires two or more people working together. Examples of the compound sentence are:

Pull the choke *and* start the engine.

Start the engine *and* release the clutch.

Press "control," "alternate," *and* "delete."

You can write every technical operation instruction with simple active sentences, and with compound sentences that consist of a subject, subject modifiers, present-tense verbs, adverbs, objects, articles, and conjunctions.

DEVELOP A STYLE SHEET

Style sheets are written mechanical rules that are created by a company to keep instructions standardized. Style sheets can be created from industry standards that set rules for what must be told and how it must be told to the consumer. If a style sheet is not provided for you by your company, you should create one so that you can use it as a checklist and guideline during the completion of your writing assignments. A sample style sheet is given below. Often style sheets may be more complex and lengthy than this example.

Basic Style Sheet for the Technical Writer

You can write technical operation instructions for every job assignment if you develop your style on the foundation of the following grammar rules. For your convenience, make sure that you have a style sheet that gives you enough guidelines to satisfy all of your company requirements for the presenting of technical material to the consumer. Develop a style sheet if one does not exist. The material listed below can help you begin the process.

1. *Use concrete nouns.* Example: A *buffer* isolates circuits.
2. *Choose the most basic, shortest, and exact word available in every word selection.*
3. *Use adjectives sparingly in order to clarify nouns.* Example: The *heaviest* sediment settles. (Most adjectives can be eliminated by using artwork.)
4. *Use nouns as predicate objects to clarify the subject.* Example: Oil is a *liquid.*
5. *Use present-tense verbs. These verbs put action into your tasks.* Example: The crankshaft *drives* the powertrain. The battery *discharges* when the current is low.
6. *Use adverbs to limit or define the parameters of action. Put them before the verb.* Example: *Slowly* turn the knob.
7. *Verbs must agree in number with the subject.* Example: Positive wires *are* red. The positive charged wire *is* red.
8. *Use one simple sentence to explain one task or idea. Put sentences in the subject, verb, object structure.* Example: Instructions are important. The engine burns gasoline.
9. *Use compound sentences to explain tasks that must be done simultaneously.* Example: Push the button, *and* adjust the pressure.
10. *Use articles to introduce concrete nouns.* Example: *The* silicon chip stores information. *An* electrode is *a* conductor.
11. *The task sentence gives a command.* Example: (You) Stop! (You) Turn the adjustment knob. Do not feel for leaks in pressurized hydraulic systems. Hold cardboard over a suspected leak. Quickly push the start button.

Use this formula to build a technical operation instruction:

WORDS = IMAGES (They help create pictures in the mind)

IMAGE + IMAGE = 1 SENTENCE

1 SENTENCE = 1 TASK OR 1 BIT OF INFORMATION.

INFORMATION + TASK A + TASK B + TASK C . . . = 1 INSTRUCTION

FAMILIARIZE YOURSELF WITH STRUCTURED VOCABULARIES

There are three basic types of vocabularies that you may be required to use as a production technical writer. They are:

The open vocabulary.
The restricted vocabulary.
The limited vocabulary.

First, the *open vocabulary* has no formal restrictions that the technical writer must follow. Every word and every sentence structure can potentially be used. The technical writer has no restrictions when an open vocabulary approach is used.

The restricted vocabulary. The opposite situation from the open vocabulary is the *restricted vocabulary.* There are commercial sources where a manufacturing company can purchase a restricted vocabulary. These word suppliers computerize a selection of a company's technical support literature for vocabulary usage. The basic vocabulary from these instructions is entered into the vocabulary that every technical writer must use. The most used words are chosen for the vocabulary.

A restricted vocabulary may consist of 600, 800, 1,200, or an arbitrarily set number of basic words. Because there are approximately a million words in the English language, the word "restricted" is a descriptive one. A specific word that the company writer or writers find necessary can usually be added to this closed list of words. This type of writing has advantages and disadvantages.

A restricted vocabulary can be edited with a computer. A computer program can scan the words of the technical writer to see if they are in the vocabulary and if they are properly spelled. There is no problem with selection of words; usually there are so few of them that the obvious word is readily recognizable.

There are some disadvantages with this type of writing. First of all, this type of vocabulary is rigid and is almost as difficult to learn as is a foreign language. The writing process usually follows this pattern:

1. Write the task in your own words.
2. Translate the words that you choose into the words found in the restricted vocabulary.
3. Run the tasks through the computer for an electronic edit.

Because of the original development cost and the time that it takes for the technical writer to learn the core words and prepare tasks, restricted vocabularies are not used by many manufacturing companies. Even though the vocabulary is strictly controlled, the meaning of a task may not be as clear as possible because of the inability to use the best possible word for the situation. If the instruction is not written clearly enough to make the meaning accurate, then the instruction will not translate accurately.

Limited vocabularies. Another approach to the problem of word selection is the *limited vocabulary.* This is a more flexible approach than the retricted vocabulary. Usually the technical writer works from a recommended list of words prepared in the company style sheet. Another method is to give the technical writer the responsibility to limit his or her personal vocabulary. Limited vocabularies are extremely compatible with the visual writing style that is explained in this chapter.

EVALUATE "FOG INDEXES" FOR TECHNICAL LITERATURE

For many years people have attempted to write "formulas" that can be used to evaluate difficult writing found in reports, newspapers, and articles. These formulas usually consist of a procedure that includes the counting of sentences or words in a passage and relating the total count to polysyllabic words that are in the passage. The assumption is that words with many syllables are harder to read than shorter words.

The noted "readability" formulas include the SMOG method that was created by C. Harry McLaughlin; the Rudolph Flesch *Reading Ease Formula;* and Robert Gunning's *Fog Index.* These formulas are readily available through any search library system.

While these formulas may be useful for the expository writer, they are not a good measure for the technical writer. First, if you choose the best word that is the shortest, most concrete, descriptive or active, then you have eliminated the need for a readability formula. Second, "meaning" is of the utmost importance to the technical writer. The formulas do not measure verbal and spatial logic. For example, these two sentences have the same readability level:

> Ignitor turn to and the "on" the fuel turn.
> Turn the ignitor to "on" and turn on the fuel.

The readability is important; however, the degree of accuracy of the technical operation instruction can mean the difference between consumer injury and safe and efficient operation of the machine. Nothing can test the quality of your technical literature as can physically acting out the tasks just as your mass-consumer audience will do when purchasing your company's machine.

Limit Your Words

The best way to take fog out of what you write is to be consistent with your selection of words. Do this by limiting the words that you use. If an artificial vocabulary is not in place for you to use on the job, then use your linguistic self-control to set your personal restricted vocabulary. You will find in time that writing becomes faster and more productive as you practice using these selected words. A vocabulary list consisting mostly of verbs and prepositions are given below. These words have been selected because they often appear in technical operation manuals and directions in both the shortest preferred form and the difficult form.

A SELECTED VOCABULARY TO HELP YOU USE THE SHORTEST AND MOST ACCURATE WORD POSSIBLE

Difficult words and phrases	Simplified word (possible replacement)
a great deal of	much
a minimum of	least
abandon	leave
abate	decrease
abbreviate	shorten
abet	help
abolish	destroy
abridge	shorten
accelerate	speed up (sometimes two short words are the most descriptive)
accept	take
accommodate	fit
accompany	go with
accomplish	do
accordingly	so
accrue	get
accurate	correct
achieve	earn
acquire	get
activate	start
actuate	turn on
actual	real
adapt	fit
additional	more
adequate	enough
adhere	stick
adjacent	next to
adjust	turn
advance	go
affix	attach
afford an opportunity	give
agitate	stir
alert	warn
align	line up
all of	all
alleviate	relieve
allow	let
alter	change
amend	change
annually	yearly
antedate	precede
anticipate	expect
antithesis	opposite
apparent	clear
appear	seem
applicable	applies
application	use
apprise	tell
appropriate	proper
ascend	climb
as of	by
assemble	put together
assert	claim
assimilate	absorb
assist	help
assume	suppose
at a later date	later
at a much greater rate than	faster
at all times	always
at the time of	during
at this point in time	now
attain	gain
attempt	try
attire	clothes
avail yourself of	use
based on the fact that	because
be cognizant of	know
benefit	help
bestow	give
biannual	twice a year
bilateral	two-sided
capable	able
care should be taken	be careful

category ....group
cease....................................stop
characterize...........................describe
close proximity ....close
cognizant of ....aware of
combine................................join
commence ....start
compel..................................drive
compensate...........................pay
component ....part
comprise ....make up
concept ....idea
conclude ....decide
conclusion.............................end
concur ....agree
conduct ....carry out
confront................................meet
considerable ....large
consolidate............................join
constitutes ....makes up
construct ....build
consult ....ask
consummate..........................complete
convey..................................carry
data ....facts
decelerate ....slow down
declare..................................state
decrease ....reduce
deed.....................................act
deficiency ....defect
defunct ....dead
delete ....take out
demonstrate ....show
depart...................................leave
deprive ....remove
determine..............................test
detrimental............................harmful
difficult ....hard
dimension ....size
diminish ....lessen
disallow ....reject
discontinue............................stop
disseminate ....send out
duration ....time
egress...................................exit
elapse...................................pass
elementary ....simple
elevate..................................raise
eliminate ....remove
emphasize ....stress
employ ....use
encounter..............................meet
end result..............................result
endeavor...............................try
entire....................................whole
enumerate ....count
equivalent ....equal
eradicate ....wipe out
evident ....plain
examine ....check
exceed..................................go beyond
excessive ....too much
exercise care ....watchout
extend ....stretch
external ....outer
fabricate ....construct
final......................................last
flammable ....burns
frequently ....often
fundamental ....basic
furnish ....send
give rise to ....raise
gravitate ....move toward
identical ....same
illustrate...............................show
immediately...........................at once
impact ....strike
impair ....harm
imparts ....gives
impede..................................block
in conjunction with................with
in favor of ....for
in order to ....to
in place.................................on
in reference to ....about
in relation to ....on
in the course of......................during
in the event of ....if
in the near future ....soon
in the vicinity of.....................near
inaccurate ....wrong
incombustible ....fireproof
incorporate...........................blend
increase ....raise
inflammable..........................burns
inherent ....basic
initial ....first
insert....................................put in
lengthy ....long
locate ....find
maintain...............................keep
make every effort..................try
manufacture ....make
maximize..............................increase
minimize ....reduce
modification ....change
monitor ....watch
observe................................see
obtain ....get
occupy ....fill
occur....................................happen
omit ....leave out
operate ....work
option..................................choice
participate ....take part in
per annum............................per year
period of time ....time
permit..................................let

pertain ....belong
pictured ....shown
place ....put
postpone...............................delay
precede................................go before
precipitate ....cause
predominantly........................mostly
prepared to............................ready
preserve................................keep
previous................................earlier
primary ....first
prior to ....before
procedure.............................way
proceed ....try
procure ....get
prohibit ....prevent
project.................................plan
prototype ....model
provide ....give
purchase ....buy
pursue ....follow
reduce ....cut
refrigerate ....cool
relocate ....move
remain ....stay
require ....need
retain...................................keep
secure ....get
separate...............................set apart
submit ....send
superfluous ....extra
supersede ....replace
supply ....give
survey ....look over
thus ....so
transcend.............................go beyond
transfer................................send
transmit ....send
transparent...........................clear
ultimate ....final
understand ....know
upgrade ....improve
upon....................................on
utilize ....use
velocity................................speed
whenever ....when
whether ....if

CHAPTER SUMMARY

- A communication model is not complete until it shows the two hemispheres of the mind where two separate thought processes occur.
- For about 90 percent of the mass-consumer audience, the left side of the brain is the center of logical verbal thought.
- The right side of the brain is usually the intuitive side where knowledge is interpreted from pictures and visual stimulus.
- Both sides of the brain must have adequate information that can transfer thought through the communication process into logical physical action.
- If information is not complete for one side of the brain, the side with the most information may take charge. This can cause confusion and frustration.
- The English language has developed in three periods. The basic language has been layered through outside influences and social and industrial advancement.
- Any language that is spoken and written is a dynamic and growing thing.
- Individuals develop their language from a shared basic survival vocabulary that includes concrete words used to deal with immediate problems.
- Words for a homogeneous audience can be selected from a wider vocabulary than can words that are written for a heterogeneous audience.
- The technical writer who writes for the homogeneous audience must use the same visual writing style as the person who writes for an audience that cannot be measured.
- A concrete writing style includes the use of concrete nouns.
- Active first-person verbs create action from written tasks.
- The simple subject-verb-object sentence is the easiest to read and understand structure for your words.
- The imperative task sentence is the type of sentence that the technical writer will use.
- The compound sentence can be used to show simultaneous action.
- There are three types of vocabularies: restricted, limited, and open.
- Fog indexes and readability formulas are not as useful for the technical writer as they are for the expository writer.

EXERCISES FOR CHAPTER 3

Choose Your Language Tools Carefully

1. Choose five words at random from a thesaurus. List the synonyms that are given in order from the most common and easily understood word to the most complex.
2. Select one short instruction. Analyze it
   a. Are the tasks written as simply and clearly as possible?
   b. Has the technical writer made the best word choice?
3. Choose a short magazine article. Rewrite it as visually and concretely as possible after you analyze each word and sentence structure. Follow the guidelines in this chapter.
4. Find as many examples of jargon as possible. Interview people from all types of jobs. Define each example of jargon that you collect.
   a. Estimate the number of people who know the meaning of your jargon sample.
5. What have been the greatest influences on your own personal language? Describe how these events have added words to your basic survival vocabulary.

4

Develop a Visual Writing Style

The information given in this chapter will enable you to:

- Communicate with pictures.
- Describe what effect artwork will have on the consumer.
- List the types of words artwork can replace.
- Systematically select the best picture.
- Describe the similarities between the visual artist and the cartoonist.
- Draw a storyboard to plan the outline of an instruction.
- Define the terms "photograph," "line art," "drawings," and "charts and diagrams."
- Explain how to use the black-and-white photograph effectively in a technical operation instruction.
- Develop techniques to take good photographs.
- Explain the views that are available for line art.
- Finish artwork and prepare it for the final printing process.

INTRODUCTION

There are two distinct skills that you must develop in order to write with an effective visual style. First, you must be able to select the *specific concrete noun* and the best *first-person active verb* that can enable your consumer audience to create a concrete mental picture of the information you are sending. You must be able to communicate with verbal logic. One well-chosen word is worth many photographs in a technical operation instruction.

Second, you must be able to use physical artwork, or photographs and drawings, to give body and form to your technical operation instruction. Artwork makes your tasks and technical information jump off the printed page and come alive for your audience. The ability to choose the correct artwork is as important as the ability to select the correct words. Pictures are a necessary part of a visual writing.

Artwork also is important because you can use it to reduce the number of words to a minimum. One piece of artwork can take the place of many abstract complex nouns and descriptive adjectives. Not only can artwork help reduce the amount of words but it can quickly show the consumer where the action must take place. Pictures can logically, efficiently, and safely move the consumer around the machine. Well-chosen artwork helps eliminate the intuition and imagination of the consumer's thought process by giving the consumer the raw material needed to create a complete concrete mental picture of the instruction.

This chapter explains how to develop the skill to choose the right piece of artwork for the writing assignment and the situation. It explains how you can integrate visual images with words to "write" a technical operation instruction that will allow you to communicate with the largest percentage of your audience as possible. In a sense, this information will describe a visual technical writing method that can be used to transfer your printed words and pictures into specific physical "actions" by the consumer.

The chapter ends with a basic set of definitions that explain the major types of artwork available to the technical writer. As a production writer or student of technical writing, you will have many options when you begin the task of selecting the pictures that you will use to complete your writing assignments. You must be able to create the best possible piece of artwork to reduce your final printing costs and produce a usable technical operation instruction. The information in this chapter will help you match the quality of your artwork with your writing assignment and your audience.

COMMUNICATE WITH ARTWORK

In Chapter 3 you saw how the "split brain" concept can be used to help explain how the technical communication process takes place. Half of the mental capabilities of the consumer is dedicated to seeing, understanding, and translating words into pictures or mental images. By selecting the best word for the situation, you can communicate with the side of the brain that deals with verbal logic and one-step-at-a-time thinking.

You also can use the split brain concept to help explain the need for artwork. The "other" side of the thought process of the consumer can communicate best with nonverbal visual images. You must use various types of drawings and pictures to give the consumer a concrete image of where the action must take place. Artwork replaces such words as:

Up, down, behind, left, and right.

Big, blue, small, and round (see Figure 4-1).

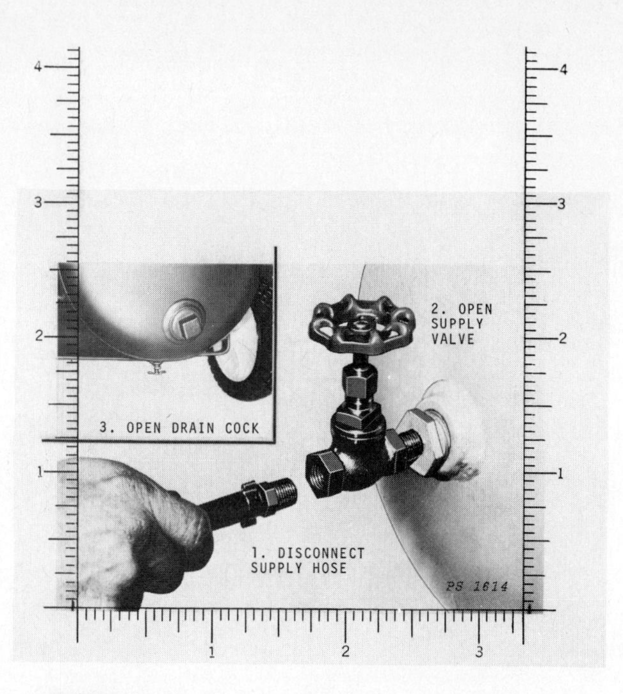

FIGURE 4-1 Pictures help reduce words to a minimum.

If you carefully select artwork it can:

Help you develop an overall sense of location for the consumer. It can show actual movement when you include in your artwork people who are doing the task that you describe (see Figure 4-2).

Show the relationships of the parts of the machine to the entire machine (see Figure 4-3). This is extremely important for large machines such as industrial robots, earth-moving equipment, and other machines that take up a large amount of space.

Locate the place where the action must take place.

Safely direct the customer's movements in the space around the machine and keep them positioned for the next task.

Artwork helps neutralize imagination. Artwork helps to prevent consumers from using their imagination or interpretation to create their own set of instructions. Words alone will not satisfy the need of the entire thinking process of the brain. The consumer also must see the machine as a visual symbol or image. By using artwork selectively, you may be able to replace the imagination and intuition of the consumer with a concrete mental picture. By using words and artwork together, you will not only be able to communicate with the largest percentage of

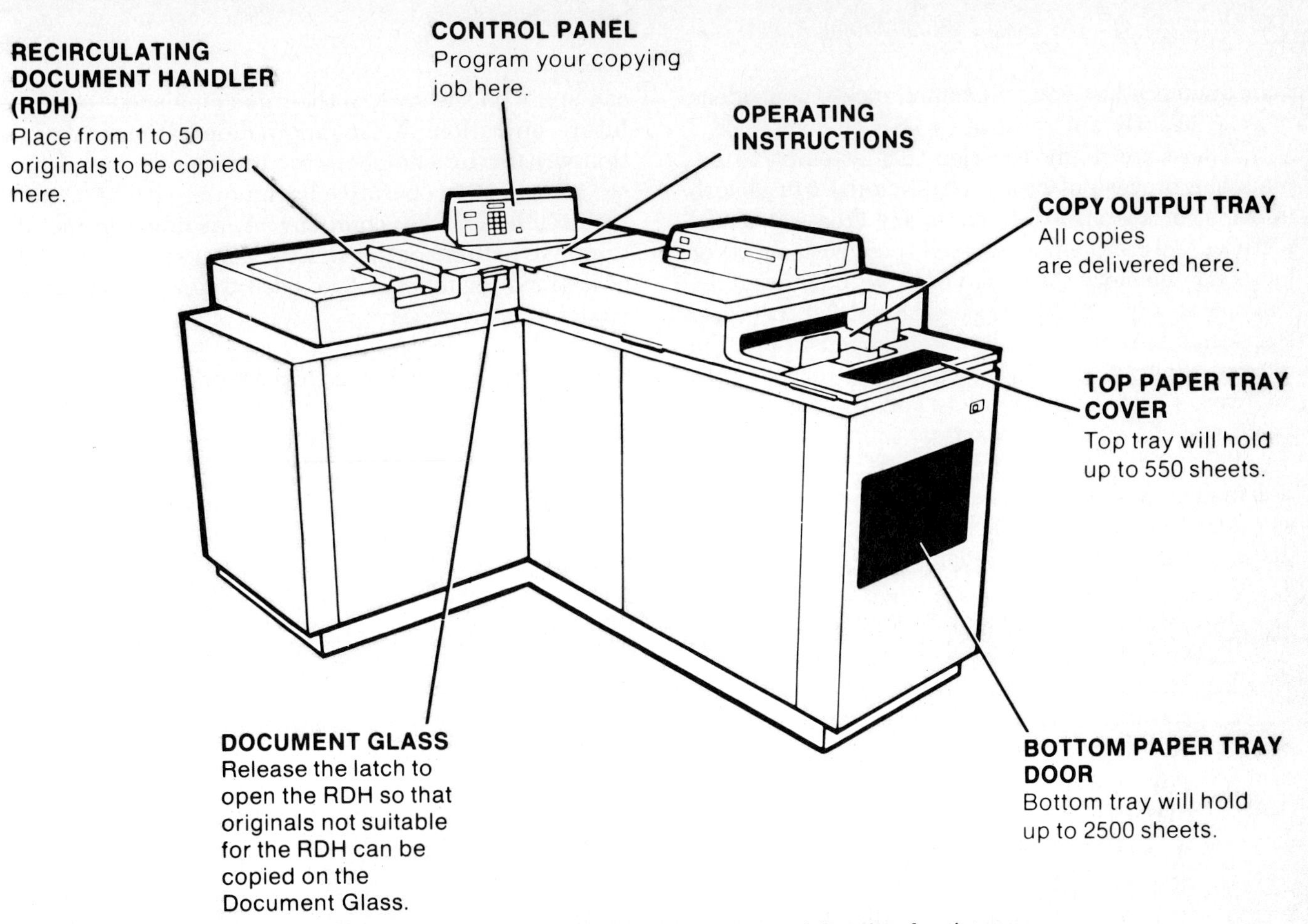

FIGURE 4–2 Pictures help develop a sense of location for the consumer. (Copyright 1981 by the Xerox Corporation, *Job Preparation and User Guide for the Xerox 8200,* 600P83889. Used with permission)

GETTING TO KNOW YOUR 1038 COPIER
ON THE OUTSIDE

Throughout this User Guide, we will refer to various components of the 1038 Copier. Those components are identified for your reference on this page and the next several pages.

Information Index

Your copier has eight different sets of messages which can be called up and viewed on the Message Display. The **Information Index**, to the left of the Control Panel, lists the subject matter for each message set.

These message sets are intended to provide you with information to help you use your copier correctly and effectively.

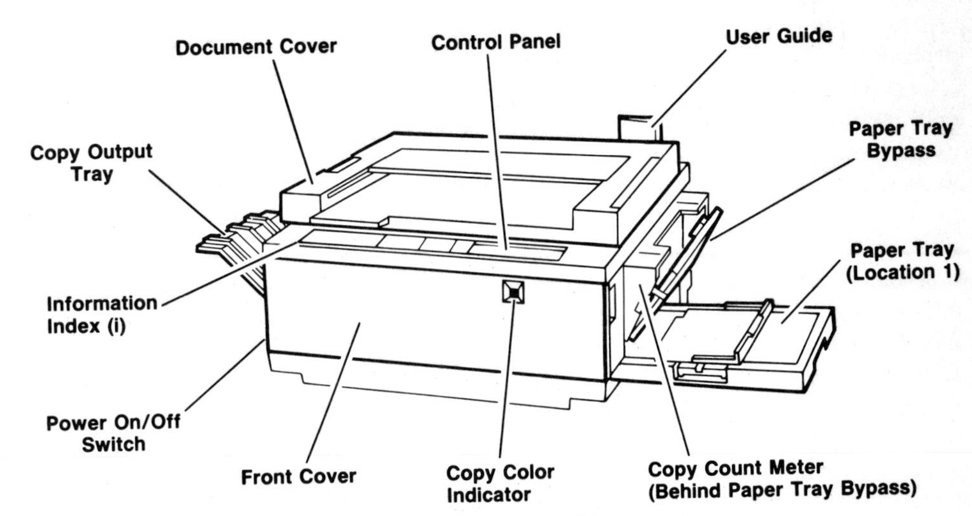

FIGURE 4–3 Pictures can show the relationship of machine parts. (Copyright 1985 by the Xerox Corporation, *The Xerox 1038 Reduction/Enlargement Copier User Guide,* 600P88316. Used with permission)

your audience but you will communicate your ideas more efficiently and accurately to each consumer.

There are many theories that attempt to explain the process of how the individual can absorb information through the senses and create "mental pictures," store them, and recall them with short- or long-term memory. Most models that have been developed to explain this process have used the basic steps that are outlined below. For the most part, the general thought is that the thinking process relies on "words" alone for abstract thinking, and mental "pictures" for concrete practical thinking. You can use this idea to see why employing a technical operation instruction to do a task is a visual process. Both words and artwork must actively create mental pictures. The basic steps include:

1. The consumer must interface with the technical writer's symbols.
2. The consumer translates the symbols according to past experiences into an understandable "code."
3. The spatial and verbal information is arranged into a logical sequence.
4. A mental picture or word definition is created and stored.
5. The knowledge is immediately used or saved for a later application.

The ingredients for a technical communication exchange. Three factors must be present before the technical communication process can be completed. The *consumer* must read the technical operation *instruction* while looking at the physical *machine.* If one factor is missing, the communication process is weakened. The instruction cannot be used effectively without the specific machine. Technical operation instructions are not self-contained works of art like a novel. They are relevant and useful only when the information in them is "applied."

In the past, instructions were written for a generic machine and could be used as a textbook to instruct the consumer or student as to how to operate every similar type of machine. Machines were basically the same. For example, a lawnmower was fundamentally like every lawnmower that was manufactured.

As a technical writer in our highly technical time, however, you must write specifically for one type of machine model because of the complexity of the machine. Each model of a machine requires a specific set of instructions. Consumers must be given specific mental pictures and tasks that they can apply on the spot to their machine and store for future operation. A lawnmower operation instruction written by one company must not be used by the consumer to operate a lawnmower that is manufactured by another company. A lawnmower operation instruction that is written for a specific *model* of lawnmower must not be used to operate another model of lawnmower.

Systematically Select the Best Picture

The consumer as an individual is able to understand even the most complex concept if it is written simply and in a small enough bit that he or she can easily understand. One picture and a few words is better than pages of words or pages of pictures. Another picture and set of words can create another small bit of information. If these digestible pieces of information follow in logical sequence, then a complex job can be done completely one step at a time without confusion and conflict of spatial and verbal logic. Knowledge and developed motor ability are more easily obtained if information is presented in *layers.*

In a sense, modern theories of communication favor the use of a "vignette" style of writing much like the ancient Egyptians used over 6,000 years ago. These vignettes are small bits of information that show a single idea or action with a drawing, and they explain it further with hieroglyphs, or words, that surround the image. Words and drawings were combined and chiselled into stone or written on papyrus to form an early type of visual communication that was basic and effective. This primal type of communication allowed ancient Egyptian technology, medicine, and society to flourish when other cultures were simply struggling for survival (see Figure 4–4).

The technical instruction vignette. You can further simplify and "modernize" this concept of communication by comparing the vignette of the ancient Egyptians with our modern comic strip. Although these two types of information transfer might seem to be incompatible, they actually obtain the same basic results. Comic strip vignettes also show action in a two-dimensional media. A comic strip character like Cathy Guisewite's heroine, for example, faces life in a two-dimensional world where action is created with images and surrounded with dialogue, sounds, and signs (see Figure 4–5).

The success of the comic strip artist depends upon how much action or how much humor the artist can convey to the audience in as few panels or drawings as possible. Unlike the ancient Egyptians,

FIGURE 4–4 Early writing depended on the relationship between words and pictures. (Courtesy Dover Publications, Inc., E. A. Wallis Budge, *Osiris & the Egyptian Resurrection, Vol. II,* 1973)

however, modern comic strip artists circulate their work on newsprint to a mass market. The action therefore is usually as simple as possible, and the words are usually as basic as possible in order to create a story line that is appealing to the mass-consumer audience who buys the newspaper.

Without the pictures that these experts create, the dialogue or words do not tell the entire story. For example, a punch line in a "Cathy" comic strip is never quite as humorous without the pictures. Remove the dialogue, sounds, and signs, and one can only guess at what is taking place in the story. The merging of artwork and technical instruction is similar to the visual artist. Ms. Guisewite tells how she merged the art and words into a humorous set of vignettes that explains a complex subject in as few words and pictures as possible:

> I created that particular strip after reading some serious articles on learning to cultivate the "right brain".... As often happens, I responded to the seriousness of the articles by imagining their theories taken to the extreme. I thought of the punch line and the picture at the same time, and then just had to figure out a scenario for the first three panels that would build to it.[1]

Balance art and words. Like the visual artist, the technical writer's "style" must be a balance between selected words and *active* artwork. The greatest mistake that can be made by technical writers who put words and artwork together for the first time is that they often rely too heavily upon artwork. You must not overly illustrate your technical operation instruction. Artwork that is not needed to keep the action moving is costly for your company. Artwork that is not needed takes up unnecessary paper and ink and adds extra cost to your budget. Above all, too much artwork can cause confusion for the consumer and slow down or even block the communication process. You must systematically evaluate and select artwork as carefully as you do the words that make up your technical vignettes (see Figure 4-6).

As a technical communicator who is objectively describing how to operate a machine, you can

[1]Cathy Guisewite, letter to author, July 5, 1987, housed in author file.

FIGURE 4–5 The comic strip is the result of a visual writing or communicating style.

CATHY

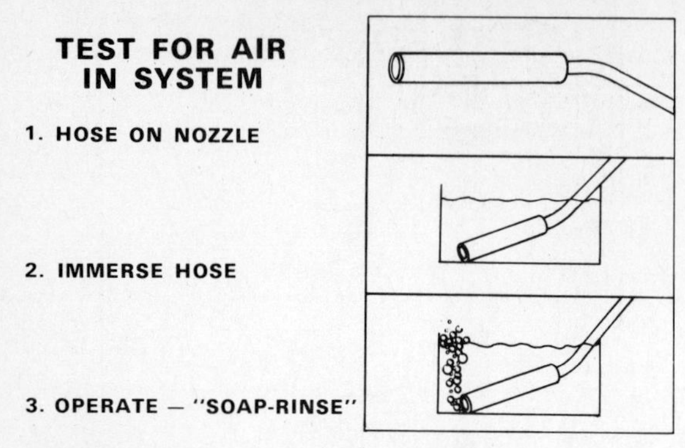

FIGURE 4–6 The modern technical writing "vignette."

use the techniques of visual communicators, including the comic strip artist, movie and television directors and script writers, and animated film cartoonists. Although your subject as a technical writer is quite different from the subject of these visual communicators, you can apply the same approach to communicating your technical information as they use to create entertainment or inform their mass audience.

These experts must share a common fundamental set of goals if they are going to be successful in reaching a mass-consumer audience. First of all, they must create as much action with as few words and pictures as possible. Second, every story, no matter what length it is, has a beginning, a middle, and a definite ending. Third, most of these people use a type of storyboard to develop the visual action systematically and unite this action with words.

A storyboard eliminates the *spontaneous* development of a story, and allows you to systematically plan every piece of artwork. For you, the technical writer, a storyboard will let you *plan* your photographs or other artwork in advance. This means that you will save time and money because you will not be creating artwork that will not be used. You will also not complete an instruction only to find that one piece of art is missing. In many writing situations, artwork is extremely difficult to obtain. By using the storyboard approach to outline the visual operation instruction, you will be able to write with an efficient, effective, and consistent style.

Use a Storyboard to Select Artwork Systematically

A storyboard is used by the visual artist to plan the contents of the comic strip, cartoon, or movie. You do not need to be a good artist to use a storyboard. Before you can write a well-constructed story you must begin with a dynamic outline that you can work with and revise as your ideas take shape. The storyboard is a simple outline of visual ideas that are quickly sketched and handwritten. Both words and sketches can be easily changed as information is added or taken away. The storyboard concept has been applied to the process of technical writing to pattern and group ideas and words together effectively. A storyboard can help you:

Eliminate the unnecessary artwork.
Eliminate the unnecessary descriptive word.
Plan the artwork that you must take or make.
Give spatial and verbal logic to the instruction.
Develop your ideas equally well both for the written technical instruction and the video instruction.

The first step in the development of a storyboard for technical communication is to use your right brain and mentally picture the task that you are going to explain. By breaking the technical operation instruction down into simple tasks, you can visualize the action with specific concrete mental images. In his essay "Politics and the English Language," George Orwell writes that, if one thinks in abstract terms, then one thinks in words from the start of the process. If you begin the thought process by thinking "wordlessly" in pictures, then you can hunt for and select the concrete word that describes the mental picture.

Write in mental moving pictures. Mentally complete the task that you must explain to the consumer. Begin at the start of the process and carefully imagine the steps that are required to operate the machine. An interesting phenomenon that is often pointed out by experimental and industrial psychologists is that it should take you the same time to do a job physically as it takes you to visualize it mentally. This can be helpful if you are describing a set of tasks that must be coordinated or timed.

The next step is to develop a method of selecting the *major* exposures of the sequence of artwork that you need. In the split brain concept of communication, the best way to begin any technical operation is to introduce the action by presenting a *general assembly* or *assembled* view of the machine that indicates where the action will occur. A popular example of how artwork can be used in this manner is the process of starting an automobile. Although this is an oversimplified example, you can see how the storyboard is developed in detail *for a large machine that requires many points of action.*

The general assembly artwork. The *general assembly* view can be used for many separate instructions (see Figure 4–7). For example, a general assembly drawing at the beginning of an automobile operator's manual can show the location of the parts that the consumer must control for these instructions:

Start-up (see Figure 4–8).
General maintenance and care.
Control of the car.
Optional equipment.
Basic safety information.

In an operator's manual, a bound publication that has many instructions, you may need to use more than one general assembly piece of artwork.

The general assembly also is usually put at the beginning or at the beginning of every major section of the operator's manual in order to introduce the tasks that will take place. The general assembly picture can be any type of artwork that realistically shows the location of the task or tasks that you explain in your technical operation instruction.

Use artwork to show location. The next piece of artwork should orient the consumer to the *general location* of the action for a specific instruction. In other words, artwork should not jump from one point to the next without showing a *transition* image. The transition image places the action in relation to other identifiable points of the machine. This artwork takes the place of the descriptive words that you would otherwise need to position the consumer properly for the task ahead.

If the transition art is not used, a specific description of how to start an automobile would begin like this:

> Open the automobile door and sit down in the seat on the passenger's side. Put your key in the ignition switch, which is located on the right side of the steering column that the steering wheel is mounted upon.

One piece of transition artwork can replace at least these 39 words that describe for the consumer the location of the action and how to get there.

Show action with artwork at a specific point. The third step is to complete the instruction with

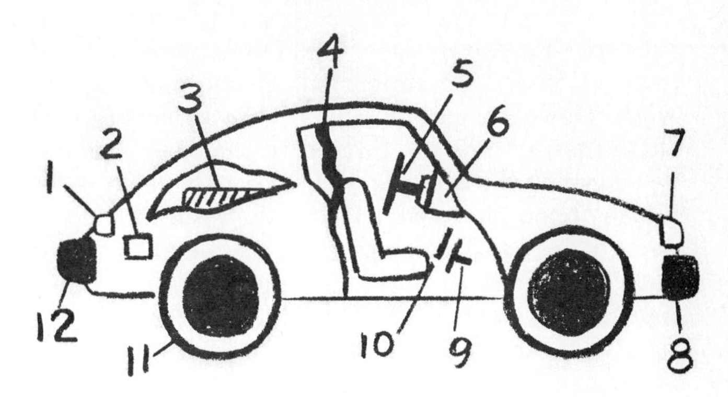

1. Tail Lamp
2. Gas Cap
3. Spare Tire
4. Seat Belt
5. Steering Column
6. Dash
7. Head Lamp
8. Front Bumper
9. Accelerator
10. Brake
11. Tire
12 Bumper

FIGURE 4–7 The general assembly sketch storyboard.

FIGURE 4–8 Think in storyboard concepts in order to write visually.

artwork that shows the action. Often more than one instruction can be supported by one piece of artwork. However, care must be taken to ensure that these instructions are limited to the ones that can be shown on the artwork without causing confusion. These action pieces of artwork show specific movement of the control points of the machine. You must keep in mind at all times that these action pictures must give the consumer all of the technical information that is needed to do the task in order to operate the machine safely and efficiently.

Artwork, then, can show three distinct situations. First, it can introduce consumers to the machine and give them a general location that shows where all of the instruction in that technical operation instruction will take place. Second, it can help you physically move the consumer around the machine. Artwork can orient the consumer to the point of action for a particular set of tasks. Third, the action is visually shown by the remainder of the artwork with symbols or by graphically depicting someone doing the task.

The storyboard. The beginning of the storyboard is "signaled" by a major heading that is larger in type than any of the other words on the page. A subhead signals the beginning of an isolated instruction. A following "third-level" subhead can effectively signal a set of tasks that allow the consumer optional information.

The following examples show how one-, two-, and three-level heads appear when they are used in printed instructions. The first example is a one-level heading. The major head on the page is:

CARE AND MAINTENANCE (see Figure 4–9)

The second example uses two-level heads to break up information. The primary heading is:

6. Copy operation

The secondary heading is:

How to make a copy (see Figure 4–10)

The third example shows how three-level headings can be used to break up information. The primary heading is:

Selections

The secondary headings are:

Water Temperatures
Agitation/Spin Speeds

The third-level headings are:

Hot Wash
Warm Wash
Cold Wash
Rinse Temperature (see Figure 4–11)

If you analyze the storyboards in Figures 4–7 and 4–8, which were created to instruct the consumer on how to start an automobile, you can see the following.

CARE AND MAINTENANCE

Your 43-207B represents a fine example of electronic engineering and construction. As such it should be treated accordingly. We offer the following suggestions so you will enjoy this product for many years to come.

If at anytime you suspect that your unit is not performing as it should, stop by your local Radio Shack store. Our personnel are there to assist you and arrange for service, if needed.

Keep it dry. If water should get on it, wipe it off immediately. Water contains minerals that can corrode electronics circuits.

Do not store in hot areas. High temperatures can shorten the life of electronic devices, damage batteries, and can even distort or melt certain plastics.

Do not drop your product. This will likely result in failure to operate. Circuit boards can crack and cases may not survive the impact. Handling your product rougly will shorten its useful life.

Do not use or store in areas of high levels of dirt or dust. The electronics may be contaminated. Any moving parts will wear prematurely.

Do not use harsh chemicals, cleaning solvents or strong detergents to keep your unit looking new. You need only wipe it with a dampened cloth from time to time.

FIGURE 4–9 The operation manual for a home intercom is basically a one-level head structure. (Copyright by the Radio Shack division of Tandy Corporation, *Plug'n Talk FM Wireless Intercom Owner's Manual,* Cat. No. 43-207B. Used with permission)

6. Copy operation

How to make a copy

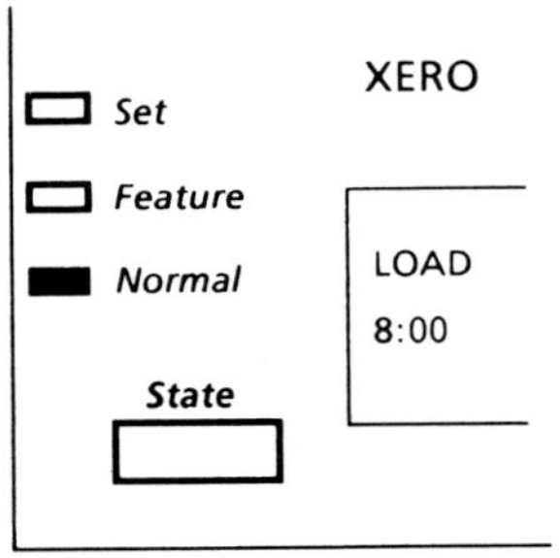

Set
Feature
Normal
State
XERO
LOAD
8:00

1. Select the normal state

Note: The normal indicator must be lit. If not, press the ***State*** *button until the normal indicator lights.*

2. Load original(s)

Note: The telephone handset must be hung up.

a. Place the original(s) face down in the ADF.

b. Adjust the document guides to the width of the original(s).

The display window will indicate:

DIAL # TO SEND OR
PUSH [START] TO COPY

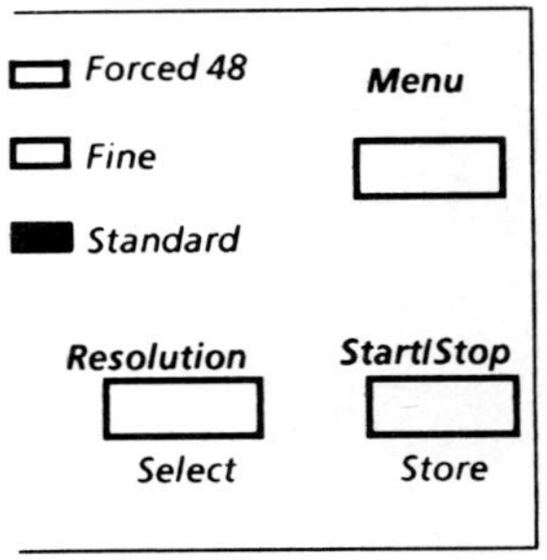

Forced 48
Fine
Standard
Menu
Resolution
Select
Start/Stop
Store

3. Start the copy operation

a. Press the **Start/Stop** Button.

The display window will indicate:

COPYING
08:00AM MANUAL RCV

Note: The copy operation automatically places the Telecopier 7010 into the Manual Receive mode and prevents the Telecopier 7010 from answering incoming calls.

FIGURE 4–10 The two-level head. (Copyright 1985 by the Xerox Corporation, *Xerox Telecopier 7010 Facsimile Terminal Operator Handbook*, 600P88404 Revision A. Used with permission)

Selections

Water Temperatures

Hot Wash Water enters the washer only from the hot water faucet. Use a hot water wash for white and color fast loads and items that are heavily soiled.

Warm Wash A mix of hot and cold water. Wash bright colors and machine washable wool in warm water. You'll also want to use warm water to wash medium to heavily soiled items.

Cold Wash The washer fills only from the cold water faucet. Wash sensitive colors (colors that may bleed or run) in cold water. Fruit and protein based stains can be successfully removed if they are soaked in cold water.

Rinse Temperature Your washer automatically provides a cold rinse. You'll save money and energy with the cold rinse. Cold rinses also help reduce wrinkling in permanent press and other synthetics.

Agitation/Spin Speeds

The agitation and spin speeds are automatically set for you.

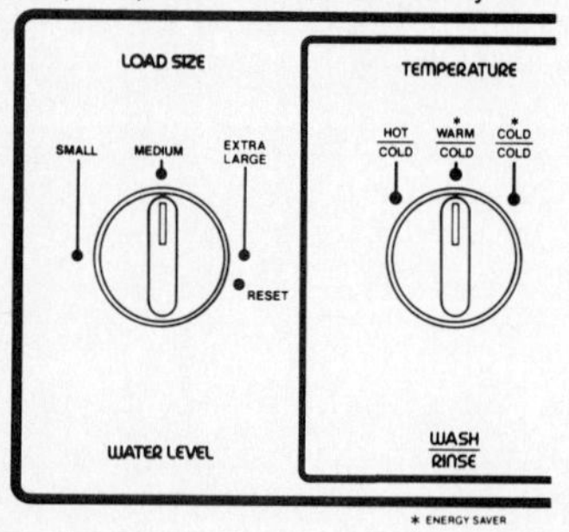

FIGURE 4–11 The three-level head. (Copyright 1987 Speed Queen, A Raytheon Company, *Operating Instructions for Home Laundry Automatic Washer Model NA 3311–3059,* Part No. 31575 4/87. Used with permission)

1. The general assembly artwork has a few cutaway sections in order to show parts of the machine that are hidden to a normal exterior view. The side of the automobile is cut away to expose the spare tire, and the door is cut away to show the interior of the dash area.
2. The general assembly artwork can be used once in order to show the consumer the correct methods of filling the gas tank, changing the spare tire, using the seat belt, changing the light bulbs in the front and rear of the automobile, and maintaining the safety bumpers.
3. The main heading for the section and general assembly picture in Figure 4-7 is "OPERATE THE CAR" because that is the information covered. The subhead announcing the task is "START THE ENGINE." It is shown in Figure 4-8. A third-level subhead can be added that further tells the consumer that there are "other" instructions for other starting situations. For example, the automobile must be started differently if the engine is hot.
4. Action is shown by people doing the task (see Figure 4–12).
5. Background material is removed from the drawing to keep the artwork as simple as possible. Pictures can be taken with a camera with the same effect.
6. Besides using pictures of people, arrows and other symbols can be used to show movement. The best way to show movement is to create the illusion of action in the artwork (see Figure 4–13).
7. Sketches and handwritten words (storyboard) can be revised easily.

The storyboard method of visual technical writing for a small machine is similar to the storyboard for a large machine. You must orient the con-

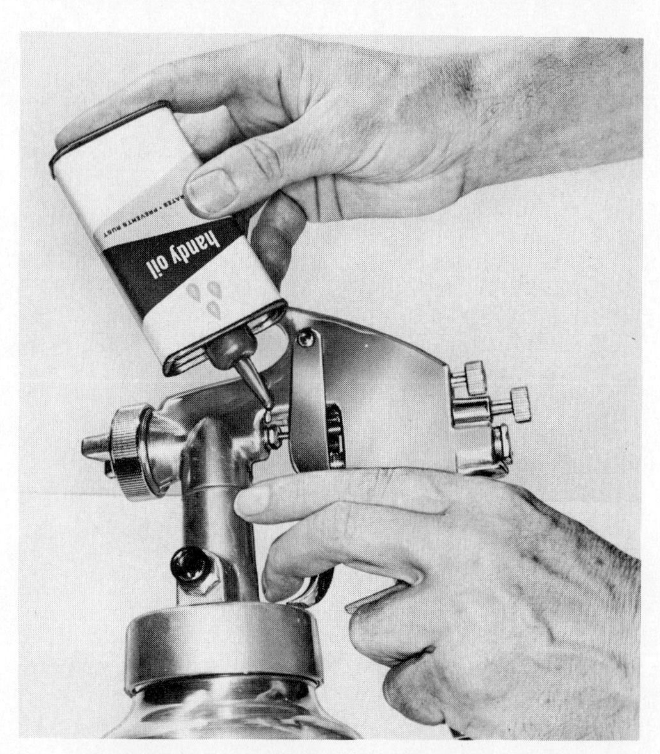

FIGURE 4–12 People doing the task create action.

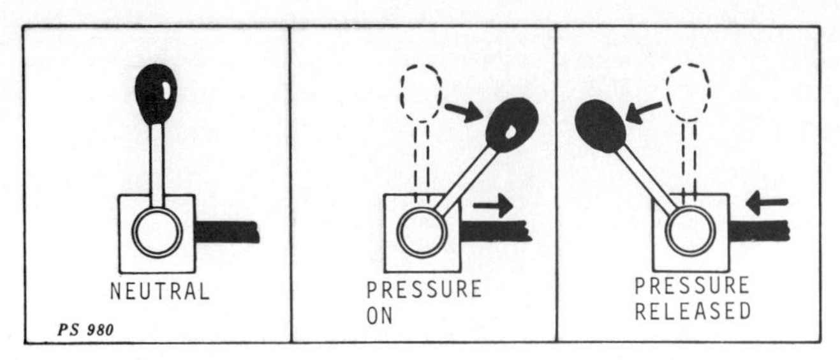

FIGURE 4–13 Arrows create an illusion of motion.

sumer to the machine before you begin the instruction or set of instructions. You must create action by showing as much as possible the tasks being completed by people. Usually small machines do not require the critical positioning of the consumer at the point of action for every instruction. The positioning of the machine is more important for correct operation for a small-machine instruction.

The storyboard can be used either to help you take the correct photographs or to have the exact artwork created that mutually supports the text of the instruction. Also, words can be eliminated or added as needed if they are shown in direct relationship to the artwork *before* the words are typeset. At this point you have the basic beginnings of either a written technical operation instruction or one that you can put on videotape. The storyboard will help you set up a shooting schedule for a videotape that allows you to get the right footage to create a complete video instruction.

When you are ready to create artwork you can reproduce the "rough" sketches that you have made. These reproductions can then either be cut out and sent to many art studios or given to the technical photographer who is "shooting" the machine. Your rough sketches have served the purpose of giving body to your technical operation instruction and helping you supervise the production of the artwork that you need to complete the writing assignment.

SELECT THE CORRECT ARTWORK FOR THE SITUATION

Part of the skill or craftsmanship that the technical writer must develop is to identify the correct type of artwork for the situation. Often the choice of artwork that is used depends upon the resourcefulness of the technical writer. You must exhaust every possible source of artwork that you can think of in order to *find* art instead of remaking it. Often artwork may be generated by many departments of a company. This artwork may be available if you can locate it.

Four *basic* types of artwork can be created for a technical operation manual. They include:

Photographs
Line art
Drawings
Charts and diagrams

Use Black-and-White Photographs Effectively

Black-and-white photographs are the most accepted and used artwork in a production technical operation instruction. Technical writing is overwhelmingly a black-and-white medium. Color and color photographs are rarely used in postsales literature. The use of color increases the overall printing cost of an instruction, and it adds time to the production process. The goal of the technical operation instruction is to *express* and not *impress* with color photographs.

Black-and-white photographs are called *continuous tone images* because they contain tones of black and white in various shades. They can be taken either with a camera that has an "instant" develop capability or with one that requires actual processing and developing of the film. A standard 35mm camera is an example of a camera that requires processing and developing of the film. Both types of film can be taken with high-quality cameras and lenses including twin lens reflex, single lens reflex, and studio cameras (see Figure 4–14).

You must look at the cost, quality, and convenience and choose the best method of photograph production that fits your situation. This is important because you are ultimately responsible for the creation of the artwork that you use in your instruction. In many writing situations in industry, the technical writer also takes the pictures. The technical writer also may supervise the taking of photographs by people inside the company or by outside photographic studios (see Figure 4–15).

Set standards for picture quality. You must ensure that each picture that you create meets your standards. Often quality of the photograph depends upon the budget that is allowed for a particular writing assignment and the environment where the photograph is taken. One important point to remember is that if the art is not understandable or identifiable, then it is as much of a hindrance to visual communication as the wrong choice of a word. Basically,

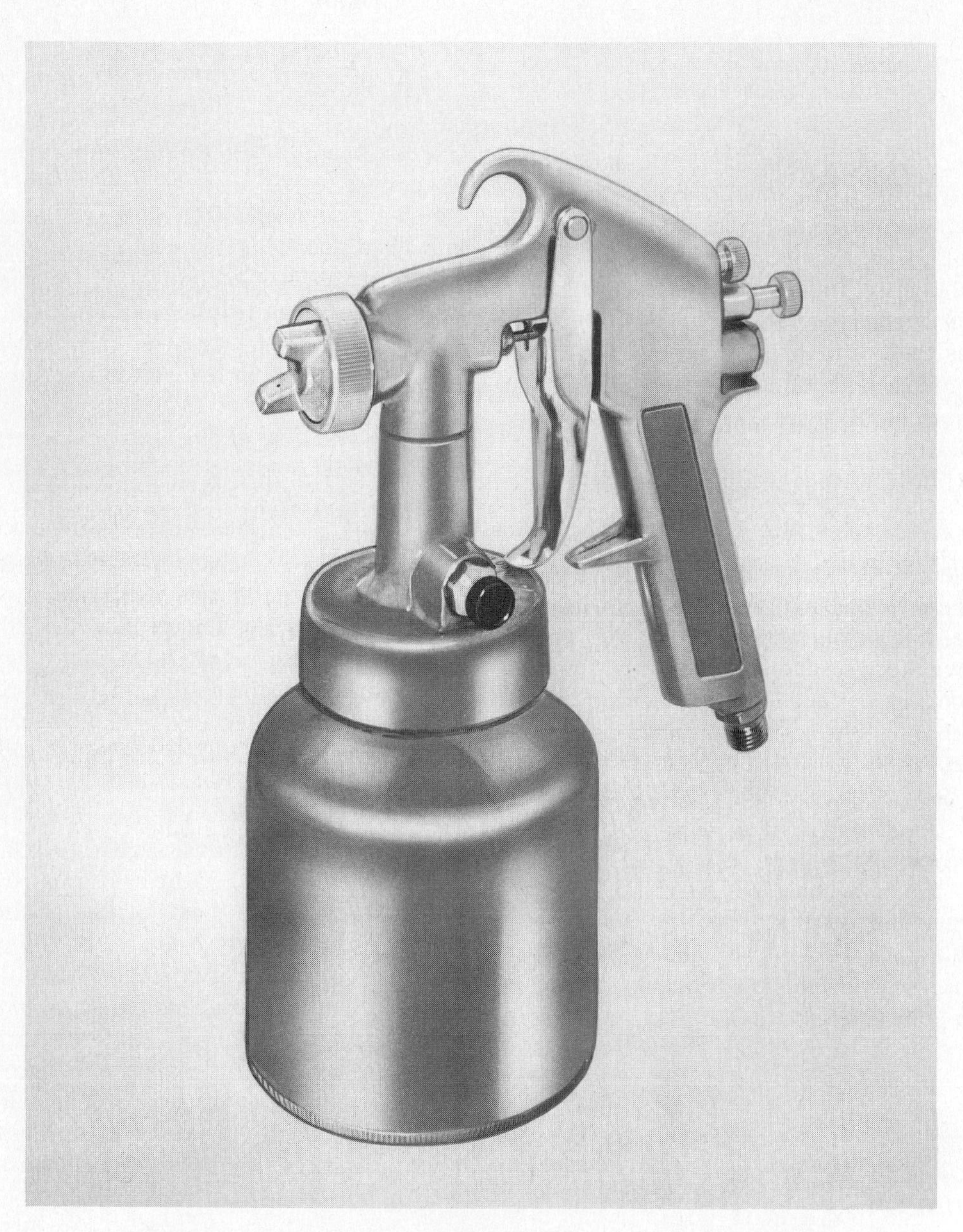

FIGURE 4–14 A continuous tone photograph—a spray canister.

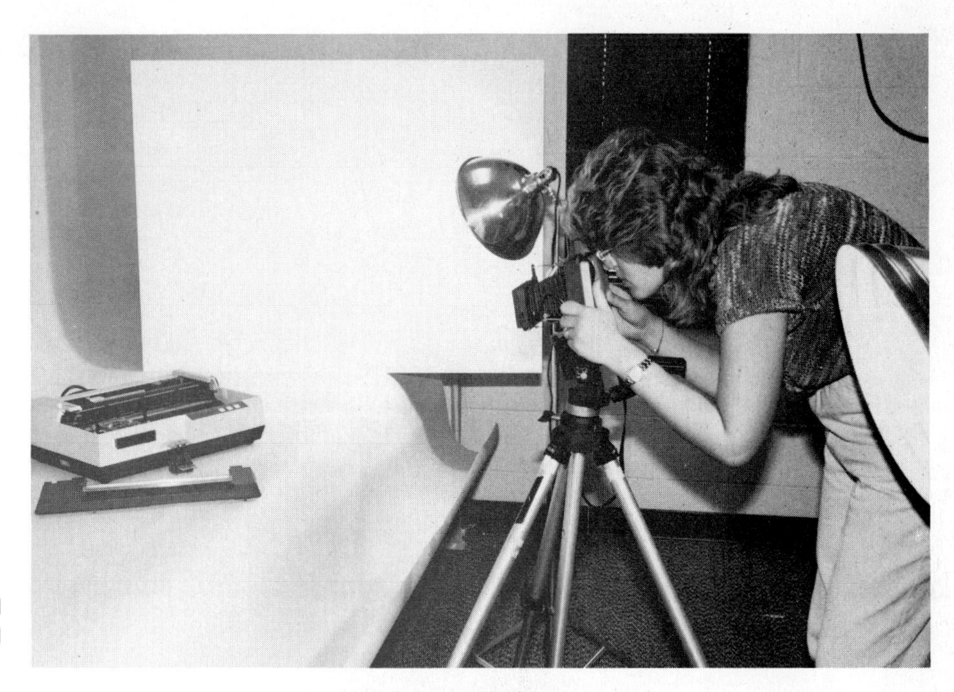

FIGURE 4–15 The technical writer also may be the technical photographer.

all "good" photographs have the following characteristics:

It must have a "glossy" or smooth finish if it is to reproduce well during the offset printing process.

It must not be taken from an unnatural angle or level.

It is correctly framed.

It has sharp contrast. This means that pictures with glare, bad exposure, or poor focus must not be used (see Figures 4–16 through 4–19).

FIGURE 4–16 An underexposed dark photograph.

FIGURE 4–17 An overexposed light photograph.

FIGURE 4-18 Shadows can destroy contrast.

FIGURE 4-19 A usable photograph with correct exposure and high contrast.

Glare is a problem when taking pictures of plastic and metal parts. An anti-glare spray or a nonglare paint can be used to highlight machine parts if necessary before the picture is taken. To eliminate the need to retouch photographs, make sure of the following:

> A clean background is behind the item that you are showing. This keeps the photograph background from being distracting (see Figures 4-20 and 4-21).
>
> A clean background is created by light gray or white paper.
>
> A sharp contrast exists between machine parts. Parts can be repainted if they are dull. If you use black-and-white film you can paint parts different colors and not confuse the operator. If a machine is filmed in color for color pictures in the instruction, then different colors may confuse the operator.
>
> Always try to *frame* the object properly within the photograph. This means that the picture of a machine part must be taken far enough away from the object to show it in relationship with other parts. It must not be taken so far away

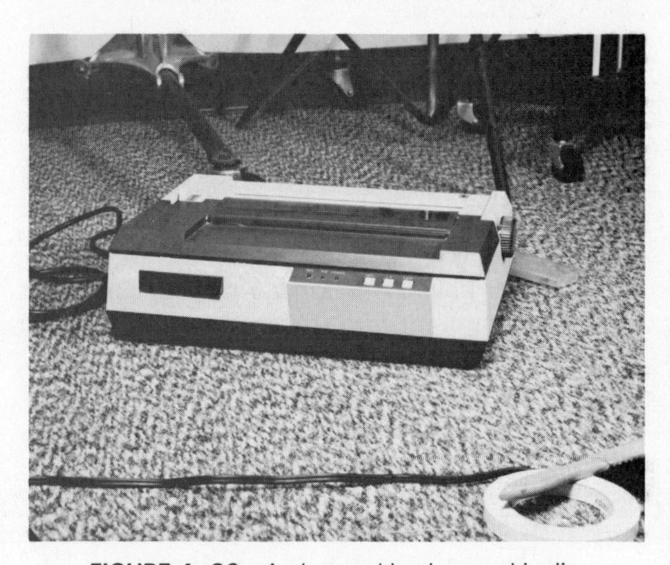

FIGURE 4-20 A cluttered background is distracting.

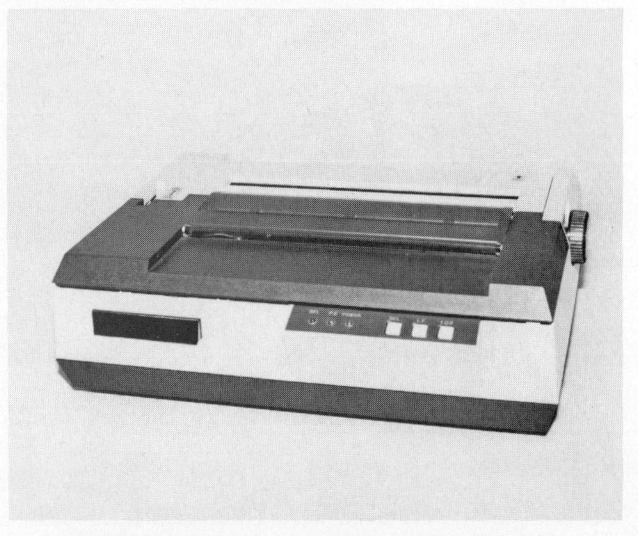

FIGURE 4-21 A good background makes the image jump off the page.

that details are omitted. The part is usually best depicted if it is in the center of the photograph.

If a poor quality photograph is the only artwork that is available, then you must try to make it usable. This can be done by *graphic artists* either inside the company or at art studios. Graphic artists, among other things, can "retouch" and "edge-up" photographs in order to sharpen lines for better contrast. These professionals can also use an "airbrush" technique to cover up the background or bring out the highlight of a section of the photograph. Most of these artwork enhancements can be done easier with photographs that have been developed remotely from the camera. "Instant" pictures are coated with a gel and are often difficult to retouch and airbrush.

Retouch techniques. Some common techniques that are used to "retouch" art when the picture-taking environment could not be controlled include airbrushing the picture with white paint to block out the background and highlight lighter surfaces. This is the most effective way to retouch most photographs; however, it is also the most expensive (see Figure 4-22). Another way is to block the background by opaquing the negative of the photograph. This also is an expensive but effective way to remove clutter. The negative should be retouched if you are going to use more than one positive print from the negative. This saves you the expense and time of retouching more than one copy of the same positive (see Figure 4-23). Two methods that are less expensive are the blocking of the background with a white film directly on the photograph positive (see Figure 4-24) and the blocking of the background

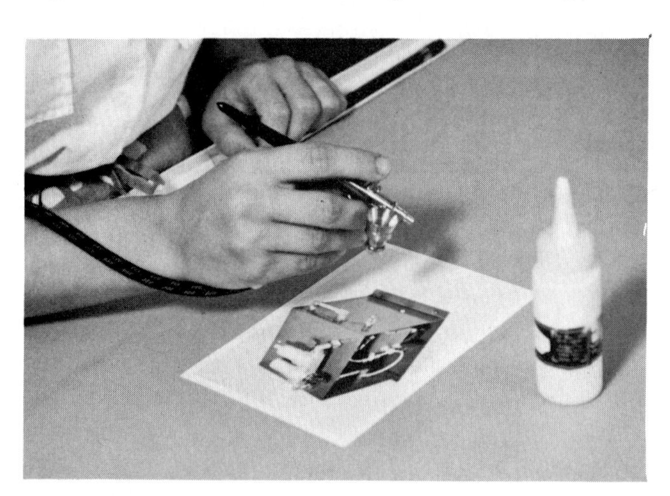

FIGURE 4-22 Opaquing technique with airbrush.

FIGURE 4-23 Opaquing technique—blocking the background.

FIGURE 4-24 Blocking out the background with a white film.

with red colored film on top of a clear plastic overlay that protects the photograph (see Figure 4-25). Red film can be used if a color is going to be added to the picture. An example is the addition of a red color to give background to a safety symbol.

Often a skilled graphic artist can outline or edge-up the photograph. This gives a slight outline to parts of the picture that otherwise may be lost. (See Figures 4-26 and 4-27). Remember, all retouching that is done to the photograph *after it has been shot* usually is going to be more expensive. It is better to create a good environment before the photograph is taken (see Figure 4-28).

When necessary, color can be used to show important contrasting information such as the color coding on the operating controls of a large combine, a robot, or other machine that requires quick identification of the parts or systems. Color in a postsale

FIGURE 4-25 Blocking out the background with a red film.

FIGURE 4-26 Edging up a photograph by outlining in pen.

technical operation instruction is used to contrast information. For the most part, color is used for the presales advertising literature to impress and persuade the potential consumer to buy the machine. Contrast can be obtained in black and white by using shading, different surface textures, or line patterns on the various types of line art drawings.

Control the Environment During Photograph Sessions

Taking the best photograph as possible can be a complicated and difficult procedure. You must first develop the ability to use the camera. You must be able to set the adjustments to correlate with the film to get as much contrast and detail as possible in the photograph. Researching a good book on basic photography is necessary if you are going to get the results that you want. However, there are a few things that you can do to increase your photograph quality in an industrial setting that most photography books will not cover.

First of all, always try to control the environment where the picture is taken. A good parallel and

FIGURE 4-27 The results of edging-up. The hand and screwdriver have been outlined.

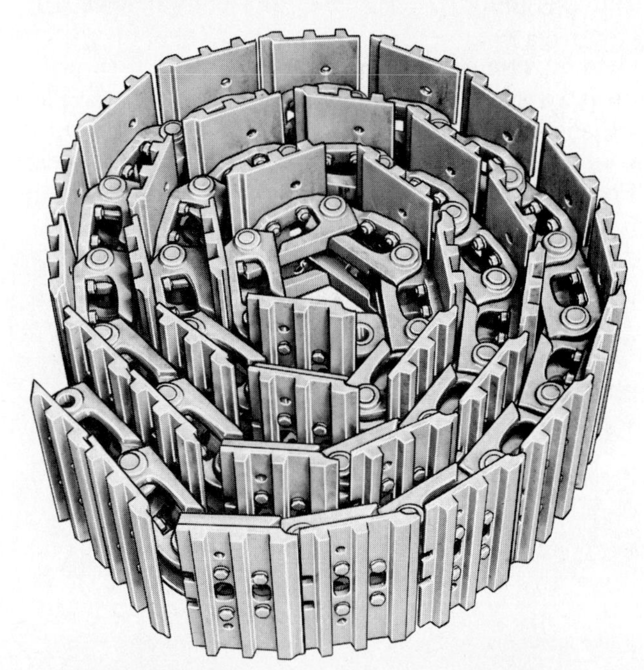

FIGURE 4-28 The result of an expensive edging-up and retouching is a sharp image full of detail.

FIGURE 4–29 A bounced light is diffused and creates a softer reflection off metal and plastic.

example of this concept is the use of a sound stage for the shooting of a movie. Although the illusion is given that the actors in films are taking a leisurely drive in the country, or are hacking their way through a jungle, the scene is usually shot on a sound stage where the sound and lighting for the camera can be controlled. By controlling the environment, interruptions and unexpected situations can be reduced or eliminated.

A controlled environment for the photographer includes a room or area that is relatively free of dust and dirt. The walls and floors should be painted light gray to create a consistent background for the machine. A storage cabinet for the photographic supplies should be used to protect the equipment when the studio is not in use. The room or area also should include an overhead crane or hoist if the machine is difficult to move.

Tripods for both the light source and the camera will greatly enhance photo quality. By supporting these staple items of the photography studio, you will be able to control your situation and make adjustments quickly and efficiently. A small table for tabletop photography of smaller machines and machine parts is also a convenient item.

A machine that has a high gloss or shiny surface must be photographed with care if the contrast of parts is going to be created. There are two basic methods you can use to photograph this type of machine. First, you can use indirect *diffused* light sources. The flash attachment or tripod-mounted light sources can be fronted with thin paper or cloth. This eliminates a direct flashback from the light during exposure. Do not put paper or cloth directly against a light source. Heat can build up and burn the material. Heat buildup can also cause premature bulb failure. You also can diffuse the light by bouncing it off a surface before it reaches the machine (see Figure 4–29).

A second method to help create contrast is to make sure the machine is painted and that there are no dull-colored areas. The beginning of a photo session for a larger machine may include many hours of painting and cleaning in order to bring out the features of the plastic and metal machine parts. Along with this enhancement of contrast between machine parts by painting are other ways to highlight the surface lines.

To reduce or eliminate the need for retouching the finished photograph, you can "pretouch" the machine parts. Lines of the machine can be highlighted with chalk, graphite, and other shading methods that will bring out detail in the finished photograph. Highlighting the machine part with a white deodorant spray is a technique that is used by Deere & Company of Moline, Illinois. By varying the amount of spray, the photographer can control the contrast of the machine before it is photographed. Another technique that can be used is to spray the machine parts that cause glare with an aerosol called "Dulling Spray" or an equivalent. This spray mutes the metal or plastic surface of the machine and helps give depth to the machine parts.

Caution: Be sure that whatever you use to pretouch and highlight the machine does not harm the surface of the

FIGURE 4–30 A horizontal light source helps highlight the image. Two lights, one on each side of the machine, are required.

machine or damage seals and gaskets. Carefully clean all the highlighting material from the machine. Sometimes machines that are used for a photo session will be used for testing or machine promotion. Sprays and aerosols that can damage the machine can be extremely costly to your company. It may be less costly in this case to simply retouch the photographs.

Machines with textured surfaces may require a horizontal light source that illuminates the area to be photographed. Direct lighting in this situation should be avoided because it will fill in the texture of the surface. Most problems encountered by the photographer and technical writer are lighting problems and not mechanically caused problems. Control the environment as much as possible and experiment with the various techniques mentioned above (see Figure 4–30).

Use Line Art Effectively

Line art is similar to the engineering drawing that you will cover in Chapter 5. It is the representation of the machine or machine part with lines. It does not have a continuous tone that helps to give features to the artwork. Lines show the physical *outline* of the machine. The graphic artists within a company and at most art studios can create line art. These drawings can be used in place of black-and-white photographs in some situations (see Figure 4–31). Line drawings have some advantages over photographs, such as:

They can be drawn without a machine. Sometimes a piece of artwork is needed and a machine is not available for photographing. By using a composite of photographs and engineering drawings, a graphic artist can create a line drawing with as much detail as is needed to communicate your task.

They can be drawn selectively to show only the parts that you are writing about (see Figure 4–32).

They can be used when retouching a photograph is not possible or practical. A line drawing can be drawn from the photograph (see Figure 4–33).

They can be used when parts must be added or taken away from a picture. For example, if an engineering change is going to be made to a part of the machine, then the change can be made on the line art to keep the instruction updated for the future. There is always a gap between the time when the engineering change is announced and the actual change. There is also a corresponding gap between the time a technical operation instruction is written and the time when it is printed for distribution.

Line art is easier and less costly to reproduce in the printing process than are photographs.

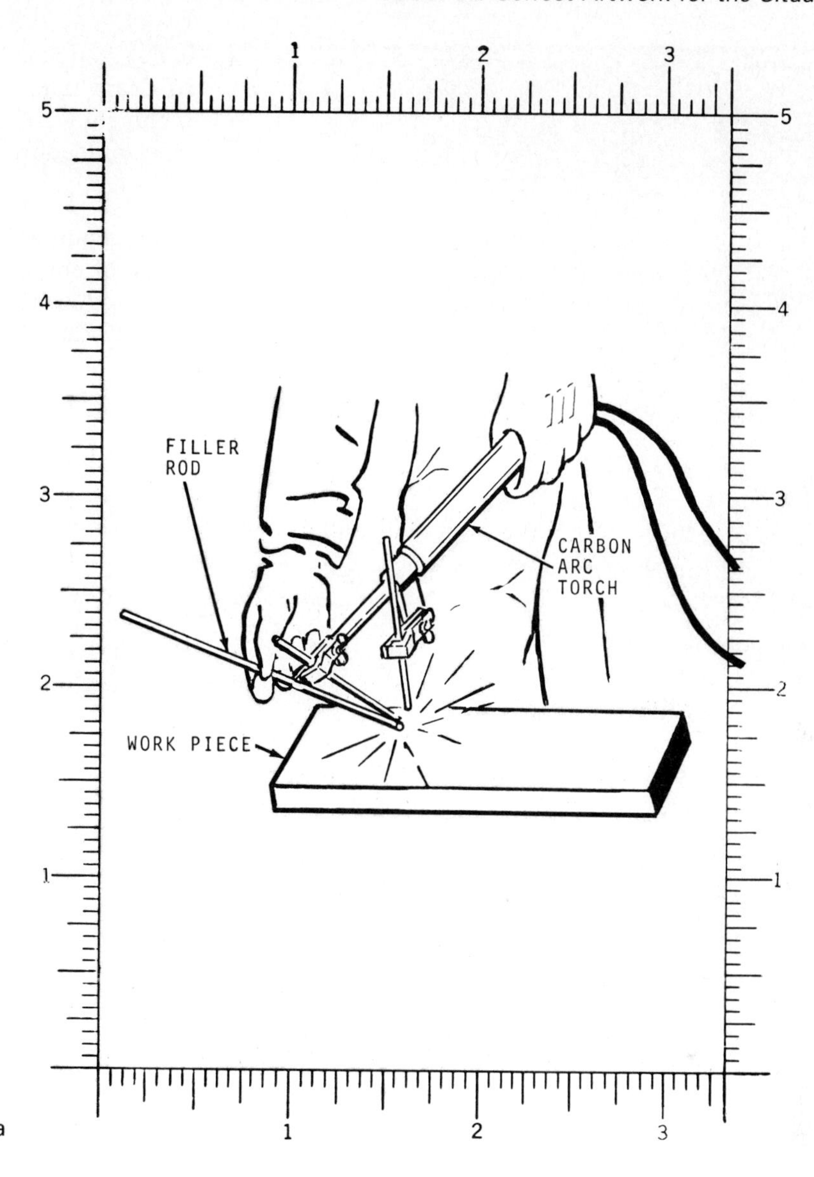

FIGURE 4–31 A line drawing of a welding unit.

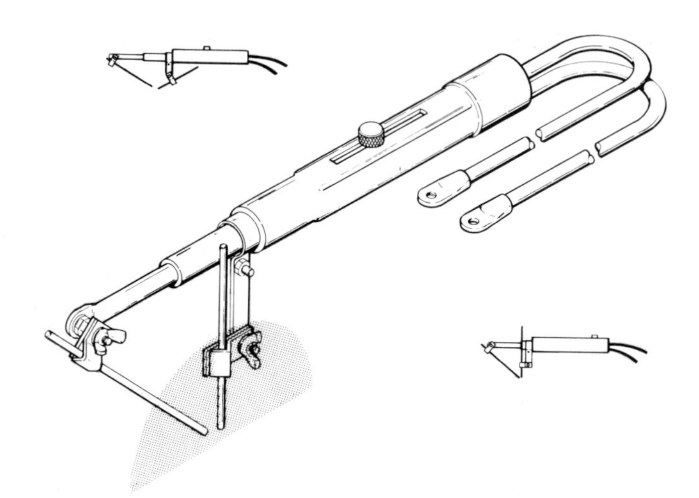

FIGURE 4–32 A line drawing can show selected information.

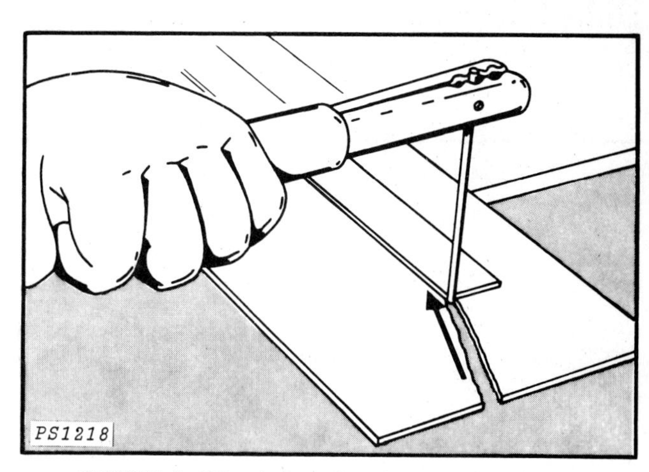

FIGURE 4–33 A line drawing can be traced from a photograph.

FIGURE 4–34 Computer software can be created to draw line art.

The disadvantages of line drawings are:

They often are not as *realistic,* and do not show as much contour and detail, as photographs. This is usually not a problem except when it comes to showing the inside of complex machines or other views that require many lines.

They can be very expensive to create, depending upon what is shown and how much detail is included in the artwork. Budget restrictions play a big part in the operation of every technical writing program in industry.

A detailed piece of line art may take many days to draw. The second major influence upon a technical writing program is the necessity to meet a deadline.

Many companies are beginning to recognize the ability of computer-aided design hardware and software, and other electronic art generating machines, to produce line drawings for use in the technical operation instruction. This technology is in the embryonic stage at this time; however, advancements and new machines are entering the marketplace every day. Consider electronic-generated artwork if you need a line drawing. It is usually readily available, and with minor revisions it can be applied to the instruction (see Figure 4–34).

A method that most graphic artists can employ to create line art is to use a copy of an engineering drawing as a base. The secondary information lines and data can be removed from the copy of the drawing with white paint. All lines can be retraced with a lead holder or ink and a straight edge. The drawing, no matter what size it is, can then be reduced. Usually a larger drawing will need to be reduced further to be usable. If this is necessary, the drawing may need to be retraced again before reduction. This method can be done by the technical writer within the company and with easily acquired tools (see Figure 4–35).

If you order the art from a graphic artist, then you must specify the *view* before a line drawing can be created. These basic views are the:

FIGURE 4–35 A line drawing can be made by tracing an engineering drawing in ink.

Pictorial
Exploded
Sectional
Phantom

Pictorial view. Pictorial drawings are created as close as possible to what is seen through the lens of a camera. They can be drawn from different angles. One of the most common angles and the most inexpensive to draw is the *orthographic projection.* Orthographic projections are drawn just as you would look perpendicular to the item (see Figure 4–36). Views that can be used in this type of graphic drawing are the front, side, rear, bottom, and top. Projections also can be drawn realistically in 30-degree angles (isometric), and in 45-degree angles (axonometric). Both the isometric and the axonometric views are more expensive and difficult to create than the orthographic view (see Figure 4–37).

Exploded view. This type of line art shows how the pieces of a part or system fit together. Pieces are drawn as if they are exploding away from each other on the same plane. This drawing is used to show relationships. In a technical operation instruction, the exploded view is usually used to show how basic maintenance items fit together. Otherwise this type of artwork is used in parts catalogues and sheets that explain how to order replacement parts, and in technical maintenance manuals to explain how to diagnose and fix the malfunctions of the machine (see Figure 4–38).

Sectional view. A sectional view shows the internal workings of the machine or machine part. The machine or machine part is shown as it would appear if it had been cut in half. In this type of artwork, the outlines of the parts that lie beneath the outer skin of the machine can be depicted (see Figure 4–39).

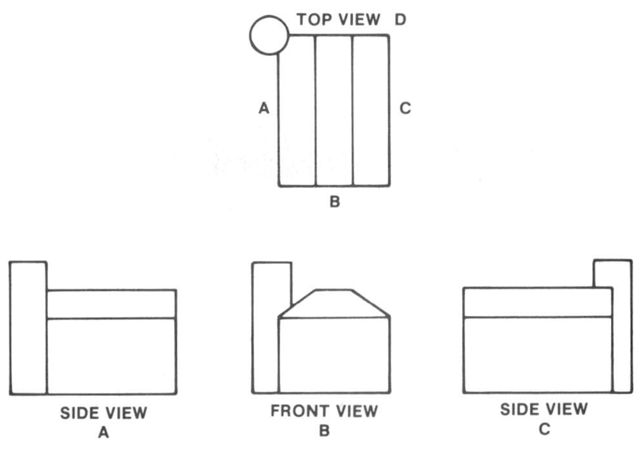

FIGURE 4–36 Orthographic views (Standard U.S.)

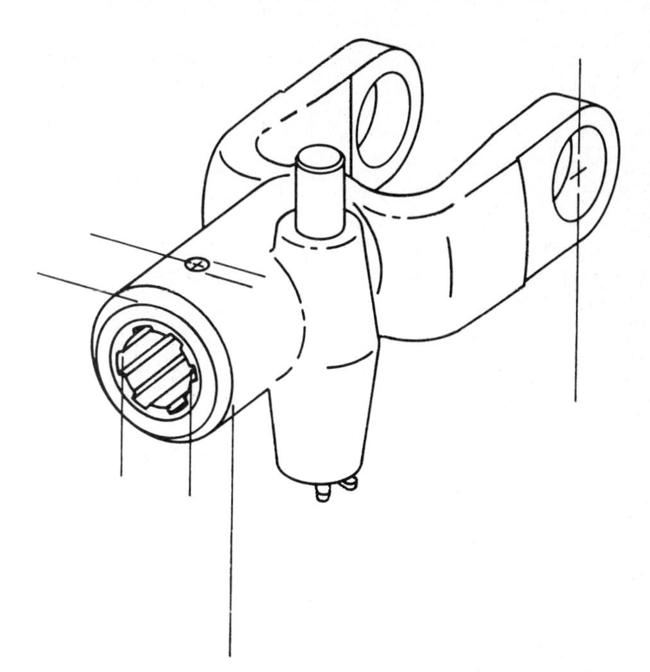

FIGURE 4–37 The 45-degree angle drawing is more realistic than orthographic views, but it is more expensive to produce.

Phantom view. In a phantom view, the machine skin or superstructure is peeled away to show some of the hidden parts that the consumer should be aware of during the operation of the machine. Phantom views help to give realism to the picture by showing the outline of the part or machine. The stripped away skin helps the consumer concentrate upon the inner workings of the item (see Figure 4–40).

Make Use of Existing Drawings

Any line art can be transformed into a realistic drawing by a skilled graphic artist. Line art can be shaded with a dot pattern, color, or texture to produce a more true-to-life piece of artwork. However, this process can be extremely expensive. Usually this type of realistic artwork appears in sales brochures and other presales literature, technical textbooks, and other informative types of technical literature. Many drawings rival the realism of the photograph. Usually this type of artwork is generated by the advertising or sales departments of the company. If the art is available and is the best piece of artwork for the situation, then you should be resourceful and attain it for your writing assignment (see Figure 4–41).

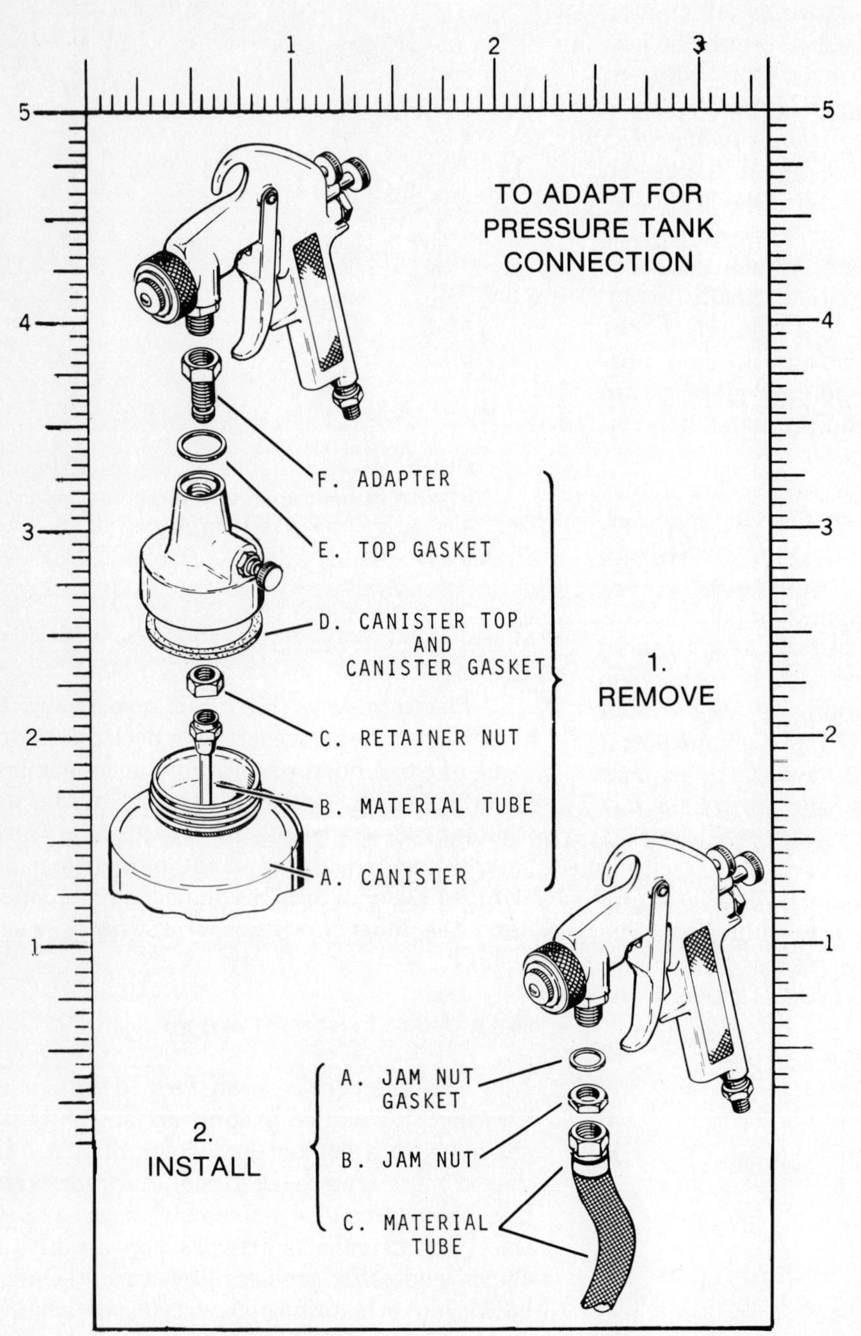

FIGURE 4–38 The exploded view drawing.

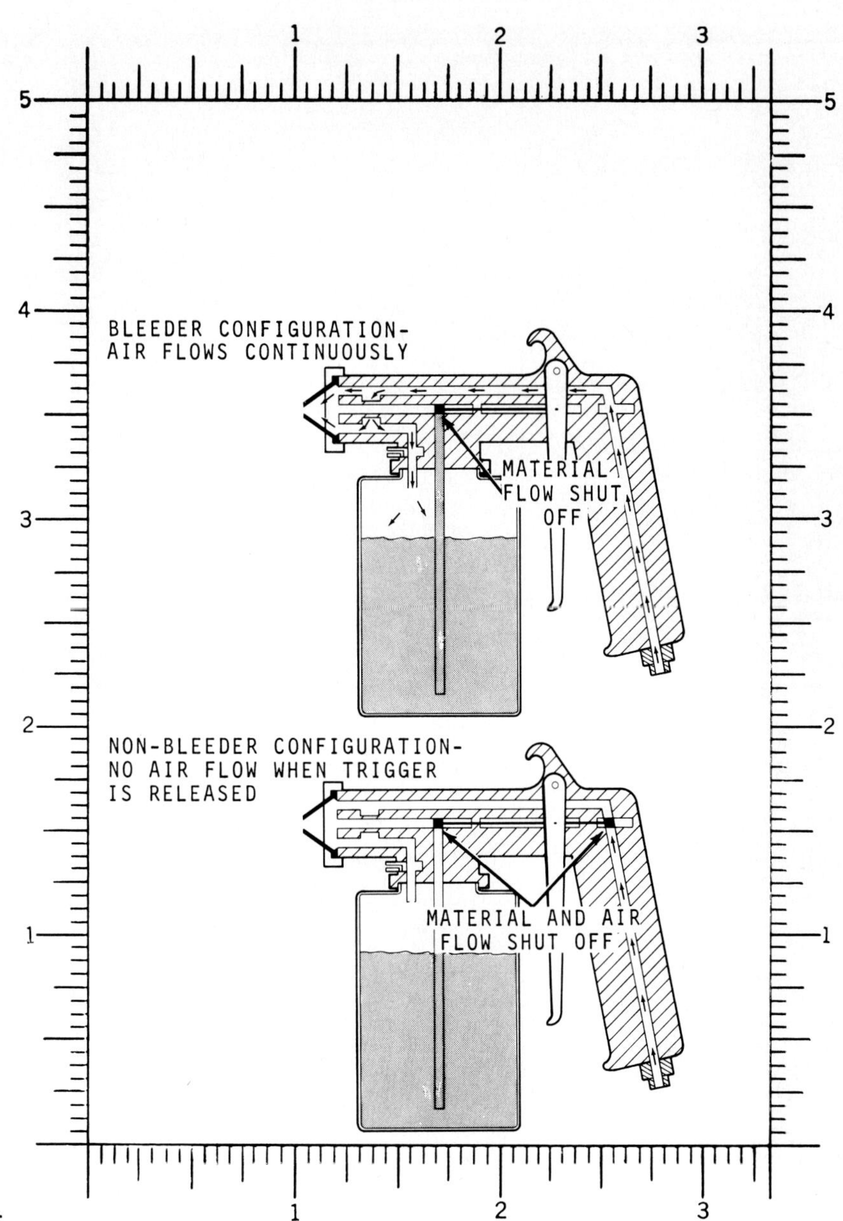

FIGURE 4–39 The sectional view.

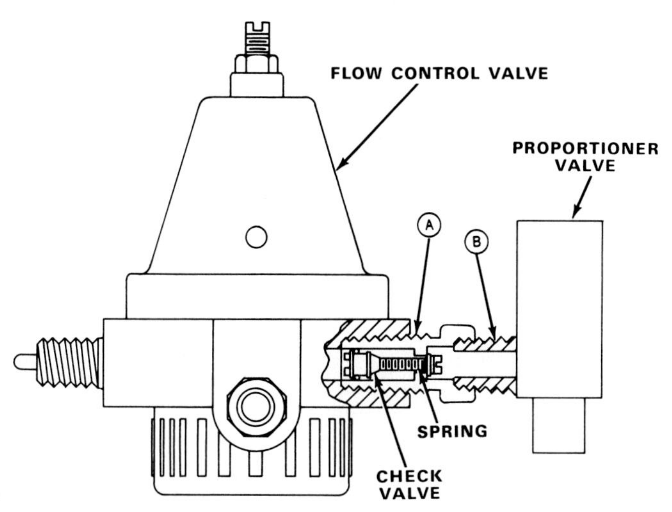

FIGURE 4–40 The phantom view.

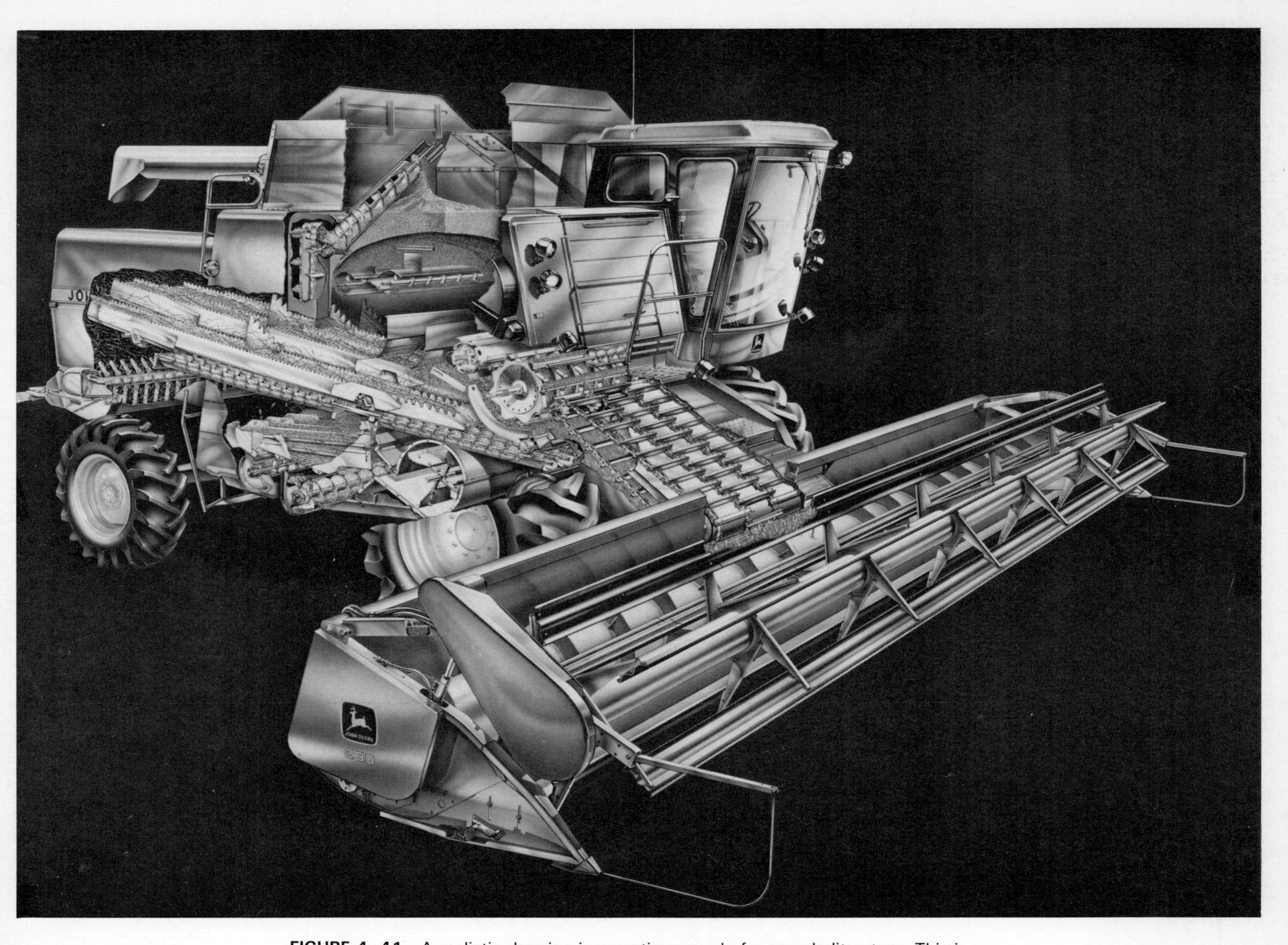

FIGURE 4–41 A realistic drawing is sometimes made for presale literature. This is a cutaway view of a modern grain combine. (Courtesy Deere & Company)

Use Charts and Diagrams Effectively

Charts and diagrams are arrangements of similar information or a comparison of different data so that the consumer can easily find information. Usually charts and diagrams are the easiest and less costly types of artwork to produce. For the price, they are an effective way to present facts in an easily accessible and usable format.

Charts are a patterning or grouping of information together that would be difficult to find separately. A common example of a chart is a *maintenance schedule* that tells the consumer exactly when to service the machine. A car operation manual, for example, usually contains a chart that quickly tells the consumer exactly what must be done and when the maintenance must take place. Charts can also be effectively used to show comparisons or contrasting information (see Figures 4-42 and 4-43).

Diagrams are another type of patterning that shows how things fit together. For example, the communication diagram that was used to support one of the major ideas in Chapter 2 is an example of a diagram. Diagrams also can show how something flows through a process or system. A diagram that graphically illustrates how power flows through a system is drawn with a type of shorthand symbolism to explain relatively complex data in one under-

COMPUTER PIN ASSIGNMENTS

Signal Name	Computer A	Computer B	Printer Port Number
DATA	1	1	1
DATA1	2	3	2
DATA2	3	2	3
DATA3	5	4	4
DATA4	4	5	5
DATA5	6	6	6
DATA6	7	7	7
DATA7	8	8	8
DATA8	9	9	9
STROBE	11	10	10
ACKNLG	10	11	11
BUSY	12	12	12
SELECT	13	13	13
FAULT	14	14	14

DIPSWITCH SETTINGS

Number	Computer A	Computer B
1	on	off
2	on	on
3	off	on
4	off	off
5	on	on
6	on	off
7	on	on

FIGURE 4-42 The chart compares and contrasts information.

MEMORY

OPTIONS

Option Adaptors and Default Setup	Switch Unit S1 Setting	
Monochrome adaptor only	#5-off	#6-off
Color adaptor only (40 × 25 display)	#5-off	#6-on
Color adaptor only (80 × 25 display)	#5-on	#6-off
Monochrome and color adaptors installed, monochrome default	#5-off	#6-off
Monochrome and color adaptors installed, color 40 × 25 default	#5-off	#6-on
Monochrome and color adaptors installed, color 80 × 25 default	#5-on	#6-off
No display adaptor installed	#5-on	#6-on
System start-up from drive A: or C:	#1-off	
System start-up from drive C: only	#1-on	
Numeric processor (8087) installed	#2-off	
Standard processor (8088) installed (coprocessor socket empty)	#2-on	
Two diskette drives installed	#7-off	
One diskette drive and fixed disk drive installed	#7-on	

FIGURE 4–43 User guides for computers and other electronic switching circuits would be more difficult to read without charts. (Copyright 1984 by the Radio Shack division of Tandy Corporation, *Tandy 1200 An Introduction & Guide,* 25-3001 9/85 SL. Used with permission)

standable piece of artwork. The main use of diagrams, however, is to show how a system is connected (see Figure 4–44).

Use Outside Sources as a Supply for Artwork

Artwork that exists outside your company or your college or university classroom may help you in explaining how to do the task in your instruction. This art may exist in vendor literature, which consists of sales brochures, operation instructions, and specification charts that another company supplies to your company. Some parts of the machine that you are writing about may be manufactured by an outside company. In this case the vendor, or the outside company salesperson, should have technical literature on the particular part or system that the vendor provides.

Be resourceful and use all sources of artwork available. Usually all that is required to use another company's artwork is a letter from you that requests a copyright release. A copyright release gives you and your company the legal right to reprint or reproduce artwork that is owned by someone else. You must remember the following points when you request a copyright release and artwork of reproducible quality from another company:

When you request artwork from other sources, make sure that you get a good negative or positive picture of the original artwork. Do not use artwork such as black-and-white pictures that are clipped directly from a printed book or brochure if negatives or positives are available. This artwork is "screened" with an overlay of little dots. These screened pictures will not reproduce clearly.

Line drawings can be clipped from printed brochures, instructions, and articles and reproduced. Screening is not a factor in the reproduction of line drawings that are clear and printed on good quality paper. Artwork taken from printed literature in this manner are called "tear sheets."

Your request letter should contain the following information:

Your name, title, and company.

The publication where you found the artwork.

The page number and location of the art.

Exactly how and where you are going to use the artwork.

Whether or not you are willing to pay for its use if it is available.

How the other company or source will be given credit in your instruction for the use of the artwork.

What form the artwork should be in. Negative? Positive?

Do you need other releases for the artwork that you are requesting?

Will you be able to use a tear sheet?

A negative of a black-and-white photograph is the coated film that shows all variations of light and dark contrast, exactly the opposite of what can be seen in the photograph. The positive, which contains a glossy finish, is usually required for two reasons. First, you can immediately judge the quality of the photograph that is sent to you. Second, you can add the necessary symbols and arrows directly to the positive photograph without processing the negative into a positive. A tear sheet is usually line art that is torn from another publication and used directly in your instruction.

Prepare the Artwork

Just as both the left hemisphere and the right hemisphere of the brain require a connection at the corpus callosum, so too must artwork have something to connect it to the words of an operation instruction. The transition between words and artwork takes place with the application of *call outs, arrows,* and other types of *nomenclature.* These symbols tightly bind the art to the words of a technical operation instruction and make both of them a single unit of thought for the consumer. Artwork cannot simply be placed next to a group of words and be expected to communicate to the consumer.

The items that you need for the finishing process are preprinted "pressdowns," a "burnishing tool" or any dull rounded tool, clear plastic acetate, an "X-acto knife" or cutting tool, spray adhesive, and art boards. All of this material can either be bought at an art supply store or ordered from a graphic arts supply house.

Artwork must be prepared by you so that your art interfaces with the particular printing process that you use to print your final instruction. Before you finish the artwork, you must make sure that the company that will print your technical operation instruction can use your finished art in its process. The standard method for handling artwork is to prepare it for an offset printing method.

The first step in the preparation of art is to protect it as well as you can. This can be done by mounting it on an art board. Adhesive spray can fix the artwork in place. Do not mount artwork if your printer can "gang-shoot" it with the graphics reproduction camera. This method is used to shoot many pieces of artwork at once in order to save on film cost and camera time. Make sure that "ganging" is not a possibility before you mount all of your art on art boards that must be shot one at a time.

A clear acetate cover can be taped over the art-

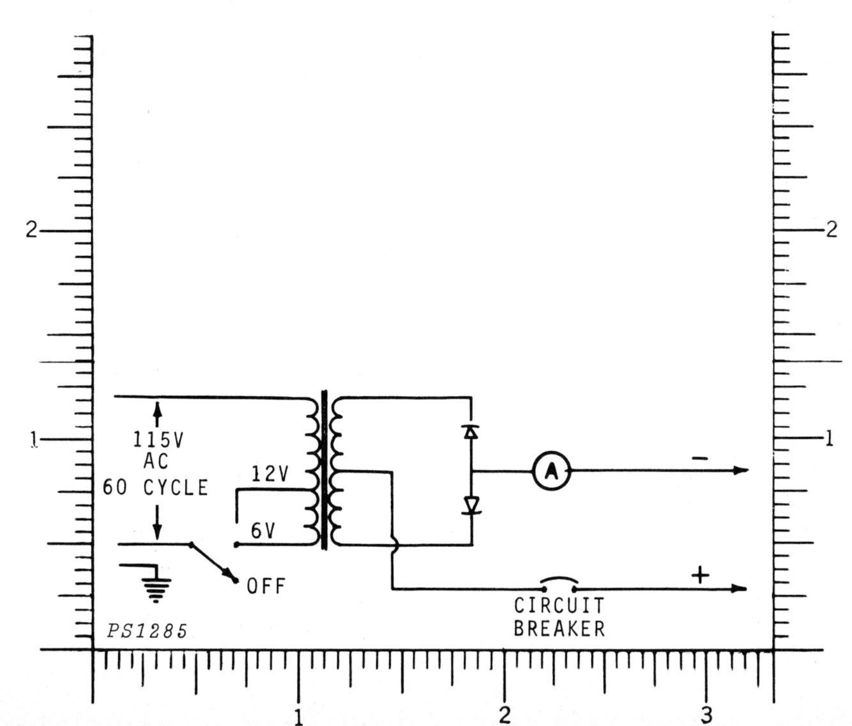

FIGURE 4–44 An electrical system diagram of a welding unit.

work to further protect it during the printing process. Pressdowns can be mounted directly on the artwork or on the acetate. A single piece of artwork can be made into many pieces of artwork if pressdowns are added to subsequent pieces of acetate.

In order to develop each layer of acetate as a separate piece of art, you must give that art a unique number. Each separate art board also must be given a unique number (cut number) so that it can be stored and retrieved in case it is needed for more than one writing assignment.

Call outs. Pressdowns complete the preparation of artwork. Pressdowns come in two main types. First is the peeloff type. Peeloffs are sticky-coated plastic that is mounted on a base sheet. As the symbols are needed, they are cut and peeled off and attached much like a bumper sticker. These adhesive-backed pressdowns are easy to handle; however, they may accidentally peel off of the artwork as easily as they are stuck down.

The second type is the burnished pressdown. They are symbols that are printed onto a plastic page. By pressing down with a burnishing tool on the paper that they are mounted on, you can transfer them to your artwork. Generic symbols such as numbers, letters, arrows, basic maintenance symbols, and some safety symbols are available on both types of pressdowns. At an extra cost, specialized pressdowns can be manufactured for your situation.

Pressdowns are used to *call out* specific information on artwork. Arrows (leaders) and numbers can point to and identify a specific item that is the central idea to your task. Bold arrows can show specific movement such as the direction of a hand that is turning an object. Arrows can help animate the artwork. Pressdowns can either be placed directly on the line art or photograph or on the acetate cover that is used to protect the artwork. Often arrows take the place of the consumer in the artwork.

Call outs are used in the task. Instead of saying "Make the adjustment at the adjustment nut located just below and to the right of the large wing nut," you can say "Adjust nut (A)." All that you need to do is to press down a letter "A" on the artwork and lead an arrow to the point of adjustment. Be sure that arrows do not cross each other and that all call outs are located off any part of the artwork that might be needed as an orientation point for the customer. Multiple call outs usually are arranged in a clockwise position around the artwork.

Burnished call outs such as arrows, lines, letters, numbers, and other symbols can be easily removed from a glossy picture with common plastic tape. Tape can also be used on some coated papers and acetate overlays.

Artwork that shows many call outs should have a "legend" close to the actual art when it is printed in your technical operation instruction. These legends help the customer find information fast and accurately. Avoid putting words or alphabet letters on the art. If the artwork is used in a translation, the words also must be translated. If words are on the art, a second piece of identical art must be made. By duplicating the process of generating artwork, you will add unnecessary expense and time to your writing assignment. Figures 4-45 through 4-52 give you a visual guide that you can use to prepare artwork for the printing process.

Figure 4-53 shows a completed piece of art. The unique number for the first acetate overlay is

FIGURE 4-45 (Optional) Mount the art on an art board.

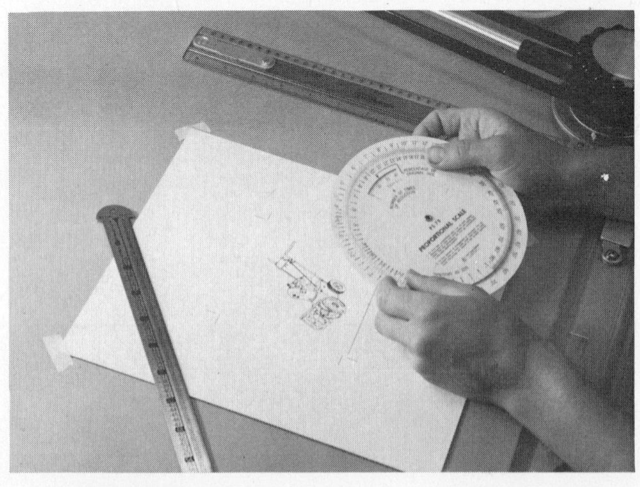

FIGURE 4-46 Size the art to fit the space that it will go into.

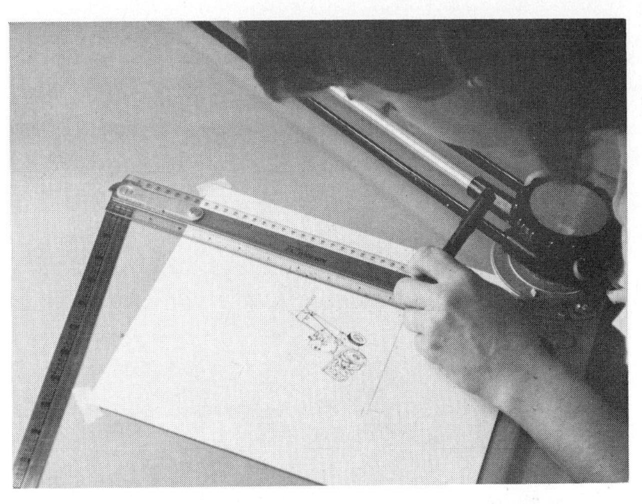

FIGURE 4-47 Draw cropmarks to show the graphic arts photographer exactly what to shoot.

FIGURE 4-48 (Optional) Add the acetate overlay. Attach it to the board at the top of the picture with a strand of tape.

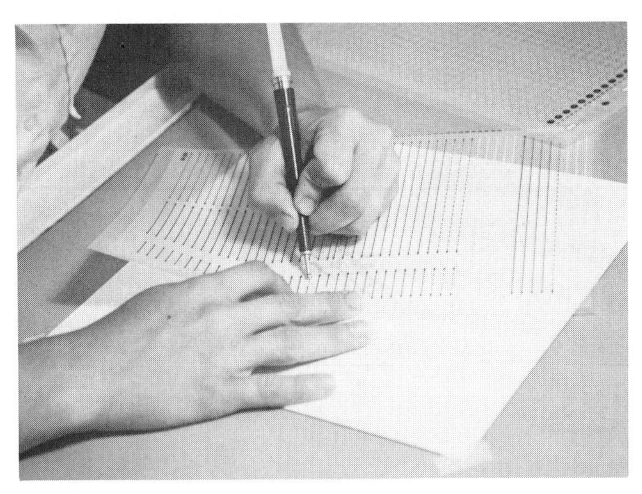

FIGURE 4-49 Add pressdown leader lines and arrows or leader lines.

FIGURE 4-50 Add call outs.

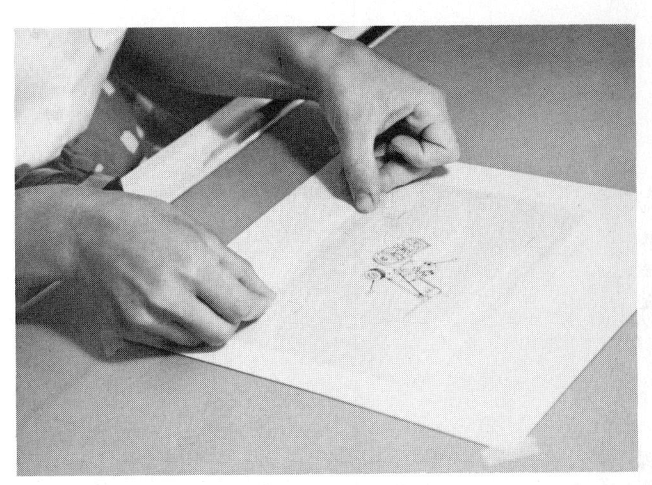

FIGURE 4-51 (Optional) Put a second overlay of acetate over the picture if more than one piece of art is made from the original picture.

FIGURE 4-52 Put on the second set of call outs. When the photographer shoots the artwork, he or she will fold back all but one piece of acetate for each shot.

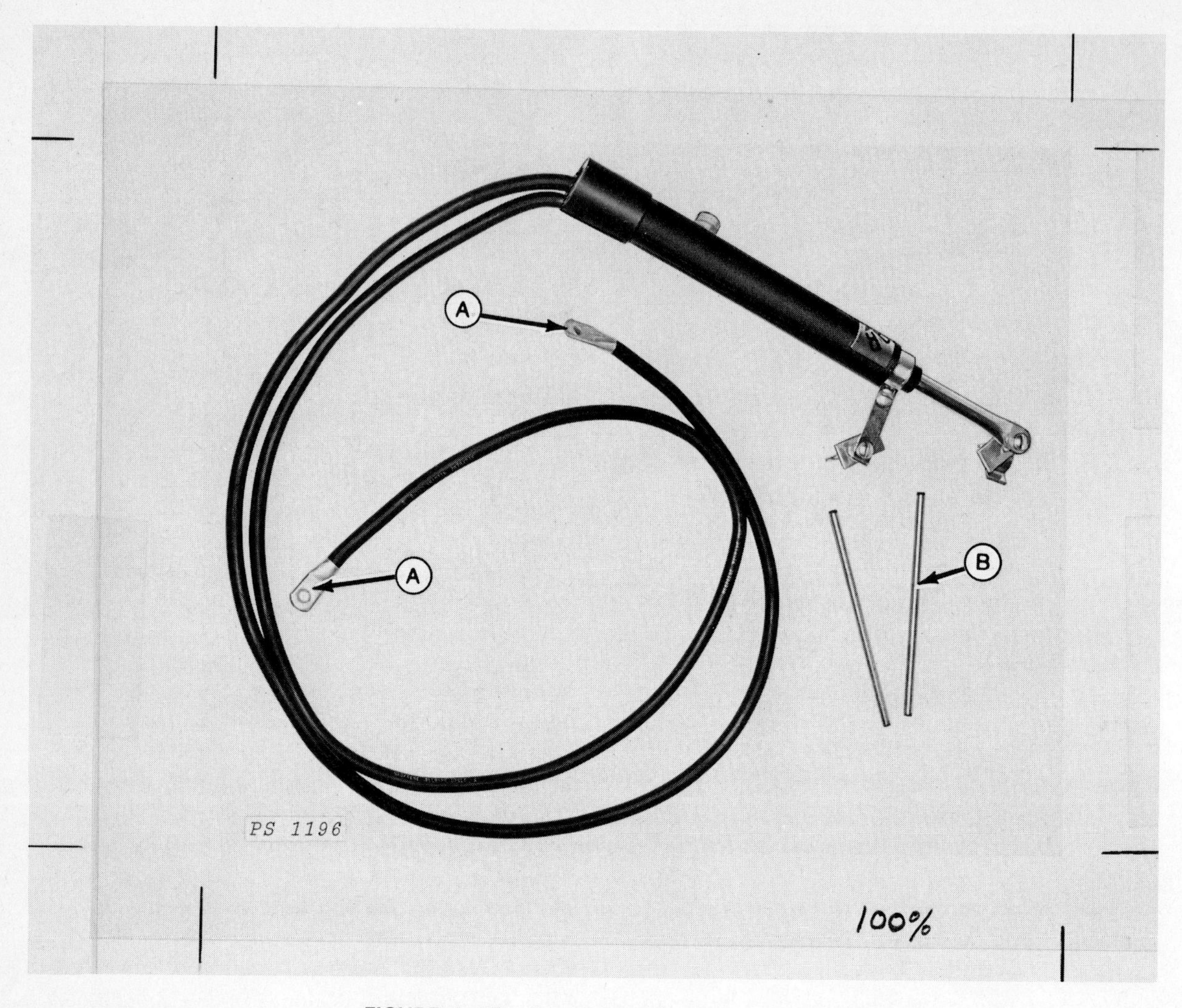

FIGURE 4–53 A complete piece of artwork.

PS 1196. With the addition of a unique number, the artwork can be stored and retrieved. The vertical and horizontal lines are the cropmarks. The call outs and leader lines and arrows point to the welding rods and electrical connectors. The welding unit artwork is ready to be sent to a printing company.

Note: Cropmarks can be used to eliminate some of the picture face if you need to focus upon a certain part of the image. Cropmarks are an effective editorial tool if they are used properly.

The art wheel or proportional scale. The final step done to prepare artwork for the layout and printing is to "size" it. This procedure must be completed if the artwork that you want to use is larger than the space that it must fit into. You can crop the art to eliminate some of the picture. If cropping is not possible because all of the artwork is necessary, then you must size it (see Figure 4–54).

An *art proportioning scale* can be obtained from a graphics art supply store. It is basically a wheel that rotates within a wheel. Both wheels rotate upon the same axis. The inner wheel contains a scale on the outer edge that is divided into 1/16-inch increments (see Figure 4–55A). Another scale on the larger wheel also is divided into 1/16-inch increments (Figure 4–55B). A window on the inner wheel shows the percentage of reduction (Figure 4–55C) and the number of times the original art must be reduced (Figure 4–55D).

Suppose that you must fit a 10-inch-wide by 14-inch-high piece of artwork into a 5-inch-by-7-inch opening or "window" on the final layout of your instruction. You must follow these steps:

1. Measure the width of the window or opening.
2. Measure the height of the window or opening.

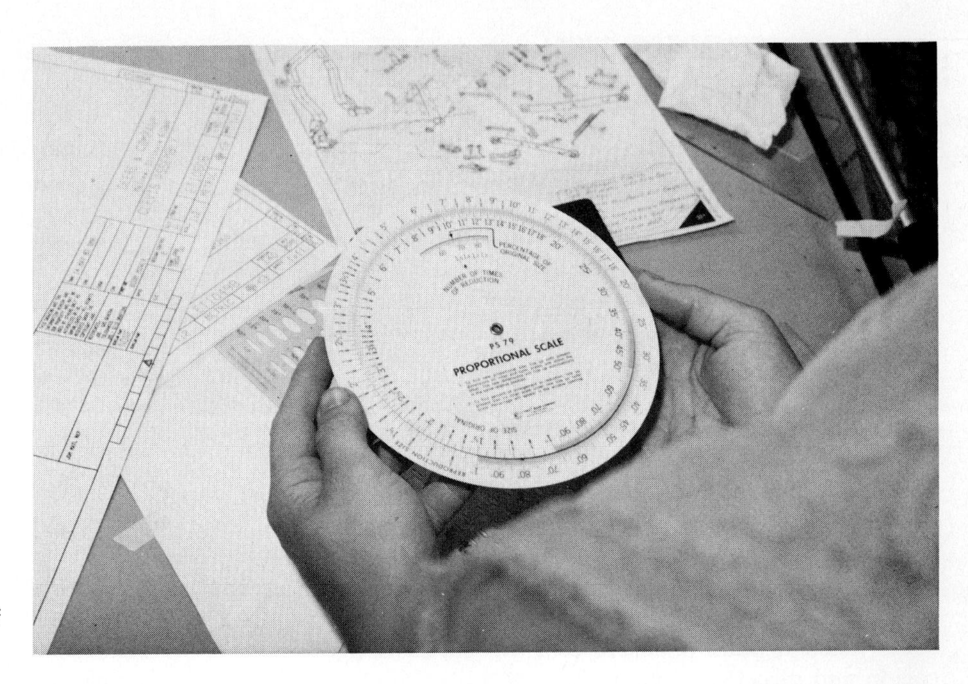

FIGURE 4–54 A general view of a proportional scale.

3. Measure the width of the art.
4. Measure the height of the art.
5. Find 10 inches on the inner wheel scale and 5 inches on the outer scale.
6. Align the 5 *over* the 10.
7. Look in the window for the percentage of reduction. Your art must be reduced 50 percent. The number of times of reduction is 2. That is, the art will be twice as small as the original art.

When you mark your artwork with the proper percentage reduction, it is then ready to be sent to a printing company. There it will be photographed with a special graphic arts camera and reduced by 50 percent. The art can then be put into the window. In Figure 4–53, the artwork is marked to be shot at 100 percent.

Both cropping of art with cropmarks and reduction can be used together to edit the artwork. By experimenting with photographs and line art, you can develop an ability to show only what is necessary in the space that is available.

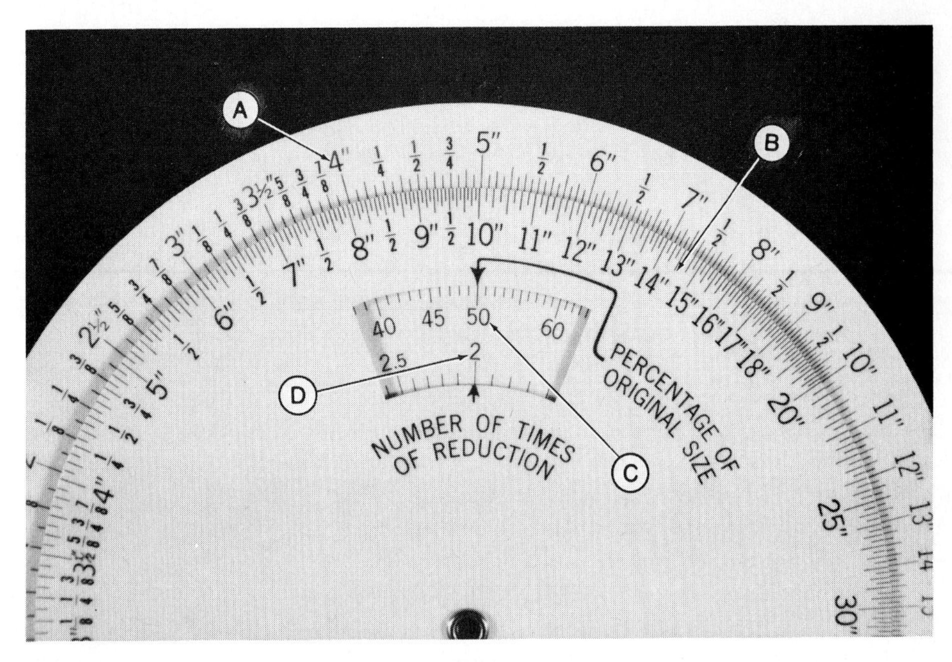

FIGURE 4–55 Detail of an art proportional scale.

DEFINE THE *VISUAL* TECHNICAL WRITING STYLE

The technical writing style is the blending of nonverbal and verbal symbols that send concrete images to the consumer. It is a visual script that details the consumer's movement around the machine and *shows* how the task is done.

For this reason the visual technical operation instruction must be put together on storyboards that show action. In a sense, the instruction must be written like a series of vignettes. This type of communication demands that the technical writer have the ability of an artist to select the words and pictures, and the ability of the craftsperson to put them both together skillfully.

The technical writer who masters the visual writing style thinks in pictures and connects pictures and language with callouts and arrows. Pictures take the place of descriptive words that tell location and direction.

CHAPTER SUMMARY

- Half of the thinking ability of the individual consumer is devoted to the one-step-at-a-time logic of words. The other half is developed by the individual to recognize and communicate by means of nonverbal artwork.
- Artwork, photographs, line drawings, charts, graphs, and diagrams can all be used to reduce the number of complex nouns and eliminate most of the descriptive adjectives in an instruction.
- Artwork helps to eliminate the tendency for consumers to interpret or imagine their own instruction.
- Three factors must be present if the technical communication process is going to be complete: the consumer, the specific technical operation instruction, and the specific machine.
- The visual technical writing style is similar in method and approach to the vignette of the ancient Egyptians and the modern comic strip artist.
- The technical visual communicator must use a storyboard to bring words and artwork together.
- Storyboards can be used to plan the photography schedule or to order line art, charts, drawings, and other visuals.
- Visual communicators must begin their thought process by thinking of a concrete picture and then adding concrete words.
- In a technical operation instruction, three basic views of artwork are usually required to communicate a task to the consumer. They are the general assembly, the transition shot that helps to place the operator for the instruction, and the detail that shows the action taking place.
- There are four basic types of artwork: photograph; line art; drawings; and charts, graphs, and diagrams.
- When taking photographs, it is necessary to control the environment as much as possible.

EXERCISES FOR CHAPTER 4

WRITE VISUALLY

1. Cover up the words on your favorite newspaper cartoon. Look at the pictures in sequence.
   a. Did the cartoon communicate without the words?
   b. Could you interpret the punch line?

   Now read only the words on another comic strip.
   a. Could you put action with the words?
   b. Could *any* words be left out to improve the dialogue in the comic strip?
2. Draw a series of rough sketches that visually explain how to do an everyday task. Be specific and cover all of the details.
3. Explain, without the aid of artwork, the task that you sketched in exercise 2.
4. Add words from exercise 3 to the series of sketches that you drew in exercise 2.
   a. Can you eliminate any of the drawings?
   b. Can you eliminate words?
5. Critique a technical operation instruction that you have collected.
   a. What types of artwork are used?
   b. Is there enough artwork to explain the tasks in the instruction?
   c. Can you improve the instruction with more artwork? Be specific.

5

Find the Facts

The information given in this chapter will allow you to:

- Explain how a manufacturing company is organized.
- Divide the areas of a manufacturing company into departments.
- Explain how the engineering, sales, accounting, and production departments in a manufacturing company cooperate to make a machine
- List the main technical documents that can be used as sources of data in the technical writing process.
- Determine the location of important documents that are stored in a manufacturing company.
- Isolate blocks of information on an engineering drawing and diagram.
- Separate the secondary lines on an engineering drawing from the main visible lines.
- Understand the importance of double-checking information.
- Interview the machine expert with confidence.

INTRODUCTION

The manufacturing company is an organization devoted to the making and selling of machines for more money than the machines cost to design, build, and distribute. Even though the modern manufacturing company is usually a highly sophisticated enterprise, it has been compared to a "tribal" organization. This is a valid comparison when one looks at the interpersonal relationships of the people who work together to design and make the machine. There are many people in a manufacturing company who are specialists in at least one part of the manufacturing process. Each "tribal" member must do his or her job as efficiently and correctly as possible in order to help the company survive the external pressures of a worldwide marketplace.

Specialists in the same area and with similar types of jobs are grouped in *departments*. Communication is the uniting bond among tribe members.

Communication unites these separate departments into an organization. Without a transfer of information from one department to another, there is lack of cooperation needed to unite the departments and produce a machine. The people in a manufacturing company must be able to create, store, and retrieve information as it is needed. An efficient information exchange enables everyone to work toward the same goal. Without communication among the departments, the company technology will not be able to advance and keep pace with the demands and tastes of a changing consumer-oriented marketplace.

The manufacturing process is a progression of steps. Each step of production adds value to the machine that is being made. For example, the raw materials of iron, carbon, nickel, and chromium are less valuable than the steel that is produced by combining these raw materials. As the steel, plastic, and textiles of a machine are formed and worked, they tend to be more valuable as each fabricated step is completed. For example, a control valve for a hydraulic system is more valuable than the steel that is used to make it.

Each major step in the manufacturing process also results in a specific type of documentation. This written communication explains what has been done to the machine. Documentation that begins during the actual birth of the machine concept is invaluable to the technical writer. The problem with company communications is that much of it is unusable as a source of data. As a technical writer, you must be able to find and select only the documentation that you need to complete your objective.

The machine design during every step of the physical manufacturing process on the shop floor is mirrored on paper. Before any step of the manufacturing process can be initiated, there must be a document that reflects that step. In this chapter you will learn where you can *find* these major sources of information. You will then be able to isolate the useful documents from the sea of paperwork that is created by the manufacture of a machine in a modern manufacturing company.

Part of your search for usable material will be from interviews. This chapter will cover the art of interviewing the machine expert. Often the manner in which you interview these people will directly affect the amount and quality of data that they give to you.

As a general introduction, this chapter will also cover the topic of how to *use* the major documents of the manufacturing process, including engineering drawings and diagrams. Engineering drawings and diagrams are the *universal language of the manufacturing process.* These specialized documents can be your best source of accurate and up-to-the-minute data. You must develop the ability to interpret the information from these drawings and other documents.

UNDERSTAND THE STRUCTURE OF THE MANUFACTURING COMPANY

The most efficient way to isolate the sources for useful technical documents is to examine the structure and the organization of the manufacturing company. Although each manufacturing company may be unique, there are certain basics that you can find in every operation. To be successful in the job of collecting data, you must know how your particular "tribe" is organized and how documents are generated in the system. You can begin your examination of the company structure by dividing the organization into two separate parts. They are:

The shop floor.
The front office.

The shop floor. The shop floor is the location of the fabricating machines and people who form and shape the raw materials into machine parts and components. It is the location for the many people who assemble the machine, test it, paint it, and package it for shipping. The shop floor is the extension of the technical documentation that is created by the machine design and manufacturing experts. For example, you may find the lathes, cutting tables, presses, drills, planers, and precision electronic part-making machines that are often initially driven by the computer-aided design machines programmed in the front office and connected to computer-aided-manufacturing systems. The machine is given its physical shape by the machines and people on the shop floor. A typical shop floor can be divided into the following main areas where the machine is made. A machine usually flows through the manufacturing company in the order that is given below: (see Figure 5-1).

The receiving and storage area where material is delivered to the company.

The tool room where special tools and equipment are stored for the fabricators and assemblers.

The fabrication area where the cutting, lathe, shaping, and press machines that make the parts for the machine are located (see Figure 5-2).

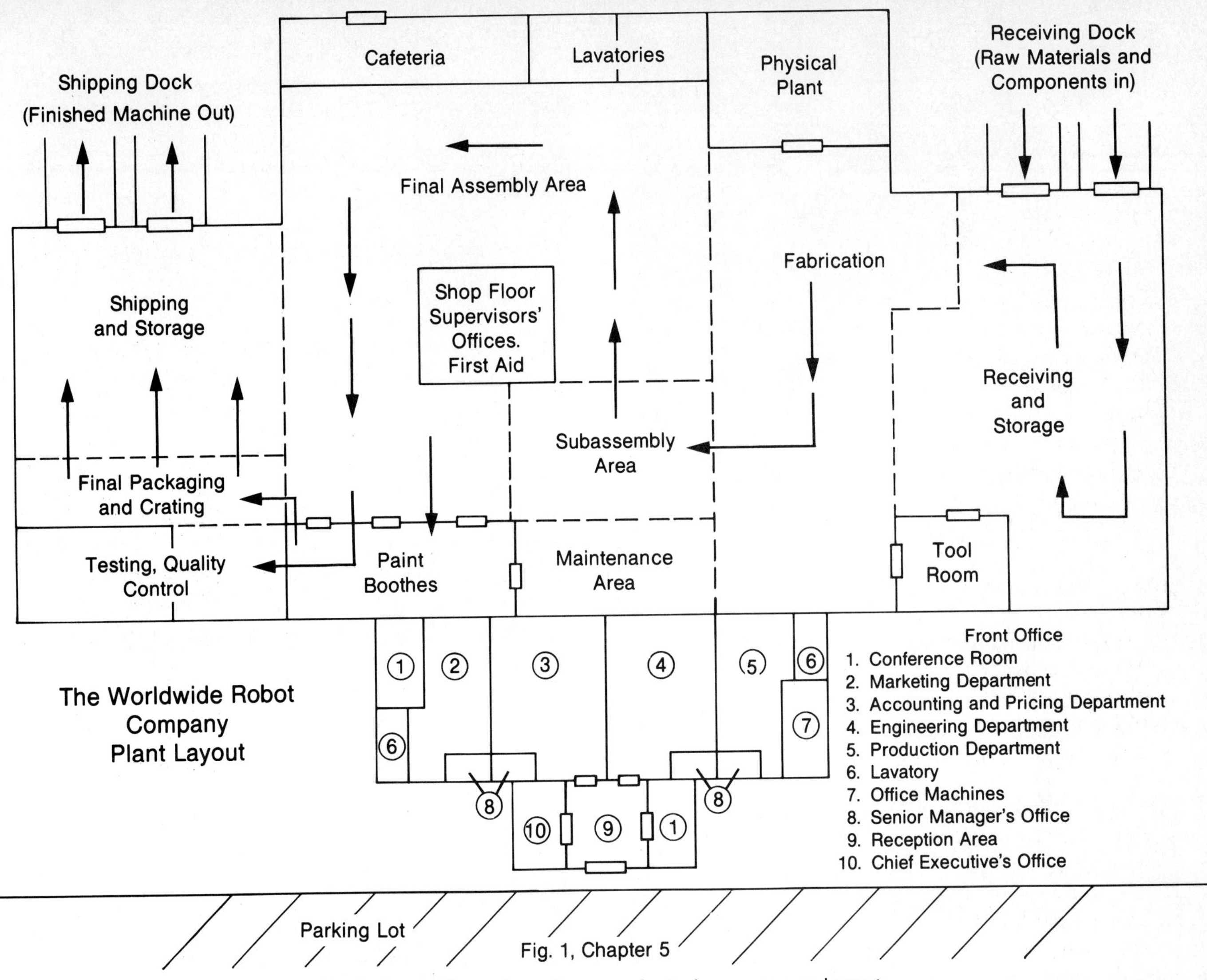

FIGURE 5–1 Floor plan of a manufacturing company layout.

FIGURE 5–2 A semiconductor is built atomic layer by atomic layer with a precision molecular beam epitoxy machine. (Reproduced with permission of AT&T)

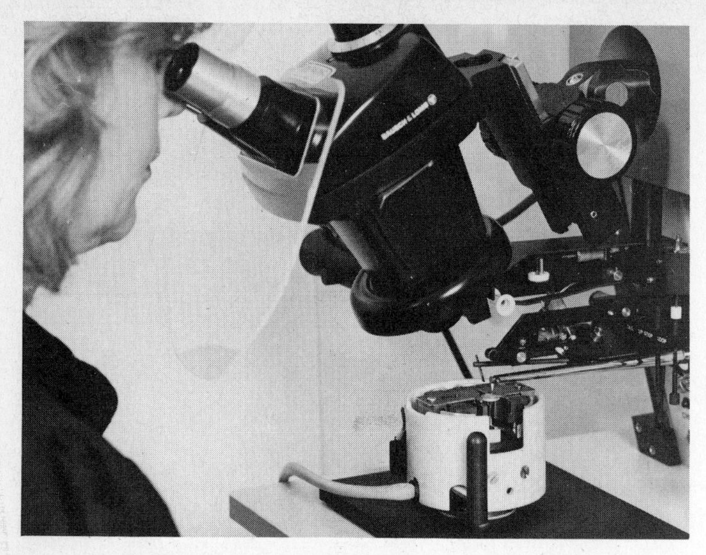

FIGURE 5-3 Assembly techniques may require more precision than can be obtained on a moving assembly line. (Courtesy Florida Department of Commerce, Division of Tourism)

The subassembly area where components and systems are put together (see Figure 5-3).

The final assembly area where all systems and components are attached to the superstructure or skeleton of the machine (see Figure 5-4).

FIGURE 5-4 Machines as large as the B-1B bomber are too large to be assembled on a moving assembly line. (Courtesy of The LTV Corporation)

FIGURE 5-5 The Radar Cross Section Laboratory developed by The LTV Corporation is one type of test facility. (Courtesy of The LTV Corporation)

FIGURE 5-6 Laboratories and testing areas include everything required to analyze the machine completely. (Courtesy Deere & Company)

The paint booths where the machines are painted.

The testing area where the machines are final-tested before being released to the marketplace (see Figures 5-5 and 5-6).

The final packaging and crating area where the machine is prepared for shipment to a world-wide market.

The shipping and storage area where the machine is held prior to distribution (see Figure 5-7).

FIGURE 5–7 Computer-generated labels, produced on Hewlett-Packard printers, make it easier to locate materials in the Ford Motor Company warehouse. (Photo courtesy of Hewlett-Packard Company)

The front office. The front office area, on the other hand, is the place where the machine begins its mechanical life. Before it is released to the shop floor, a machine design is developed and refined by the experts in the front office. While the shop floor is the place where the machine is physically made, the front office is the place that creates the instructions and gives directions that tell the people on the shop floor exactly how it is to be made (see Figure 5–8).

The people in the front office have the job of developing sales and after-sales programs for the machine (see Figure 5–9). Purchasing agents who buy parts and components, marketing experts who interpret marketplace trends, and engineers are a few of the specialized people in the front office. As an example of what is meant by the front office, look at an organization chart for a typical small or mid-sized company. The Worldwide Robot Company is an example.

FIGURE 5–8 The front office is where the machine is designed. (Courtesy Deere & Company)

FIGURE 5–9 There are many specialists in a variety of jobs in the front office. (Reproduced with permission of AT&T)

Construct an Organization Chart of Your Company

Your first exposure to the shop floor of a manufacturing company may be a confusing experience. People on forklift trucks may be driving haphazardly through the aisles, some employees may be busy at their work stations, and others may seem to be moving aimlessly from one place to another. With some materials being constantly moved on conveyors or assembly lines, and others stacked undisturbed in the corner of the factory, the manufacturing process may seem very unorganized.

You may think that it is impossible for the employees to produce the finished machine that you see going out the back door of the shipping dock. What seems to be in a state of confusion is exactly the opposite. *Successful* manufacturing companies, which seem to be in mass confusion, are actually well organized and efficient. Every movement of employees and materials on the shop floor is intended to have a definite purpose and goal.

The first thing that you as a technical writer must do is to obtain or create an accurate organization chart of the company that you are writing for. The seeming confusion and disorganization of your manufacturing company will become logical with an organization chart. Without this guide, you will also find that obtaining information is often a more difficult job. As an example of how complex the manufacturing process is, look at the structure of the The Worldwide Robot Company that is outlined below. Its organization is like many new companies that have developed in the past few years. The Worldwide Robot Company contains all of the major parts or divisions that you can find in a "typical" modern manufacturing company.

Case Study: The Worldwide Robot Company

The Worldwide Robot Company is a small-sized manufacturing organization if it is compared to the larger multinational corporations that dominate the business world. Five years ago it was created to make high-tech robots. The company's major customer until now has been the larger parent company and the other subsidiary companies that comprise the corporation to which The Worldwide Robot Company belongs. As an offshoot of the much larger corporation, the first five years of its life were tightly controlled by the corporate officers who run Worldwide Robot from the corporate headquarters.

The company is led by the chief executive officer, who has been appointed to the position by the board of directors of the parent corporation. Because it is part of a much larger organization, Worldwide Robot Company officers can use some of the services of the parent corporation. The company, for example, can employ the expertise of the well-staffed corporate legal and safety department.

The Worldwide Robot Company branches out from the chief executive's office like the limbs of a symmetrical tree. Four senior managers control four different areas or departments directly under the chief executive officer (see Figure 5-10). These areas are:

Engineering and product support.
Sales and marketing.
Accounting and pricing.
Production.

Examine these areas closer to familiarize yourself with the division of responsibility found in a manufacturing company.

The Engineering and Product Support Department

The engineering and product support department at Worldwide Robot is responsible for creating the original machine concepts and parameters. Design engineers and their staff develop the machine from basic specifications or measurements and ideas. The machine actually begins to take shape on paper in the engineering section of the department. Another responsibility of the department is to give final approval to the overall design of the machine and create detailed engineering drawings that can be used to manufacture the machine (see Figure 5-11). Other important functions of the engineering department are:

To do research and development to improve machines, materials, and manufacturing techniques.

To send members of the department to industrywide meetings where standards of safety, design, and other machine parameters are created.

Because Worldwide Robot is not a large multinational company, this department is also staffed by the service people who help ensure that any mechanical trouble with the machine in the field (or in the hands of the customer) is handled quickly and efficiently. This includes the filling out and filing of

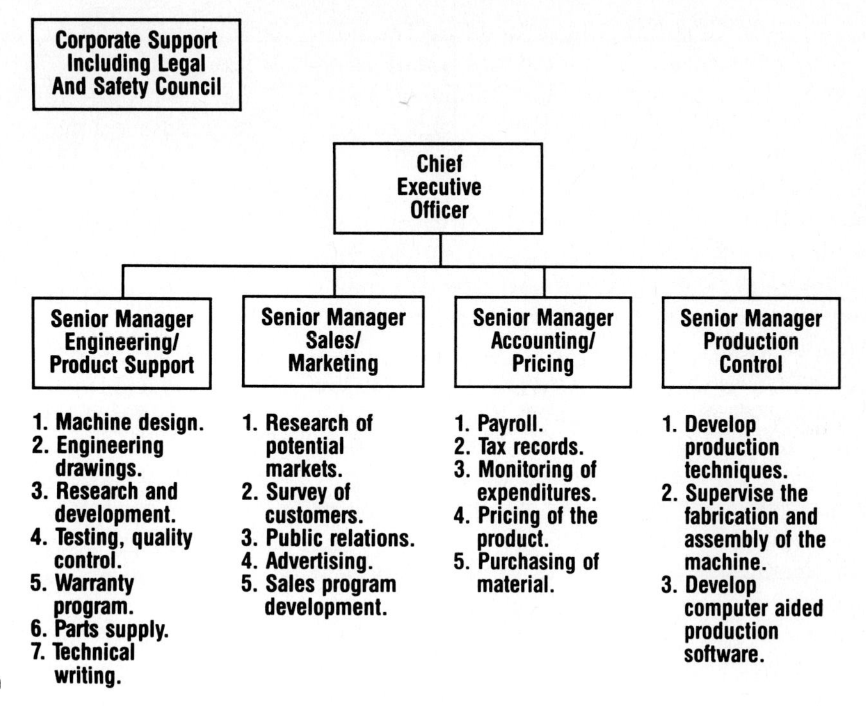

FIGURE 5–10 In an organization chart, the departments of a company are like the branches of a tree.

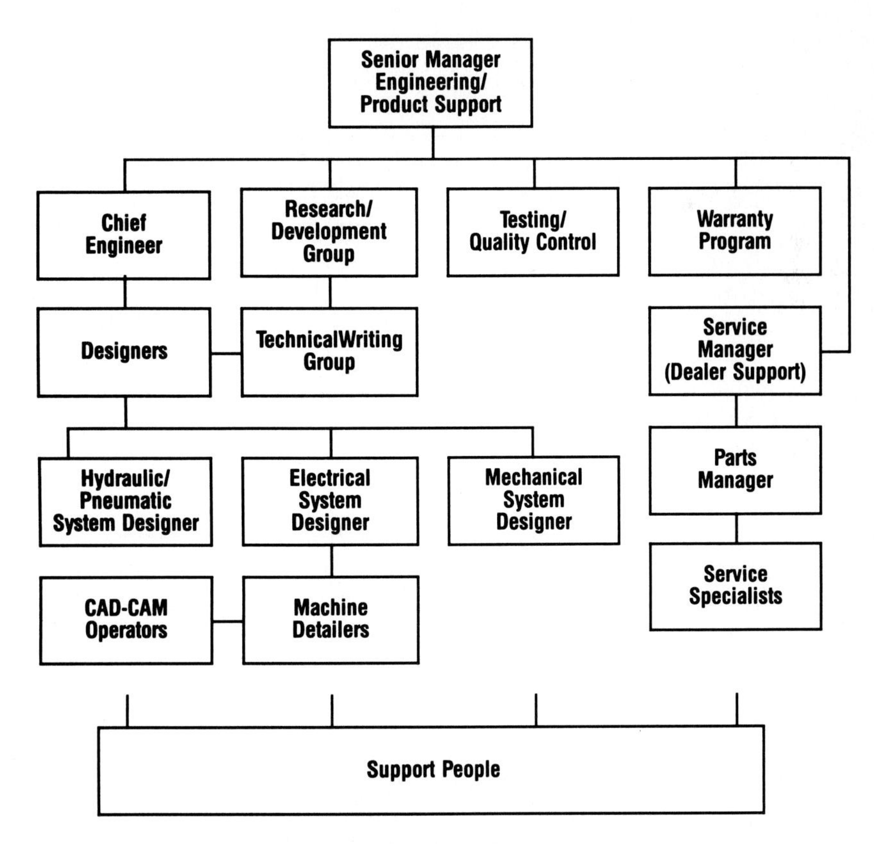

FIGURE 5–11 The Engineering Department structure.

warranty claims that have been made by the consumer. Along with warranty satisfaction and parts supply, the engineering and product support department sees that parts are available to the service technicians who fix and maintain the machine.

In case there is a design or manufacturing problem with the machine type, the warranty experts must be able to trace the product to the consumer, who must be informed of any safety or maintenance programs available for the machine.

Although the testing program is usually done by a separate department in larger companies, the engineering and product support department at Worldwide Robot has the responsibility to assure that the machine performs correctly. There are two testing programs that have been set up on the shop floor. One is the testing of the "prototype" machine. In every manufacturing process, a prototype, or first machine, is manufactured to test the methods and processes of the manufacturing company. The prototype machine is checked to ensure that both the quality and the design meet the parameters originally set by the conceptual design engineers.

A second testing program will continually check and evaluate the machine throughout the manufactured life of the machine. Every step in a manufacturing process must be continually tested to assure that the quality demanded by the company is being maintained. A machine also may be reengineered, changed, or fine-tuned throughout its manufactured life. A machine design must be dynamic and change as market demands or the manufacturing process changes. Every change in a machine must be thoroughly tested before it is allowed to enter the marketplace and be sold.

The technical writers. Most technical writing programs are either attached to the marketing and sales department or the engineering department of a manufacturing company. Because Worldwide Robot Company is an engineering-oriented company that relies on extensive research and development by the engineering department, the technical writers are assigned to this area. They must work closely with the engineering and product-support people in order to keep track of new and developing company technology so that they can get updated information to the customer as quickly as possible.

Because the company has limited operating capital, the technical writers assigned to the engineering department have more job responsibilities than just writing instructions. They also are writers, editors, and publishers who are in charge of supervising the final printing process of the instruction. In a sense, the technical writers of Worldwide Robot have complete *control* and complete *responsibility* for the production of the company technical operation instructions.

A trend that is developing in industry is to make technical writers responsible for their own editing and production. This trend eliminates the technical editor and publisher from the technical writing staff and directly reduces overhead costs of the department that must create technical operation instructions.

The Sales and Marketing Department

Because Worldwide Robot Company was formed to supply machines primarily to the corporation, the marketing department remained small for the first two years of the company's existence (see Figure 5-12).

Today, Worldwide Robot has an extremely active marketing department. The company began its manufacturing and marketing cautiously. Now the members of the department aggressively research the marketplace and attempt to analyze what new machines the consumer wants or will want in the future. The decision to produce new products that will sell to the mass-consumer market has forced company executives to expand their marketing functions.

Not only do the marketing experts analyze the buying habits of the consumer but they also often make recommendations that are the stimulus for engineering redesigns in the machines manufactured by the company. The marketing experts analyze the potential for a machine before it is designed, and they continue an ongoing survey of the consumers who purchase the company's machines. A manufacturing company will not survive if the machines that are made cannot be sold in the marketplace for a profit.

A company cannot produce a machine and expect the machine to sell itself to the customer. A sales staff must set up distributors and sales outlets that provide a convenient market location for the consumer. Salespeople work closely with the service people to ensure that the company's postsales and presales activities fulfill the consumer's needs. If the consumer cannot conveniently buy a machine and get the machine fixed and maintained, they they might buy their machine from a better-organized company.

Because Worldwide Robot is marketing new machinery in many countries, a foreign sales staff must be recruited and trained to handle the new

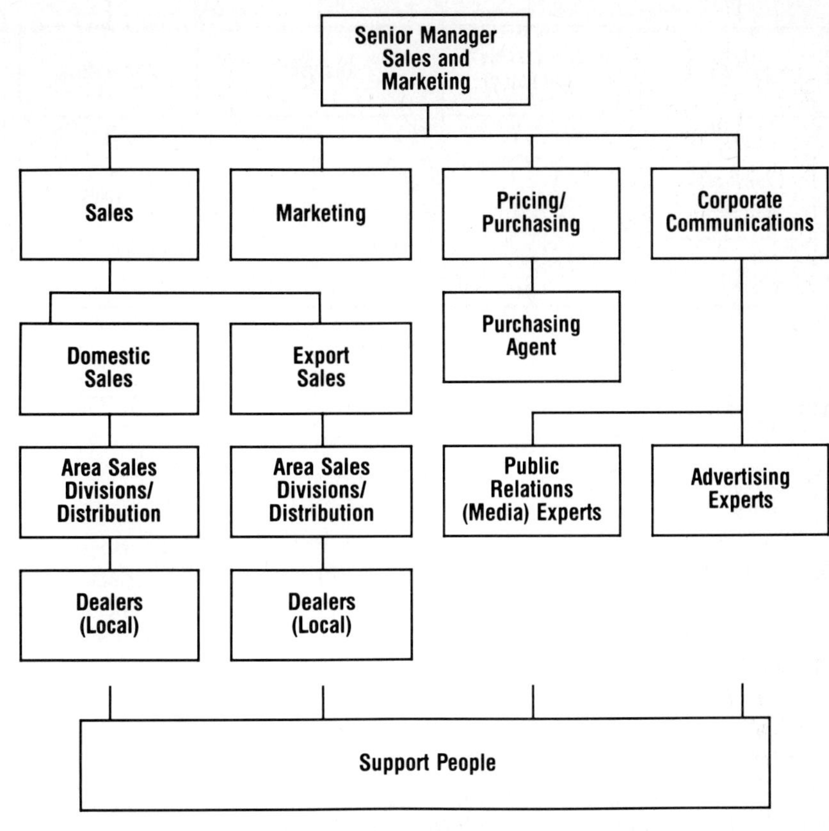

FIGURE 5–12 The Sales and Marketing Department structure.

product. Worldwide Robot Company has divided the world map into five basic sales areas and has given five people responsibility to ensure the machine can be supplied and supported in every market. New sales offices are built in every country that the company executives believe will be a potentially strong market for the new machine.

A major division of the company's sales and marketing department is the communications office and the advertising group. Combined, these people make sure that the consumer is informed of new machines and machine developments that will be available in the marketplace. The public relations writers work with the mass media through news releases, while the advertising people develop and place paid advertisements in newspapers and magazines, and on radio and television.

The Accounting and Pricing Department

In order to streamline Worldwide Robot's organization, the accounting and pricing groups have been combined into one major department. Most of the employees in this area handle payroll and tax matters. Without a payroll office, the company employees would not receive their paychecks on a regular schedule. Without strong tax records, the company may be forced to pay more taxes than is necessary (see Figure 5–13).

The accountants also monitor the other departments of the company to ensure that these areas are getting the most value for the money they spend. The accountants and other employees in this group are involved in every phase of the manufacturing process whenever capital is borrowed or money is spent and whenever a return on the investment is realized.

The pricing experts work to determine the lowest possible price that will be charged for the finished machine. They must take into consideration the cost of raw materials, design costs, labor costs, and overhead, such as rent, utilities, and upkeep of the plant, when they set a selling price for a new machine. Pricing experts must be aware of the total cost of all phases of the manufacturing process.

The company purchasing manager finds the best possible price for the raw materials and the components that the company needs to make the machine. To do this, bids are issued to other suppliers and manufacturers. The company with the best raw

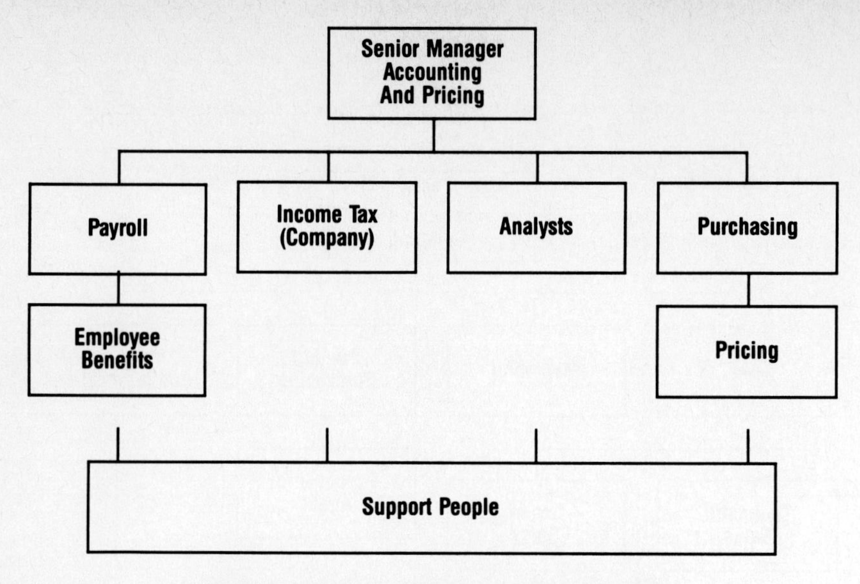

FIGURE 5–13 The Accounting and Pricing Department structure.

materials and components at the best price are contracted by the purchasing agents to supply their product to Worldwide Robot.

The Production Department

The production department acts as a bridge between the front office and the shop floor. The machine experts in this area translate the theoretical ideas and machine designs into concrete fabrication and assembly instructions that can be used to make the machine as cost-efficient and reliable as possible (see Figure 5–14).

The production department is concerned with developing the techniques of production. This also is a form of Worldwide Robot's technology. New and innovative methods and techniques are constantly tried. Production experts must advance their meth-

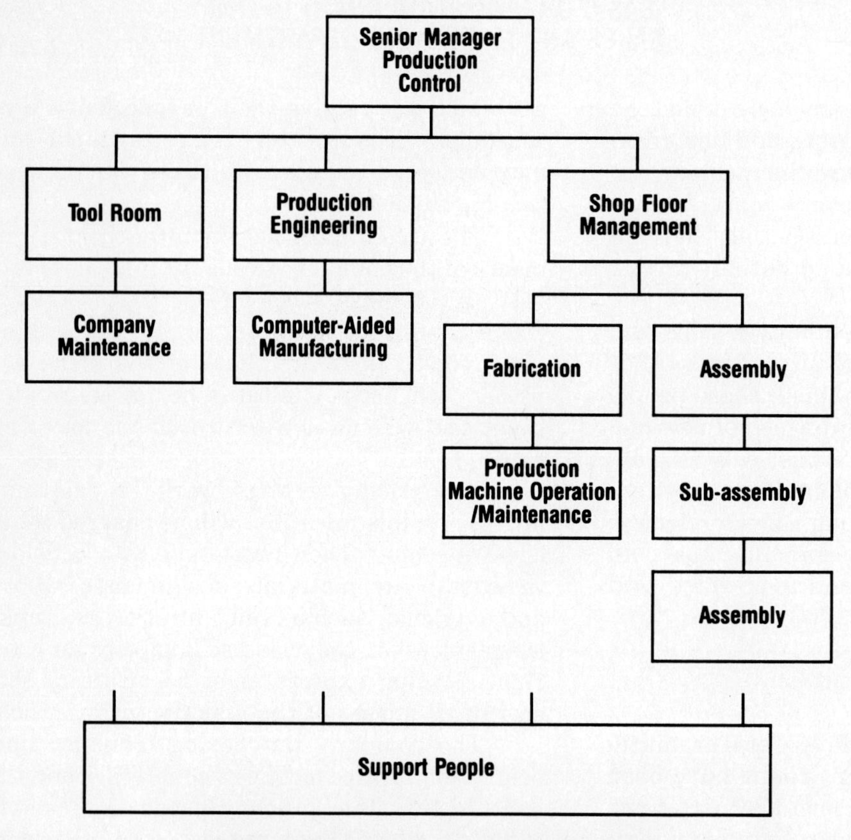

FIGURE 5–14 The Production Department structure.

ods in order to keep pace with the machine that is being manufactured. The development of a new machine is often stopped because the machines and the techniques of production are unable to produce the correct results that are necessary.

Not only is this department concerned with developing and structuring the manufacturing process but the experts who work here must also see that the right components and raw materials are at the correct place on the assembly line. Along with this, the production specialists must assure that the tools and manufacturing machines are available and in working condition on a daily basis.

There is a wide variety of skilled people in this department. They include computer-aided manufacturing specialists who work with the sophisticated electronic fabrication machines on the shop floor. The department also is the base of operation for the production engineers who develop the process of manufacturing that is the most economical and productive for Worldwide Robot Company machines and shop employees.

Support Personnel

Each department is staffed by secretaries, file clerks, mail room workers, and other non-technical support personnel who keep the paperwork and communication smoothly flowing among all the machine experts. Support personnel make up approximately 25 percent of all the employees at Worldwide Robot Company. Their jobs are extremely important to the day-to-day operation of the company.

ISOLATE THE SOURCES OF DOCUMENTS

The model of the company outlined above is a typical type of organization that you will find in most small and medium-sized companies. In the smaller company, one person may have more than one responsibility. Larger companies tend to have experts who concentrate on a single phase of the manufacturing process.

No matter how big or small the company is, the manufacturing process *is a complex interaction of machine experts and support people.* A manufacturing company may be a mixture of technical and non-technically trained people who must work together and communicate with each other in order to manufacture a machine and make a profit. Everyone must cooperate to produce a machine that can compete in the worldwide marketplace. Each person in the manufacturing process must do his or her job *and document what is done* before the next level of manufacturing can begin.

Each manufacturing step is recorded. Documents that tell exactly what each person does to advance the manufacturing process is recorded at each step. The physical machine on the shop floor of the manufacturing company is mirrored in computer files or on paper in the files of the various departments of the manufacturing company. This fragmented information is useless to the consumer if you—the technical writer—are not able to collect and organize it into a technical operation instruction.

In the manufacturing process, the basic steps that produce your best resources are:

The creation of the idea of the machine.

The development of the basic design of the machine.

The creation of the specifications of the machine.

The development of the steps and techniques of the manufacturing process.

The manufacture and testing of the prototype of the machine. (For some high-cost machine designs, a working model may be built before the prototype is manufactured.)

Warranty claims and liability cases that give you valuable negative feedback. This documentation is not available to you when you write an instruction for a new machine, but it becomes a major part of the manufacturing process and your resources as the machine penetrates and is sold in the marketplace.

Often these steps overlap as the people at Worldwide Robot go about the business of manufacturing machines. In the following case study you can see how two technical writers research the documents of Worldwide Robot Company. These technical writers will concern themselves only with the stages of the manufacturing process that can help them develop their technical operation instruction. Their goal is to find the information and translate it into technical operation instructions before the new machine is mass-produced, packaged, and shipped out the company's back door.

A Case Study

Background: The Birth of the Machine

A year ago, the New Products Committee of Worldwide Robot Company met in its quarterly meeting to discuss possible new enterprises (see Figure 5–15). Bob Rops, the company's chief executive officer, outlined the future strategy of the corporation.

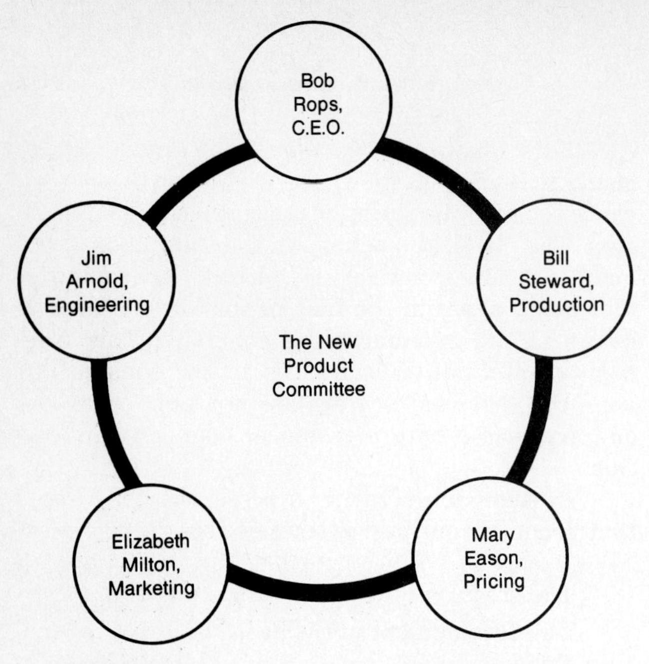

FIGURE 5-15 The "New Products Committee."

Worldwide was instructed by the corporate directors to expand the product line from industrial robots to include a machine for the domestic mass-consumer market.

Elizabeth Milton, the representative of the marketing department, was the first to give input on the topic. Her department had been researching the general market for this type of machine for many months. With corporate information, surveys, and customer questionnaires, the marketing department of Worldwide Robot had found a strong potential market for two basic types of machines. They were:

A preprogrammed robot for an educational market.

A computer-programmed robot that could be used as a teaching device for anyone with a computer.

The conceptual design. In the three months between the next New Products Committee meeting, Jim Arnold of the engineering department worked closely with Elizabeth Milton on the project. He also asked the new machine concept designer to put together four designs for each of the proposed machines. These concepts included simple operation and maintenance parameters that explained in basic nontechnical terms how each proposed machine would work. The concept design engineer approached the project from three points of view:

Design an affordable machine.

Design a machine that could be manufactured by the machines and production technology of the company.

Design a machine that will produce a profit.

The Machine Design Is Chosen

Out of the eight basic designs that Jim Arnold presented to the New Products Committee, one was chosen as the machine that had the most market potential. The new product, although it was designed for a radically new market, still enhanced the company image. It could also put the company name and reputation in a market that had as yet been unexplored.

Each member of the committee assumed his or her part in the project. Each of them went away from the meeting with a specific responsibility:

Jim Arnold of the engineering department had to prepare a more detailed design of the product. He had to work with all of the departments to make sure that the machine was the best possible design that could be manufactured at an acceptable cost.

Elizabeth Milton, in sales and marketing, had to explore the market possibilities in greater detail. The marketing experts were also responsible for making sure the company marketing organization was able to properly promote and sell the product on a worldwide basis.

Mary Eason's accounting and pricing department geared up to find and purchase the material needed for the machine. The experts in her area must be sure that the machine components and parts selected by the design engineer were the best possible products on the market for the machine.

Bill Steward, in production control, asked the people in the production area to explore the manufacturing technology that might be available for producing the machine. These people had to work closely with the engineering documents to assure that what was being designed could in fact be made.

Bob Rops, the chief executive officer, had to develop the final figure estimates for the cost of the machine in order to approach the corporation officers for the capital needed to build and market the machine.

The Machine Is Released for Production

Six months ago, the machine was given the name of the "McGuffey Project" and released for production. In that time all of the final preparations were made. The preproduction detail drawings, systems diagrams, subassembly drawings, and assembly drawings were finished. The production process was developed. Every step of the machine assembly was documented by the production engineer. Copies of the new machine design and safety signs and warnings that had to appear on the machine were sent to the legal and safety department. Existing patent rights were checked to determine that Worldwide Robot was not unknowingly using technology that was developed by another company.

Every new part or component that was bought especially for this project was given a unique number. Every nut, bolt and component that was purchased or made by Worldwide Robot was also given this unique identification number. All of the purchasing and production preliminaries were completed last month. The schedulers, who give the production people a schedule of what to manufacture and when to manufacture it, set a date for a trial run. Only 20 of the new machine were made in this special trial. These *prototype* machines are in the testing and quality control lab for analysis.

The Prototype Is Tested

There are two people in the testing lab besides the usual technicians, and they have recently become extremely interested in the McGuffey Project. They are Ted Smith and Kathy Wentworth—the company's technical writers. They must produce technical operation instructions to support the machine when it is mass-produced.

Ted has been a technical writer for two years. He moved into the position from his entry-level job as a testing lab technician. Kathy is a new hire. She majored in technical writing and technical business communication at the state university. She is in the process of learning as much as possible about Worldwide Robot machinery.

Kathy is also learning what the people at Worldwide Robot expect from the company technical writers. She is getting this specialized training firsthand as she works with Ted and the machine experts.

The Technical Writers Collect Technical Data

Ted knows from experience that he and Kathy will collect technical data more efficiently if they develop an overall plan of action. Without a general idea of how the data will be collected, Ted realizes that the two writers will waste valuable time because of redundant and useless research. His general plan looks like this:

Examine the prototype machine.

List the most up-to-date standard operation procedures (SOP) and get the parameters upon which the idea of the machine was constructed.

Research the safety documentation.

Research the systems diagrams for the machine. Research the subassembly and final assembly drawings.

Interview the machine experts.

Research vendor literature.

Ted and Kathy first learn about the machine by watching it work as it is being tested. They observe how it is controlled. They make notes about the safety precautions the technicians follow as the machine is being "cycled" through a normal sequence or standard operation procedure. They carefully list the operation procedure in order to compare it to the parameters that Jim Arnold and the conceptual designers of the McGuffey Project developed in the engineering department a year ago.

From testing checklists that the technicians use to evaluate the machine, the technical writers can learn useful information. They can find information that at a later date will help them double-check their collected information on such things as:

The correct operation procedure for the completed machine.

Adjustment points where the machine can be fine-tuned to work properly.

Maintenance points, such as lubrication and other daily-care procedures.

Safety points and safety information.

Fast-wear parts that may be added to the specification section in their instruction.

Problems in the design or the standard operation procedure.

Maximum and minimum power inputs for the machine.

Before they leave the prototype test area, Ted and Kathy make a videotape of the actual operation of the machine so they will have a reference to refer to when they create the technical operation instructions for the machine. Kathy also takes "instant" developing photographs of other special areas that they must write about. These photographs can be conveniently filed with the other information that they collect in the "hanging file" that they maintain.

The engineering drawing. After their session of firsthand observation, Ted goes to the engineering department and collects copies of the early-concept drawings and other basic information that

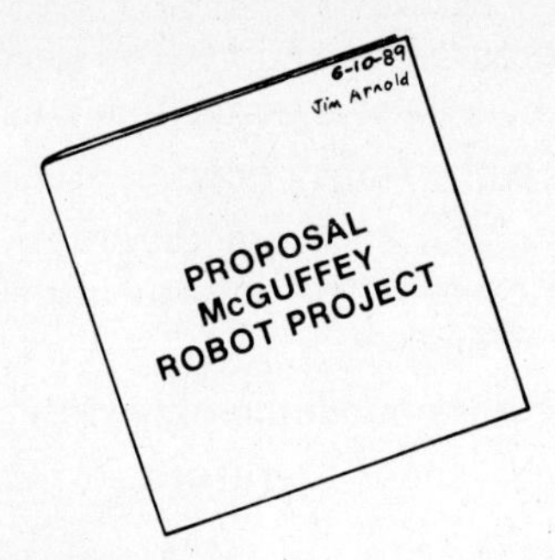

FIGURE 5–16 Proposals for machines are often written in nontechnical language.

Jim Arnold developed. This basic information is a good way to learn about the machine quickly. Ted knows that at some point the machine had to be put into basic terms and sequences of operation that could be used to sell the idea to the nontechnical company executives who release the funds for the project. The collection of this nontechnical explanation of the machine and the observation of the machine prototype in the test lab are the first and second steps that Ted and Kathy take when they collect information about the McGuffey project (see Figure 5–16).

The Technical Writers Research the Legal Documents

The lawyers and safety experts for the corporation that Worldwide Robot belongs to must review the machine for any possible dangers that may be inherent in the machine design. They inform the designers of the current industry requirements for safety and warning signs and instructions that must be put on the machine. With this feedback from the corporation legal and safety advisors, Ted and Kathy can design technical instructions that conform to both federal and industry safety standards.

The safety messages that are placed on the machine in the form of "warnings" and "cautions" usually are the direct result of the interaction of the safety and legal department with the manufacturing process. Consumer safety is a big concern for technical writers, and the signs and warning stickers that are part of the machine design are good resources for Ted and Kathy when they begin to develop safety messages for their technical operation instructions.

The Technical Writers Research the Drawings and Diagrams

The next step in the collection of technical data is the research of the drawing files that are located in the engineering department. Drawings that show each part, subassembly, and assembly of the machine must be created by the detailers or draftspeople. Diagrams that show how the power flows through a system and how the system converts it to work are also made.

Because Worldwide Robot Company is a recently established manufacturing plant, the detailers are furnished with the latest computer-aided design programs and computers to help supplement the standard method of producing engineering drawings and diagrams, which is by lead holder and paper. Because of the computer system, some drawings and diagrams are stored and retrieved electronically. Others that are drawn by hand are filed in the drawing file.

Ted and Kathy begin with the *general assembly* drawing. This document shows exactly how the machine outline must look after it has been manufactured. A *bill of material* completes all drawings and diagrams. The technical writers use this bill of material to get the drawing and diagram numbers of the main subassemblies and systems configurations (see Figure 5–17).

No matter what type of drawing is produced, it is not a complete drawing until the bill of material is made. While bills of material differ among companies, they usually contain the following useful information:

1. The revision block. This may appear on the bill of material and not the face of the drawing for complex views.
2. The company logo.
3. The engineer responsible for the design.
4. The machine name and model number.
5. The number of sheets and sheet identifier.
6. The title of the part or assembly shown on the drawing, and the drawing number that the bill of material corresponds to.
7. The banner that heads the columns for the part number (each part has a unique number), and the quantity of the parts required to make the part that the drawing depicts.
8. The information block.
9. The corresponding drawing number is usually listed in more than one location on the bill of material to aid in the retrieval process.

Each machine is made of a system or many systems. These include the hydraulic, pneumatic, electrical, and mechanical systems. Diagrams explain how these systems are connected in the machine.

Date	Revision to Part #	By	COMPANY LOGO	Engineer: Jim Arnold
10/8/	1534	EL		
			Title: Hydraulic sub assembly	Machine Name: The McGuffey Prototype 25
			Drawing No.: D-596	Sheet 1 of 2

Part No.	Item No.	Part Name	Quantity
PE-5475	20	Elector cylinder	2
PC-25751	21	Stabilizer assembly	1
PE-1534	22	Accumulator valve assembly	1

FIGURE 5–17 A bill of material is a necessary piece of paperwork for the manufacturing process.

Subassembly drawings on the other hand show how parts of the machine are actually assembled. To streamline the production process, some machines are broken down into subassemblies. These subassemblies are made in separate locations and brought together for final assembly in one location.

Detail drawings further explain how the machine is constructed. They show the parts of the machine that are used to make up the subassemblies. Kathy and Ted use the following three levels of drawings and diagrams to help them understand how the machine is put together and how the systems are connected.

The general assembly drawing and systems diagrams.
The subassembly drawings.
The detail drawings.

Ted and Kathy Use the Engineering Drawings and Diagrams

Ted and Kathy know that there is a great deal of useful information on the drawings and diagrams that have been made for the machine. When they examine drawings and diagrams they isolate information so that these sources are easier to interpret. They look for the following "blocks" of information:

The revision block.
The drawing number.
The title block.
The view block (see Figure 5–18).

Ted and Kathy know that their company has standardized drawing and diagram formats. When they use material from other companies, these blocks may be located in different areas of the drawing and diagram.

The revision block. The revision block tells the technical writers the history of the revisions that have been made to the drawing or diagram (see Figure 5–19). The information usually contains the date of the revision and the initials of the person who made the change on the drawing. The part or section that is changed in the drawing block is also marked with the date and initial of the person who made the correction. *Revised and up-to-date engineering drawings show the machine as it is and not how it was.*

The drawing number. Each drawing is given a unique number so that it can be stored and retrieved. If it is a drawing of one specific piece, or of a part that is purchased from a vendor, then the drawing number may be the *part number.* Every part that is purchased or made by the company is given a unique part number. A description of the item is usually referenced in the company's engineering files. This may be usable if Ted or Kathy need specific information about a certain piece or purchased part. A *bill of material,* which contains information about every item, is on the drawing.

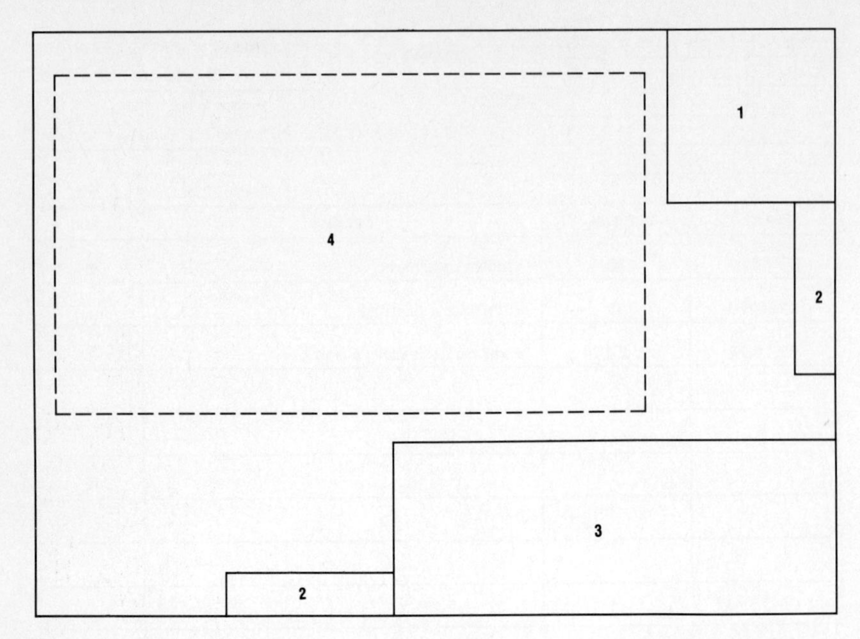

FIGURE 5-18 Information is given in "blocks" on a drawing or diagram.

Date	Drawer	Description	Approval
7-10-88	L.R.	Hand relief valve added	J.F.
3-5-89	B.F.	Filter added	J.F.

FIGURE 5-19 The revision block tells what changes are made and who made them.

The title block. A wide range of information is usually contained in the title block (see Figure 5-20). This depends entirely upon what information the company needs to manufacture the machine. The title block also states who owns the drawing or diagram. A title block on a Worldwide Robot engineering drawing includes the following.

1. The engineer who designed the system, part, or chose the piece originally for the machine design is listed on the drawing. The person who made the drawing may not be the same person as the engineer. If this is so, their initials, as well as their supervisor who checked their work, are given (see Figure 5-20-A).
2. The title of the part, system, or piece is written on the drawing. This "terminology" should be used consistently throughout the company (see Figure 5-20-B).
3. The *scale* gives the ratio of the line length to the actual item that is represented on the drawing. It is impossible to draw a machine that is 24 feet long on a 2-foot-wide area without a scale. If each inch equals 1 foot, then the draftsperson could draw a 24-foot-long machine in a 24-inch area. This scale is 1" = 1' (see Figure 5-20-C).
4. The "logo" or company symbol is put on each drawing to show ownership. Other companies cannot legally use the drawing without obtaining permission (see Figure 5-20-D).
5. The drawing number is usually placed on

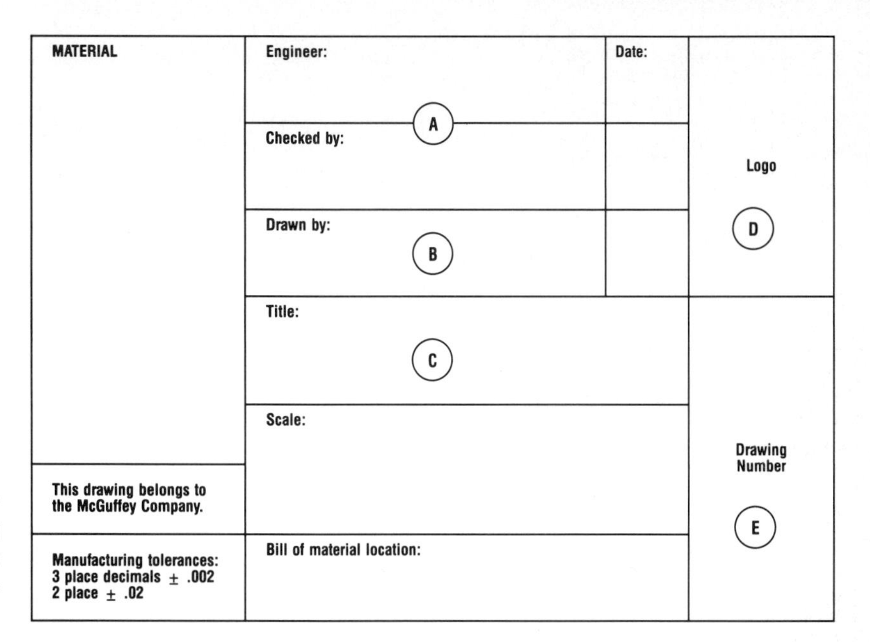

FIGURE 5–20 The title block establishes ownership of the drawing and gives specific information needed to interpret the information.

this angle to help those who are retrieving drawings from files (see Figure 5–20–E).

The view block. The view block is a visual line drawing representation of the assembly, subassembly, detail, or diagram. The physical dimensions of the things represented also are given in the view block. Revisions to the view block are usually initialed and dated.

The most common view that is used on an engineering drawing is the orthographic view. This is a view that shows the side, front end, top, and sometimes the rear end of the part. If you remove the dashed, broken, and light phantom lines from the drawing, you will get a line drawing of the physical outside of the part. Some drawings give two of the most informative views of the assembly, subassembly or part if they are needed for clarity (see Figure 5-21).

Diagrams are usually drawn with schematic symbols. Electrical circuit, hydraulic, and pneumatic schematic drawings are a shorthand method of visually linking parts together for such things as hydraulic, electrical, and pneumatic systems (see Figure 5-22). Each part of the system is represented by an abstract symbol. A list of schematics used by an engineering department to construct its diagrams was obtained by Kathy and Ted from their

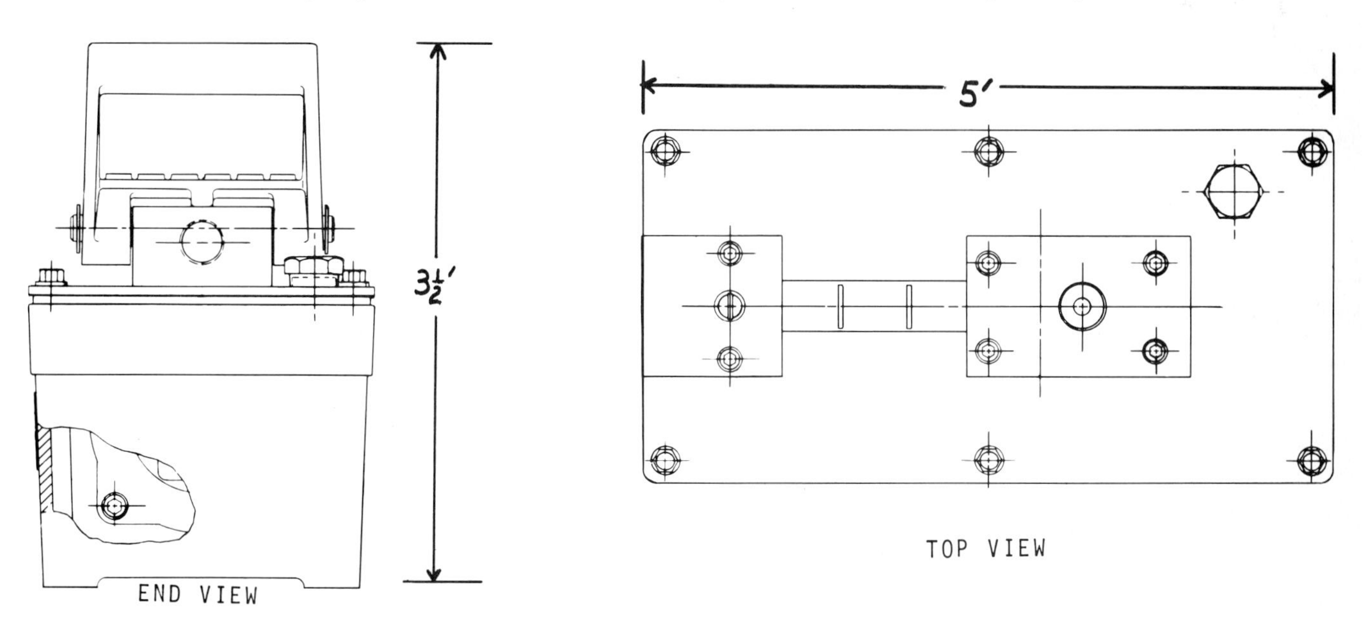

FIGURE 5–21 Two orthographic views with dimensions.

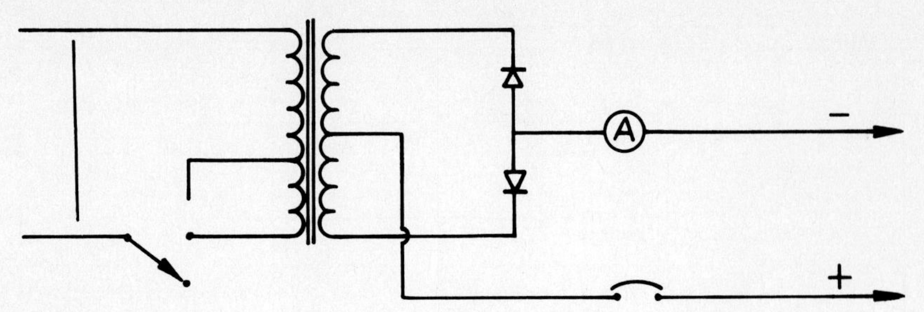

FIGURE 5-22 An electrical system schematic diagram of an arc welding unit.

own engineering department. They know that each company may use their own modified set of symbols, or one of the national or international sets of symbols that are common in industry (see Figures 5-23 and 5-24).

The other types of views correspond to the ones we covered in the section on artwork in Chapter 4. They are the:

Pictorial. Shading and shaded pressdown applications are added to give contour and contrast to the part represented on the drawing. This is not a common type of drawing because it is more expensive to produce than the orthographic line view.

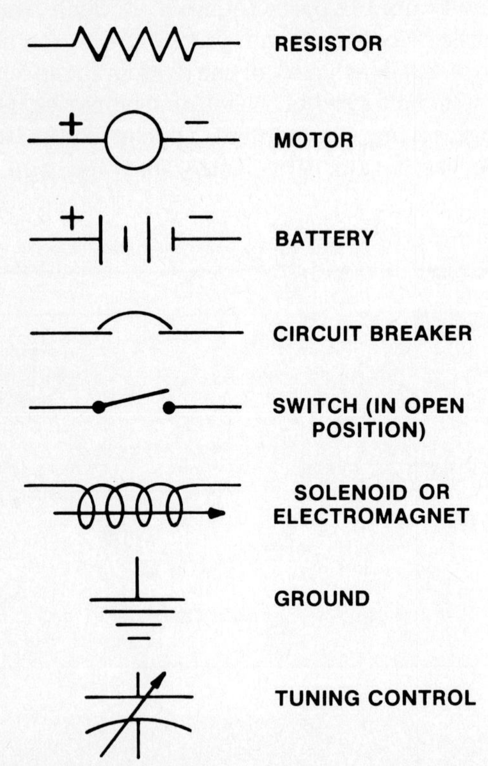

FIGURE 5-23 Basic electrical symbols used on schematic drawings.

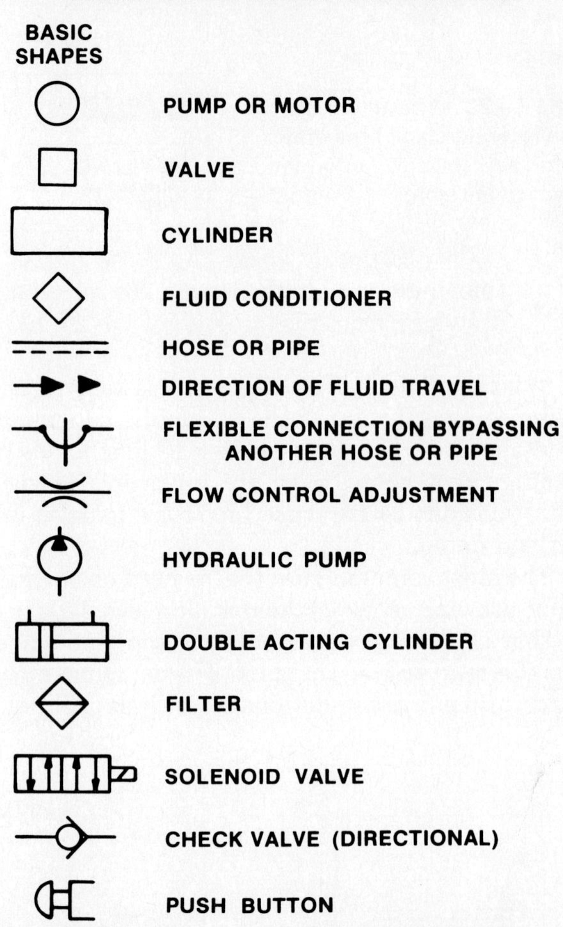

FIGURE 5-24 Basic hydraulic and pneumatic symbols used on schematic drawings.

Exploded. This is a common type of drawing that shows exactly how to put the pieces of an assembly together.

Cross section. A cross section shows the outline of the pieces that are inside a part. The part is shown as it would look if it were cut apart down the middle. This also is not a common view because it is usually expensive to produce.

Wiring diagrams. Like schematic drawings,

the wiring diagram shows how the parts of an electrical system are connected. Symbols or pictures of the actual connecting points can be used.

Lines. Ted and Kathy used the following chart to help them decipher what lines mean on an engineering drawing or diagram. In an engineering document, subtle changes in lines can mean different things (see Figure 5–25).

The Technical Writers Interview the Machine Experts

Ted and Kathy have been informally asking questions of the lab technicians who are testing the machine and other employees who are involved with the McGuffey Project. They are always asking questions and formulating pictures in their mind of how the machine works. They know that these informal interviews are going to help them prepare for more formal interviews with the machine experts who are instrumental in the development and completion of the project (see Figure 5–26).

The accuracy of this information depends upon who is interviewed. Machine experts are walking storehouses of both useful and sometimes useless information. They are sources of knowledge that Ted and Kathy must constantly tap. Machine experts involved with the logic of making the machine may not be the easiest type of person to interview. First

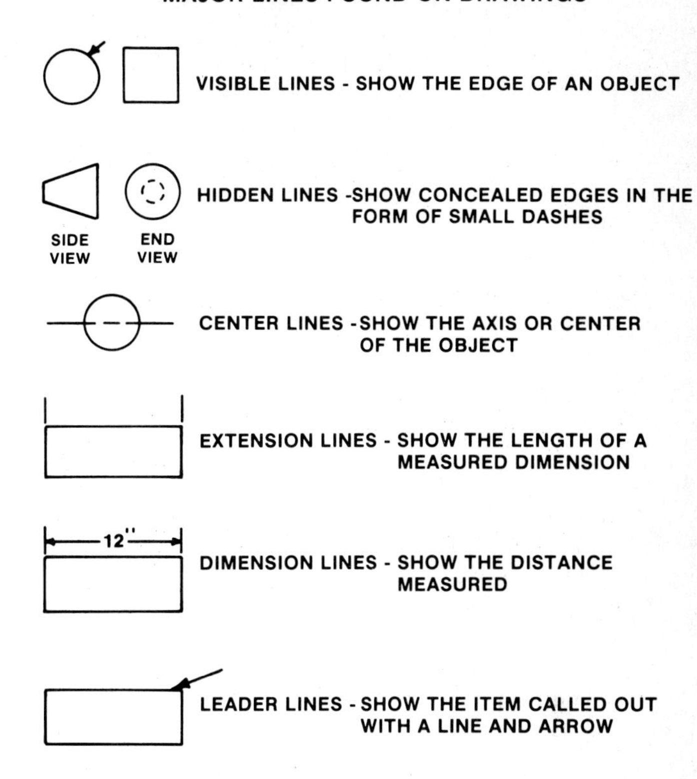

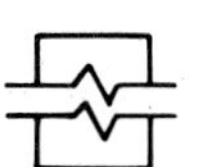

FIGURE 5–25 Line variation is the key to communication on engineering drawings and diagrams.

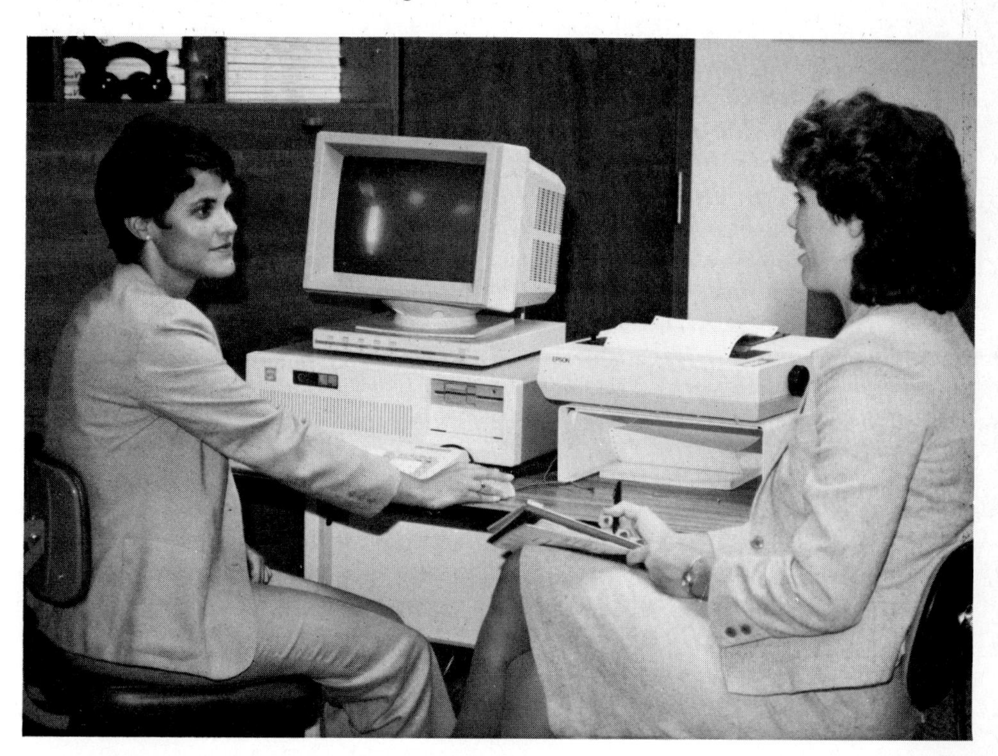

FIGURE 5–26 The interview is a communication art that must be developed.

of all, they may feel that an interview is interfering both with their time and with the logical flow of information to the people below them in the manufacturing process. Second, the machine experts may take for granted that Ted and Kathy know more about the machine than they actually do. Below are the guidelines that Ted gave Kathy on her first day on the job.

GUIDELINES FOR CONDUCTING AN INTERVIEW

1. Be patient and methodical. Keep the interview as short as possible.
2. If you arrange a formal interview, schedule the meeting in your source's work area. You can then obtain documents quickly from your source if you need them.
3. An expert is usually knowledgeable about only one area of the manufacturing process. Organize your questions to fit the expert's area of concentration.
4. Ask permission if you record the interview. Even though you record the interview, *always* take written notes.
5. Ask concrete questions that require concrete answers.
6. Do not let your source answer you in technical words and jargon that you don't understand. Politely challenge the meaning of these words and ask your source to define them.
7. Before you leave, ask the source for another interview at an unspecific time in the future. This will prepare the source for another interview request in case you need one.
8. Your job is to get an accurate picture of the machine. If you find that your source's information does not coincide with other sources, then you must return to your source with the inconsistencies.
9. Organize your notes as quickly as possible after the interview.

Kathy's Interview

Kathy's first interview with a hydraulics expert was successful because she followed the guidelines listed above:

Kathy: I am writing the technical operation instruction that supports your new machine, the McGuffey. In my research, I found that there are two operator controls on the hydraulic system. They are the direction control valve and the pressure regulator. Are there other controls for this system?

Answer: Yes. A PC26A valve has been recently added.

Kathy: What is a PC26A valve?

Answer: It's a simple manual shutoff valve that the operator must control by hand.

Kathy: What does the valve control?

Answer: It relieves the hydraulic system of hydraulic fluid pressure when it is in the "off" position.

Kathy: Why is it in the system?

Answer: Before the operator cleans, adjusts, or maintains the system, the hydraulic pressure must be relieved to help prevent a sudden movement of one of the parts. Sudden movement is a major cause of accidents for this type of system, according to our safety department.

Kathy: Will a safety caution sign be put on the machine for this valve? Should this safety message be added to the instruction?

Answer: Yes. A sign will be added to the machine. Our industry standards require that you put the message in your instruction. Please send your written warning to the safety and legal people for their review.

Kathy: Thanks for the information. I'm researching all sources of information. Can I rely on you for more information in the future?

In a short time, Kathy answers the questions of who, what, where, and why of the valve story. She goes to the primary source of the story to get her basic information. Now she is in a position to double-check the accuracy of the interview and of the documents that she collects.

Kathy sets the tone of the interview by telling the engineer at the beginning of the interview exactly what she knows, and exactly what the interview is going to cover. This allows both people to save time and concentrate on the one topic that interests Kathy. The interviewee doesn't have to guess about the reason or goal of the interview.

The Writers Double-Check All Sources

In order to be accurate, Kathy and Ted must double- and even triple-check the accuracy of their information during the creation of operation instructions. They must use their resources carefully. Drawings may not be corrected in some companies until days after the change is made. By carefully blending interviews and document searches, the two technical writers can obtain the most accurate data possible.

Note: Always double-check information that you collect from documentation with an interview. Always double-check information from an interview with documentation.

The Writers Research Vendor Information

Vendors, or the sales representatives of companies that supply *components* and *parts* for the machine, can also furnish Kathy and Ted with instructions and specifications that explain the product. This literature is usually on file in the engineering department. Each vendor product, as with every other product used in the machine, has a unique number that is given to it when it is purchased.

By finding the component or part on a drawing or diagram, the technical writers can use the bill of material to locate the unique number. The parts index file, with all parts listed by number, can be used to identify the vendor and the component or part number. With this information, Kathy and Ted can go to specific catalogues and brochures that the vendor has supplied.

Ted and Kathy Continue Their Research for Data

Ted and Kathy continue to find information and develop sources of information for as long as they can. At some point they know that they must set a deadline for accumulating data. They must give themselves plenty of time to write, edit, and produce their technical operation instruction in order to meet the set deadline. They must create their instruction so that it can be easily revised to keep up with the McGuffey Project as it develops and matures during its manufactured life on the shop floor.

CHAPTER SUMMARY

- The results of manufacturing companies are often taken for granted. The advancement of technology rests upon the hard work, intelligence, and organization of our manufacturing companies (see Figure 5-27).
- The manufacturing company can be divided into two separate areas: (1) the shop floor where the machine parts are fabricated and the components are assembled and (2) the front office.
- A successful manufacturing company is a well-organized group of people dedicated to making and selling a machine for a profit. Success is often built upon how well the front office and the shop floor people communicate.
- Every manufacturing company is divided into departments where people who do a similar type of job work.
- The basic divisions of a manufacturing company are the engineering, the sales, the accounting, and the production departments.

FIGURE 5-27 New technology such as this remote-controlled excavator is the result of the hard work of many machine experts. (Courtesy Deere & Company)

- The engineering department develops the machine concept and details the machine for production in engineering drawings and diagrams.
- The sales department must market and distribute the machine to all profitable markets that the manufacturing company can supply with machines.
- The accounting department works on securing capital and applying the capital to the manufacture of the machine. Other accounting department responsibilities include payroll, taxes, and employee benefits.
- The production department physically oversees the manufacturing process. Experts in this department develop the production methods and processes and issue schedules that are critical to the people who make the machine.
- The main sources of information for the technical writer are the drawings and diagrams. A basic engineering drawing and diagram include blocks of information. A bill of material completes the drawing and diagram.

EXERCISES FOR CHAPTER 5

Search Out the Sources of Information

1. Find examples of the following views. These can be either drawings or photographs from instructions or the popular media. Are they used effectively?
   a. Exploded view.
   b. Orthographic view (more than one view of the same object).
   c. Pictorial view.
2. Interview a person from the front office of a company. If possible, interview someone in the same company who fabricates or assembles parts of the machine.
   a. Can the people you interview tell you all of the basic steps that are needed to make a machine at their company?
   b. Do they use engineering drawings in their job?
3. Get samples of engineering drawings and diagrams. (See the drafting or engineering instructor for examples if you are in a classroom situation. If drawings and diagrams are not available, use engineering drawing and design books from your local library.)
   a. Block out the sections of information that you can isolate, including the title block and drawing number.
   b. Erase or write-out with acrylic paint all of the lines on a drawing except the visible ones. Can this drawing be used as artwork?.

6

Know Machine System Concepts

The information given in this chapter will enable you to:

- Explain the "system" concept for machines.
- Give a definition for the term "machine-powered system."
- List the benefits and uses for "feedback" in a machine system.
- Follow a basic hydraulic and pneumatic system from the time fuel is transferred into energy until the energy is transformed into work.
- Define the term "prime mover."
- Explain the difference between the reciprocating and rotary power source.
- Give examples of how the operator interacts with machine systems.
- List the basic characteristics of fluids under pressure.
- List and explain the uses for the major components in a hydraulic system.
- Give an example of a pneumatic system and explain how it works in system terminology.
- List the major parts of a pneumatic system and explain how they work.

INTRODUCTION

In the first five chapters you were given guidelines that explain the skills that you can use to develop a clear and concise writing style. You were also given a basic foundation upon which you can build a visual technical writing style that enables you to communicate with as many people in a mass-consumer audience as possible. In this chapter you will be given an overview of the skill that you must develop in order to write "technically." You can use this information to understand technology from the unique point of view of the technical writer in industry. With a basic understanding of machine system concepts presented in this chapter, you will be able to explain technology more clearly and accurately in your technical operation instruction.

This chapter also includes some of the basic logic that design engineers use when they create machines for industry. Their machines can be as big as a space shuttle, or as small as a hand-held radio. No matter how big, small, complex, or simple the machine is, design engineers use basic systems concepts to transform power into mechanical life. By studying machine systems, you can develop a better understanding of how energy is transformed into useful work. By studying the systems illustrated and explained in this book, you will know the basic fundamentals that modern technology is built upon.

The backbone of every machine is one or more of the five "machine-powered" systems. The remainder of this chapter explains the basic parts used in three of these systems. By examining them piece by piece, you can see how each part affects the whole system. You can also see how the machine operator interfaces with the machine parts. The three systems covered in this chapter are:

The power system (internal combustion engines).
The hydraulic system (fluid under pressure).
The pneumatic system (air under pressure).

The remaining systems will be covered in Chapter 7:

The electrical system (electrons under pressure).
The mechanical system (torque).

You can compare machine-powered systems to more familiar systems such as the human body. Biological systems work together in the animal organism to do the day-to-day work of living. Within each human, there are circulatory, nervous, digestive, reproductive, and respiratory systems that are arranged together on a frame or skeleton. For example, the parts of the circulatory system, such as the heart, blood, and blood vessels, are connected in an *unbroken* line from the heart to the points of work at the muscles where the blood delivers oxygen and picks up impurities.

Another example is the ecological system. Complex animal and plant interdependence evolves in specific geographical areas. The survival of these living organisms in the ecological system is directly dependent upon the weather and the physical well-being of the land upon which they live. Certain animals are a food supply for other animals, and certain vegetation is a food supply for the animals that are eaten. In order to function properly, every plant and animal has a role to play. Every plant and animal is a valuable member of the ecological system.

By comparing the power, hydraulic, and pneumatic systems to these more common types of systems, you can see how the consumer or machine operator must control or adapt the machine to do work. You can see how essential it is to give the consumer enough information to judge how well the machine is converting power into work. Using simple systems concepts also enables you to find the major control points, internal machine protection devices, and maintenance and adjustment areas that are essential topics that must be covered in the technical operation instruction.

USE SYSTEM THEORY TO UNDERSTAND TECHNOLOGY

As you examine your everyday world, you can find many types of systems. Each one is characterized by an interconnection of parts, a control that affects the output of the system, such as the weather in an ecological system, and necessary internal controls that ensure the system survives. Machine-powered systems have the same characteristics. Machine-powered systems are designed from parts that are connected together, mounted on a machine frame, and controlled by an operator in order to produce a *specific* type of work (see Figure 6-1).

Understand Machine-Powered Systems

Feedback. Machine-powered systems use power that has been created from the breakdown of a fuel to do work. *Work* is the result of power that

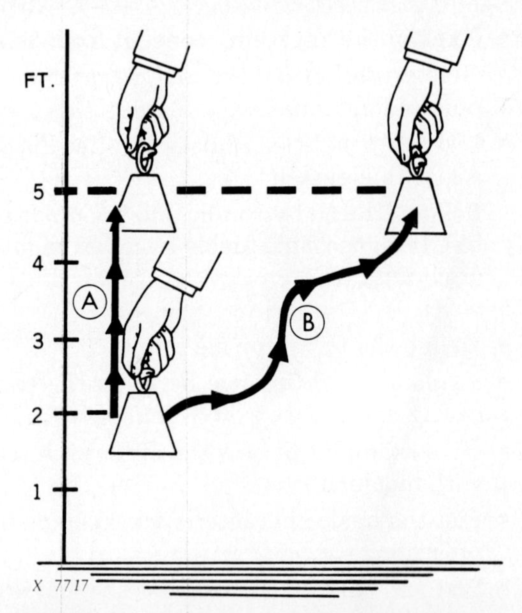

FIGURE 6–1 Work can be measured. If you lift a weight 5 feet (A), or lift it 5 feet as you carry it (B), then you have done 60 foot-pounds of work. (Courtesy Deere & Company)

is transformed into a force that can overcome a resistance. In machine systems, for example, work can be a pushing or pulling movement at the mechanical fingers of an industrial robot. The amount that is done, and the quality of the work produced, can be measured against an expected outcome. This comparison between what is expected and the actual outcome is called *feedback.* Without feedback, the machine operator cannot properly *adjust* and *control* the systems in a machine. For this reason, feedback must be considered an integral part of the machine system, and the operator must be considered to be the "mind" of the machine.

Without the correct operation instructions that explain how to use feedback to adjust the machine, the consumer will be similar to someone who is behind the wheel of an automobile for the first time. Consumers will either be awkward and unsure of their actions or they will not know how to turn over the ignition key. A technical operation instruction informs the operator of a machine how to evaluate and control the work that is done by the machine. The operator must be told how to start, stop, operate, and keep the machine running properly on a daily basis. Without this information, the operator will not be able to keep the machine systems working in harmony.

System control points. Operator-controlled systems are designed with basic fundamental concepts. A well-designed system has these characteristics. First, there must be an external control point. For example, the lighting system in your home has an on and off switch in each room. A computer monitor may have control knobs that allow the user to adjust the contrast. A refrigerator has a temperature control that is adjusted to keep the food inside at a preferred temperature. Without an external control point, the operator cannot make the necessary adjustments to fine-tune the system to produce maximum work.

Internal protection. Second, there is an internal protection that keeps the system from self-destructing. Internal system protection can be compared to a hot water heater safety valve that blows open if the internal pressure of the heater is too high. A fuse in a home electrical system disconnects a radio from the electric current if the current is not under control. A shear pin is used to protect a mechanical gear from being destroyed if a machine transmission "freezes" and becomes blocked. Without internal protection points that may or may not be set by the operator, the system can be damaged internally.

Power source. Third, every system in every machine needs a power source, or prime mover, such as an internal combustion engine. Power is created by the spending of fuel in the prime mover. This created power is mechanically transmitted to the system. The machine-powered system transfers this power through the system parts and ultimately transforms it again into mechanical work at the end of the system. For example, when you ride a bicycle, you are converting food fuel to muscle power. This power is transferred to the pedals; the pedals turn the drive sprocket; the chain transfers power to the rear sprocket; and the rear sprocket turns the rear wheel. The wheel turns and converts friction at the tire treads into forward motion.

Note: Read this chapter and the chapter on safety before you attempt physically to trace the flow of power through a machine. Be aware of the safety and warning signs on each machine before you operate or work on it. Every machine is a potential hazard if you mishandle or misuse it.

BEGIN THE STUDY OF MACHINE SYSTEMS WITH POWER SOURCES

Prime mover. Every machine-powered system is driven by a prime mover that converts fuel into energy. A prime mover power source might be mounted on the machine, like the engine of an automobile. Or it might be located hundreds of miles away from the machine, like the nuclear power plant that supplies us with electricity. A stationary power system can be located near a machine and can be connected to it by a belt, chain, or pully, much like the gasoline engine that powers a carnival Ferris wheel.

Manufactured prime movers also are systems. They are designed to convert a fuel supply into a mechanical rotating action that is measured by torque. Through combustion, heat and force convert fuel into energy. Prime movers must be supplied with fuel and kept cool if fuel conversion is to be as efficient as possible. Most of our manufactured prime movers are extremely inefficient. A gasoline engine, for example, loses one-third of the heat or energy produced during combustion through the exhaust. One-third is lost through necessary cooling of the combustion chamber. Usually this is done with a cooling subsystem of water and antifreeze. The final one-third of the fuel is converted to useful mechanical energy.

The natural prime mover. The most abundant natural prime mover power sources are the sun, the wind, and water. A chemical breakdown of fuel on the sun supplies a light source that we on Earth can convert into fuel with a photoelectric cell and store in a solar battery. Solar power is thus transferred into electric current. Although it is abundant, solar power has not been efficiently employed in our machines. Wind power, on the other hand, has been harnessed by windmills for many years. Similarly, waterwheels have been used to mill corn, wheat, and other grains. All of these sources require bulky machinery and equipment to transfer natural energy into machine power efficiently.

Another power system is the conversion of food to power in humans and animals. Both human power and animal power were the dominant prime movers until the Industrial Revolution in England and Europe in the late eighteenth century. Human and animal power built the Pyramids of Egypt, the Roman Empire's highway system, Stonehenge, and the Great Wall of China. Like natural energy power sources, animal power is less efficient, practical, and reliable than manufactured ones (see Figure 6-2).

A—First mill. B—Wheel turned by goats. C—Second mill. D—Disc of upright axle. E—Its toothed drum. F—Third mill. G—Shape of lower millstone. H—Small upright axle of the same. I—Its opening. K—Lever of the upper millstone. L—Its opening.

FIGURE 6-2 The main types of nonmanufactured energy sources are illustrated in this sixteenth-century woodcut from a book on metallurgy.

Analyze Common Manufactured Power Sources

The most common source of abundant power is the internal combustion engine. Along with the diesel engine, these power systems drive most of our modern machines. To appreciate the wide use of these fossil-fuel-burning power systems, imagine a world without automobiles, without heavy construction equipment, recreational equipment, or outdoor power equipment. Gasoline and diesel engines power much of our technology (see Figure 6-3).

There are two basic designs of these fossil-fuel-burning engines. They are:

Reciprocating
Rotary

The reciprocating engine. A reciprocating engine is the most common fossil-fuel-burning power source. It is called a reciprocating engine because gasoline, diesel, LP (liquid petroleum) gas, kerosene, or other fuel is vaporized and exploded in a cylinder chamber to produce an *up and down* motion of a movable cylinder (see Figure 6-4). For example, in an internal combustion engine, gasoline is vaporized and pulled into a cylinder chamber where it is ignited by a spark from a spark plug. The cylinders are connected by a rod to a crankshaft. The crankshaft is turned by the reciprocating movement of the piston. The reciprocating engine piston creates torque or circular motion on the crankshaft. A drive shaft or other type of linkage connected to the crankshaft transfers this mechanical circular motion to machine-powered systems (see Figures 6-5 through 6-7).

Some machines, such as a lawnmower, may be powered by a one-cylinder engine. Some machines are designed with engines that use many cylinders. The most popular arrangement of multicylinder engines is in a horizontal row like an automobile engine. No matter how many cylinders are mounted in the engine and attached to the crankshaft, they must be timed by the engine manufacturer so that they work together.

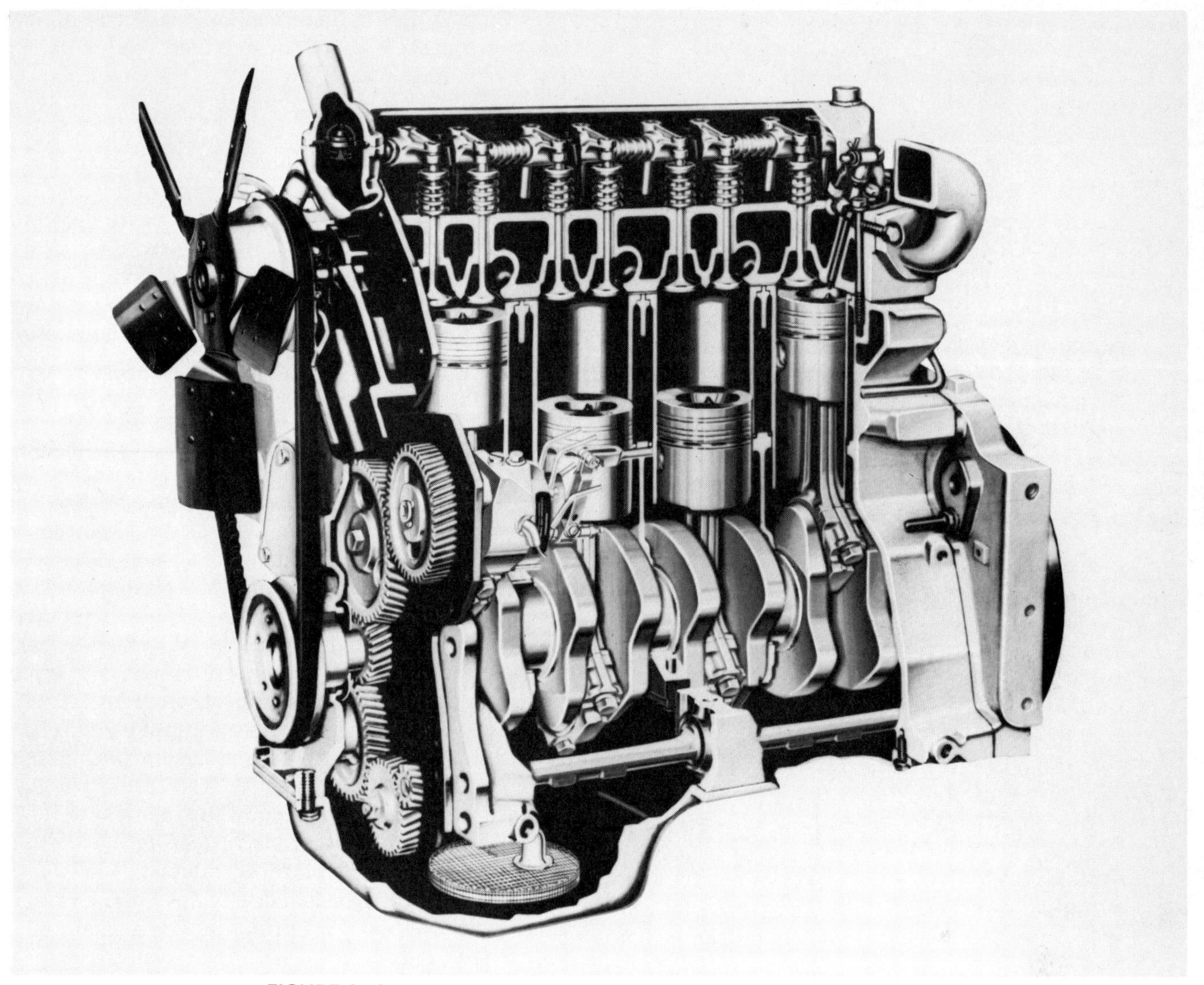

FIGURE 6–3 A typical reciprocating engine. (Courtesy Deere & Company)

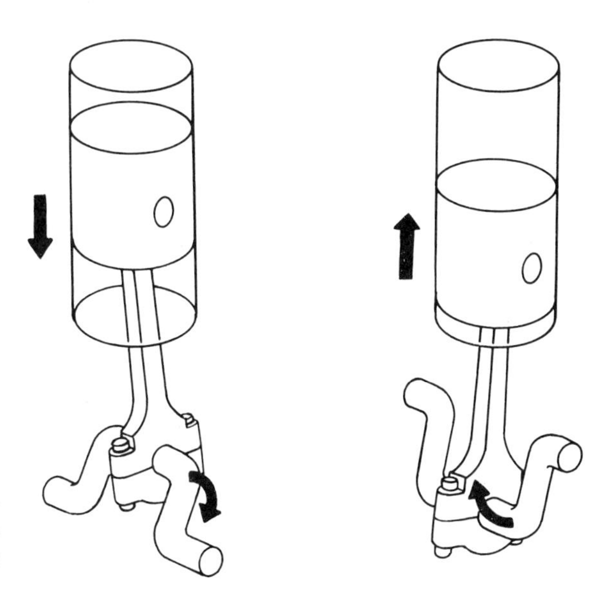

FIGURE 6–4 Reciprocating motion turns the drive shaft of an engine. (Courtesy Deere & Company)

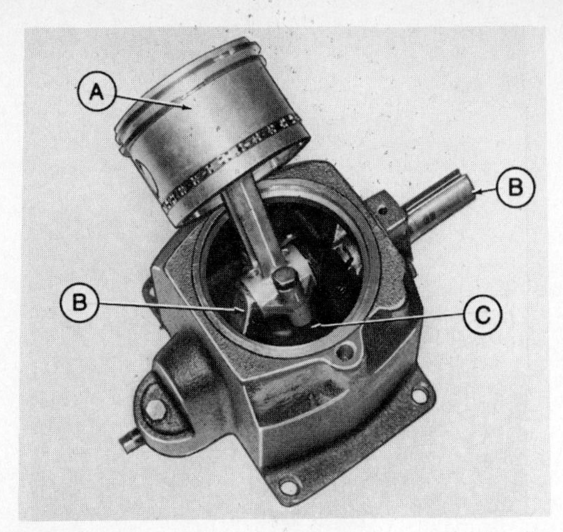

FIGURE 6–5 With the head removed, the cylinder (A) is shown attached to the drive shaft (B) in the cylinder chamber (C).

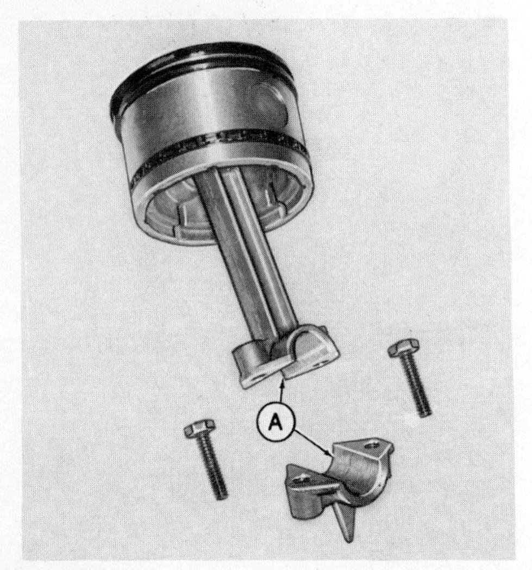

FIGURE 6–6 A cylinder assembly. The drive shaft connects to the cylinder rods at the bearings (A).

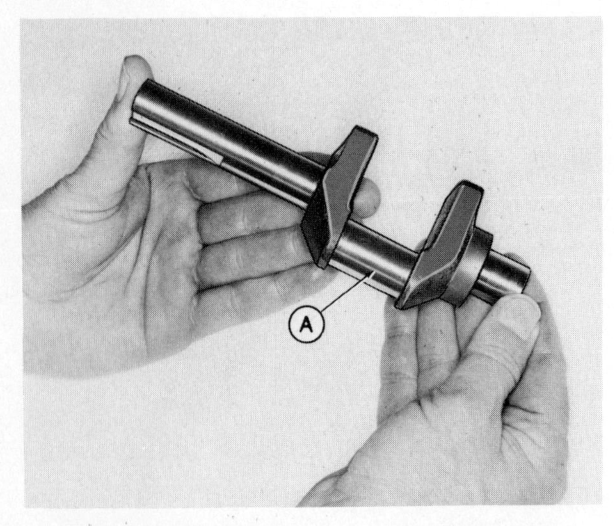

FIGURE 6–7 A drive shaft for a one-cylinder engine. The cylinder bearings attach at point (A).

Some single-cylinder engines have one up and one down movement of the cylinder every combustion cycle. In this case, the force of the weight on the drive shaft pushes the cylinder head up and compresses the fuel mixture to the proper ratio. Combustion takes place and forces the drive shaft down as the cylinder chamber is exhausted of gasses. The fuel mixture is again drawn into the cylinder combustion chamber, and the cycle begins again. Most multiple-cylinder machines and some single-engine machines have four cycles or two up and two down movements of the cylinder every time combustion occurs (see Figure 6–8).

The rotary engine. The rotary engine does not have reciprocating motion, because it does not have cylinders that move up and down. In a rotary engine, a series of chambers are located around the crankshaft. The piston or pistons are mounted on the crankshaft. As the fuel ignites, the pistons spin in a circular motion and turn the crankshaft. Like the reciprocating engine, valves are used to control the flow of fuel into, and exhaust out of, the rotary cylinder area. Because there are no reciprocating pistons, the number of moving parts is drastically reduced in the rotary engine (see Figure 6–9).

Rotary engines are used in automobiles and in some stand-alone engine units. The future use and acceptance of the rotary engine may extend to more areas because of the extensive testing and design modifications taking place in industry. Versatility and the ease of maintenance of the rotary engine may make it a feasible prime mover in many machines that are equipped with reciprocating engines today.

List the Common Characteristics of Power Systems

The internal combustion engine physically transforms a fossil fuel source into thermal energy through combustion and then into mechanical energy at the crankshaft. Although heat is a necessary step in the combustion process, heat in an engine outside of the cylinder combustion chamber means friction. Friction is a sign of inefficiency and wasted energy in an engine. The goal of all internal combustion engine designers is to reduce heat or friction to a minimum and deliver as much torque to the point of work as possible.

The point of work. The point of work for a power system such as the reciprocating and rotary engine is the drive shaft that is put into circular mo-

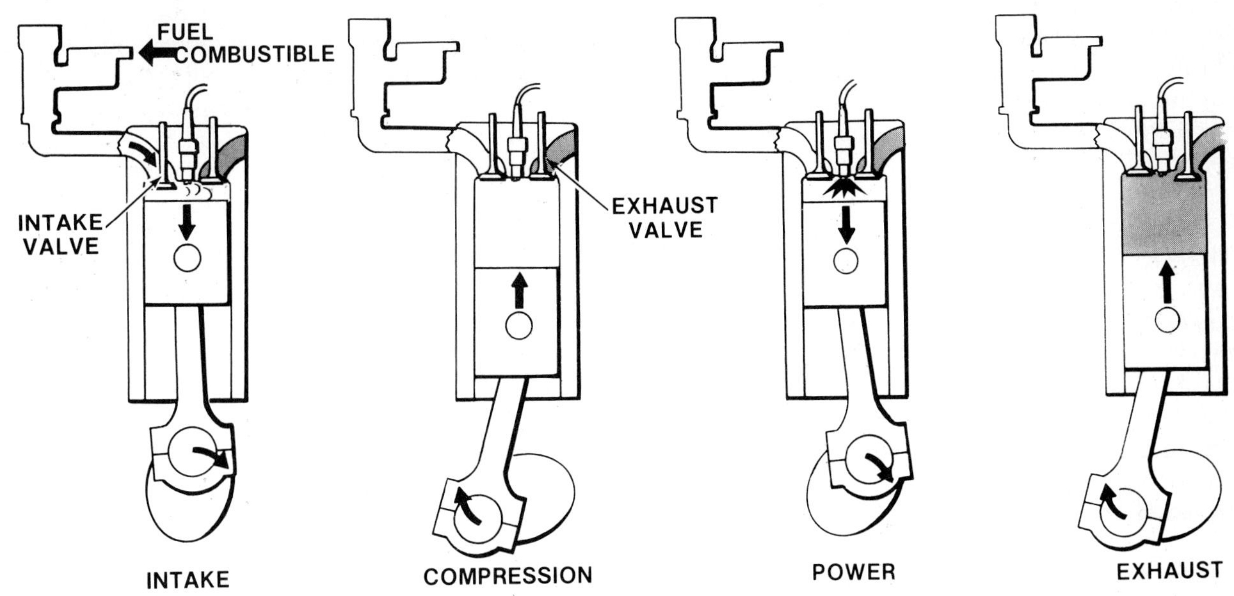

FIGURE 6-8 The four-stroke engine has four distinct actions during one cycle. (Courtesy Deere & Company)

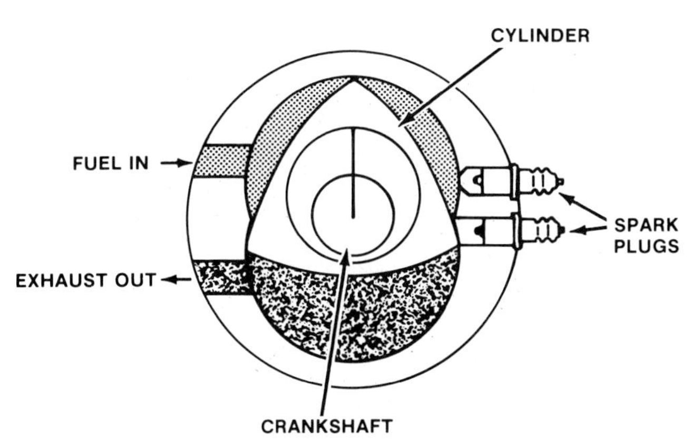

FIGURE 6-9 The rotary engine combustion chamber rotates and turns an eccentric, or off-center, crankshaft.

tion by the crankshaft when the engine is running. The size and strength of engine components determine the load that can be put on the power system when it is connected to another system. Engines are rated by how much "horsepower" they can produce. One horsepower (hp) equals the work required to raise a 150-pound weight a distance of 220 feet in one minute.

Points of control. Engines are generally started by turning the crankshaft. A flywheel, or large gear, is attached to the crankshaft. A starter motor, usually an electric motor, turns this flywheel and creates circular motion on the flywheel. This circular movement creates piston and valve motion, which begins the process of combustion in the cylinder chambers. The cylinder head is forced up and compresses the fuel mixture so that it can be burned. The driver of an automobile, for example, controls the startup at the ignition key. By turning the key, the operator activates a small electric motor that turns the flywheel (see Figure 6-10).

The consumer controls the speed of the crank-

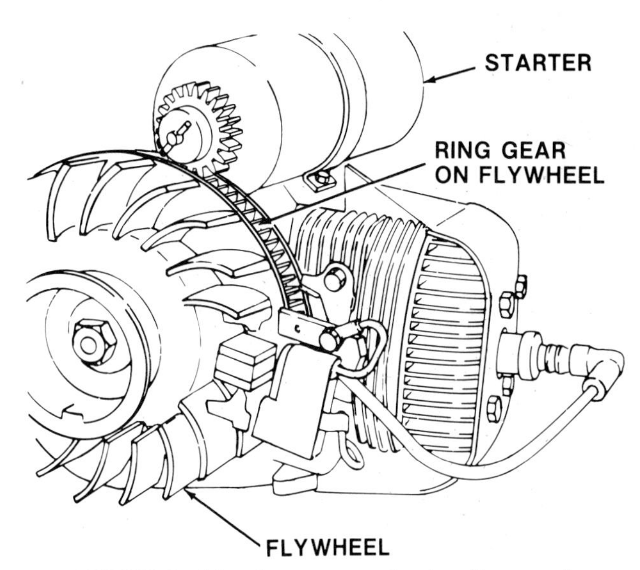

FIGURE 6-10 A starter motor is often used to begin the combustion cycle of an engine. (Courtesy Deere & Company)

shaft movement by regulating the fuel and air intake in the cylinder chambers. An increase in fuel and air increase the speed of the combustion cycle. In a non-fuel injected gasoline reciprocating or rotary engine, a carburetor air intake opening is increased or decreased by the operator at the accelerator pedal, which in turn draws more fuel into the cylinder.

Conversely, diesel engines do not have carburetors. The fuel is injected directly into the cylinder chamber where it combusts because of the pressure put upon it by the upward movement of the cylinder. Fuel injection into the diesel-fuel-burning cylinder is timed and controlled to produce the maximum output of the combustion and exhaust cycle. In both types of engines, the rate of fuel delivery to the cylinder chambers regulates the speed and the torque of the crankshaft.

Analyze the Role of the Operator in the Operation of Power Systems

General maintenance. Engines are internally lubricated by oil that is pumped or splashed on the moving parts. Internal oil quantity and quality are probably the greatest factors in keeping an engine working efficiently. Consumers must be given a schedule that they can follow to ensure that the oil used as internal lubrication is clean and that it meets the engine manufacturer's specifications.

Filters, such as the internal lubrication oil filter, must be changed at regular intervals. Fuel filters or fuel conditioners also must be maintained to ensure that impurities in the fuel or lubricant do not cause excessive wear on internal power system parts. Every moving part made out of metal, plastic, or rubber will eventually wear down and deposit impurities into the system. Filters are designed to remove this grit and protect the system from internal wear.

Gasoline engines must be regularly checked and often adjusted at the carburetor or fuel injectors. Also, spark plugs do not last indefinitely, and they must be replaced on a regular basis. Diesel engines, although they do not have spark plugs, need to be monitored because of the possibility of fuel injector clogging because of impurities in the fuel. In general, the fuel, and the parts that supply fuel to the cylinder combustion chambers, must be maintained *on a regular schedule.*

There are two basic types of coolants that the consumer must know about for reciprocating and rotary engines. First is a coolant used mainly for cooling aluminum parts. Aluminum has become a primary metal used in modern over-the-road vehicles like automobiles and trucks. Most of these vehicles are made with a mixture of aluminum and cast iron components. A second type of coolant is for an engine that is heavier and made of cast iron. This heavier engine is used in such vehicles as off-the-road construction, forestry, and farming machines. If an antifreeze that is blended with additives for an aluminum engine is used in a cast iron cooling subsystem, the subsystem may clog. The consumer must be told exactly what cooling fluid is to be used in the engine. If the wrong fluid is used, the heat that is produced during the combustion process will build up and damage engine parts.

ANALYZE HYDRAULIC SYSTEMS (FLUID UNDER PRESSURE)

Hydraulic systems are common on larger machines. Hydraulics are found in:

- Earthmoving machines (hydrostatic drives, movement of grader blades, and loader shovels). (See Figure 6–11.)
- Outdoor power equipment like lawn and garden tractors (hydrostatic drives, steering, brakes, and equipment control). (See Figure 6–12.)
- Robots (basic movement).
- Machines that are used in the manufacturing process.
- Automobiles (brakes).
- Large agricultural machines. (See Figure 6–13.)

Isolate the Basic Properties of Hydraulics

The science of hydraulics includes the study of how liquids, such as hydraulic oil, react under pressure. Hydraulics also includes the study of how specially designed machine parts can control and transfer high-pressure fluids in the hydraulic system. In order to help us understand how these parts work together to make a system, let us look at a few basic principles that are the foundation for the study of the science of hydraulics.

Liquids are shapeless. Liquids are free flowing and take the shape of the container or reservoir in which they are kept. For this reason, liquids will flow through hollow transfer connectors, such as reinforced rubber hose, or metal pipes. This property makes hydraulic systems extremely flexible. Hy-

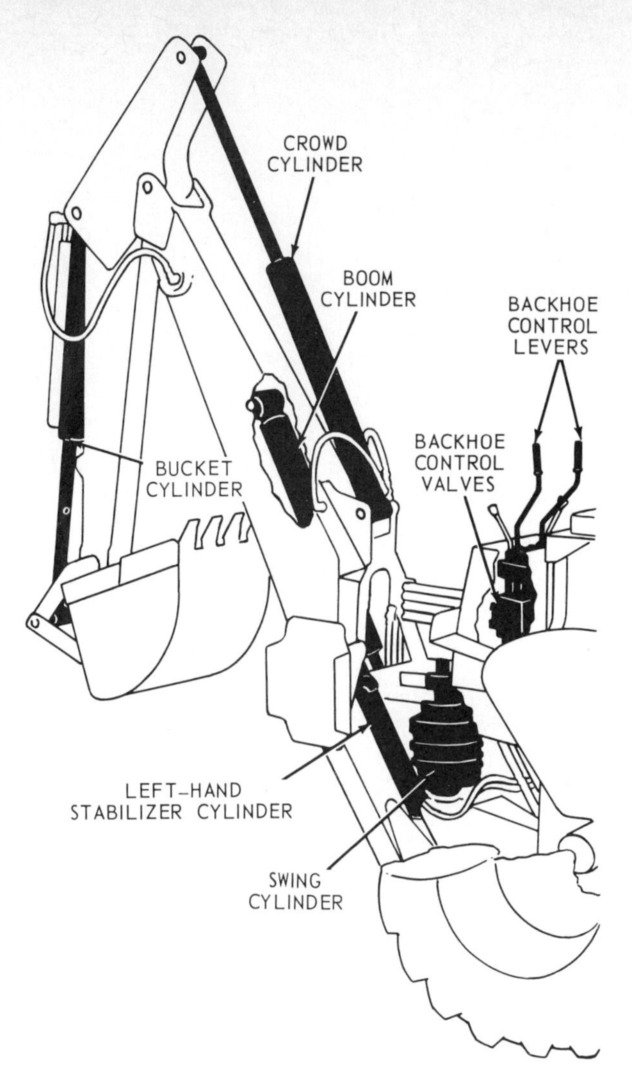

FIGURE 6–12 Small outdoor power equipment may be powered by a hydraulic system. (Courtesy Deere & Company)

FIGURE 6–11 A backhoe attached to a construction machine turns hydraulic power into work such as digging a foundation for a house or a ditch for laying pipes. (Courtesy Deere & Company)

FIGURE 6–13 Large machines are often designed with an extensive hydraulic system. (Courtesy Deere & Company)

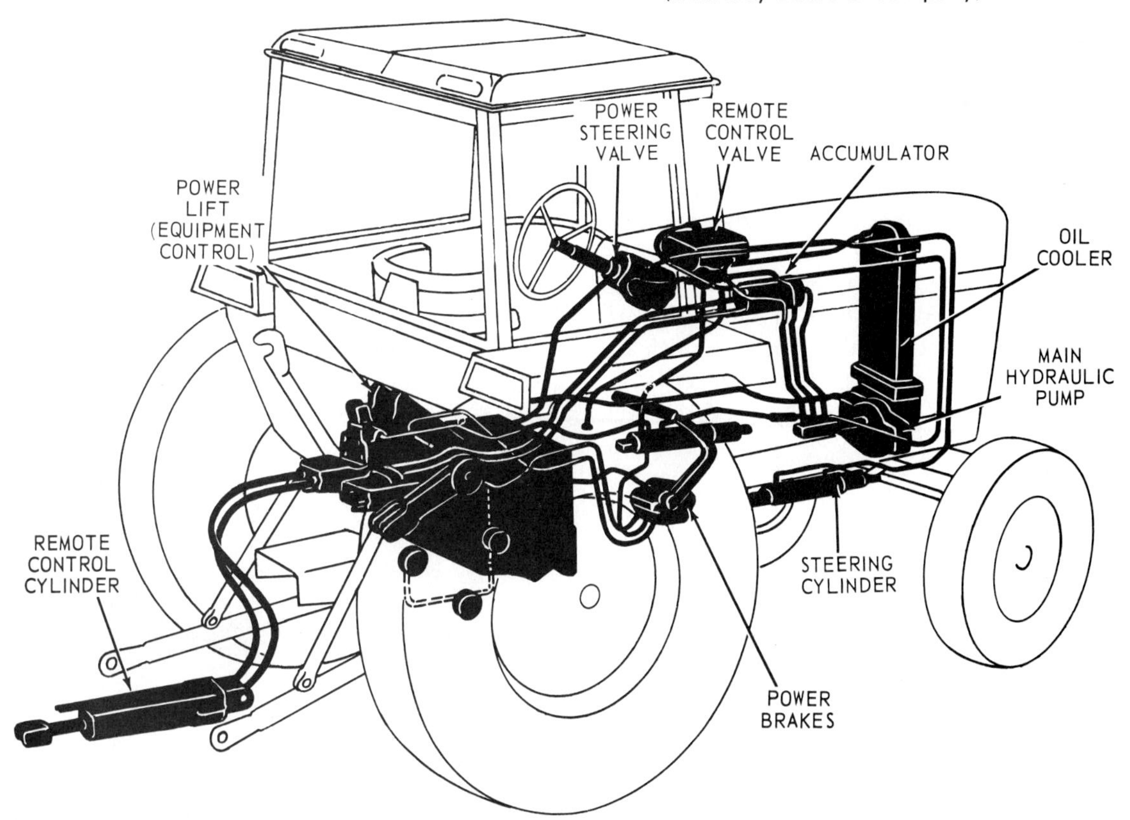

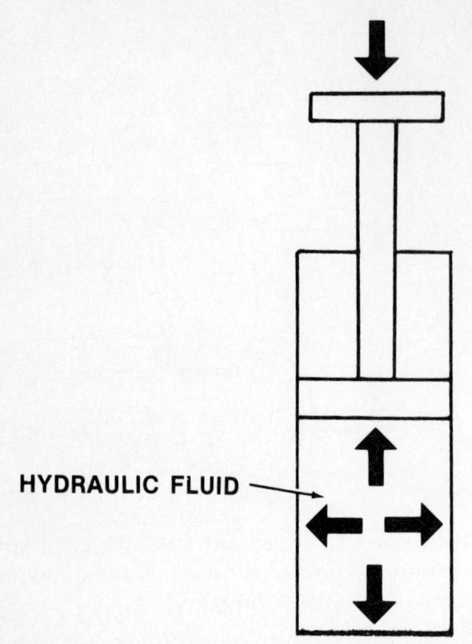

FIGURE 6–14 Liquids are shapeless and almost incompressible. They transmit pressure equally in every direction.

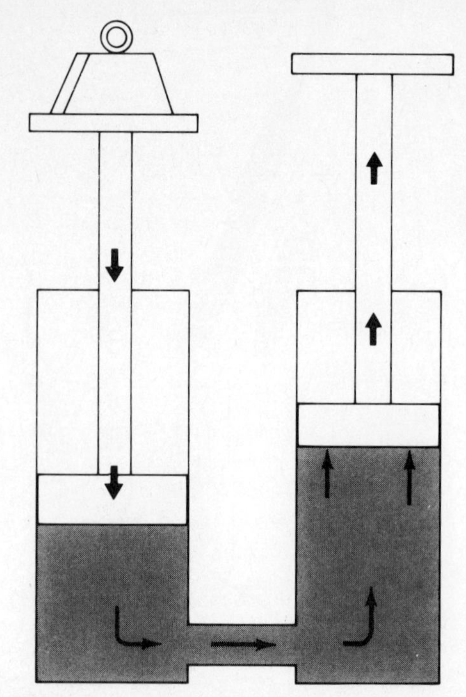

FIGURE 6–15 Downward pressure on the left cylinder transfers the same pressure to the right cylinder. (Courtesy Deere & Company)

draulics can supply power to a point of work that would be impossible for other power systems. Pipes and hydraulic hose can be bent and routed to points of work that other machine power systems cannot reach. Hydraulic pressure does not occur until there is resistance or a blockage of the hydraulic fluid flow.

Liquids transfer pressure equally in all directions. Unlike gases, liquids are almost incompressible. If you put a liquid in a container and put pressure on top of it with a sealed lid, the amount of pressure that you put on top of the lid will be the same on all sides of the container. Liquids under pressure transmit power equally in all directions (see Figure 6–14). For this reason, pressure can be applied at the power source to a hydraulic fluid and transmitted by hose and pipes to a point of work many feet away and not usually lose a significant amount of the original pressure (see Figure 6–15). Fluid travels to the point of least resistance.

Understand the Points of Work in a Hydraulic System

Hydraulic systems are used to do two basic types of work. First, hydraulic systems can be used to push or pull something. Second, the hydraulic pressure can be returned to torque or circular motion at the drive shaft of a hydraulic motor.

The hydraulic cylinder. The hydraulic cylinder is used to push and pull. The cylinder is a hollow tube that is sealed and has a plunger inside of it (see Figure 6–16). As fluid flow is applied to one side or the other of the plunger, the plunger moves back and forth in the cylinder tube. A rod that sticks out of the end of the cylinder is connected to the plunger. The rod end is connected to something that needs to be pushed or pulled. If hydraulic pressure is used to push the rod out, and a spring or some other nonhydraulic method is used to push the rod back, then the cylinder is "single acting" and needs only one control for operation. If the rod is pushed and pulled by hydraulic pressure, then it is a double-acting cylinder and needs two controls (Figure 6–17).

As an example of how a cylinder is used in a machine, look at the design of a hydraulic-operated automobile brake. When the brake pedal on the

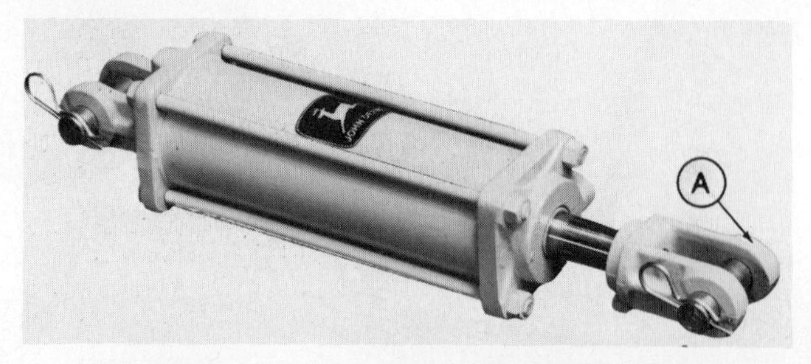

FIGURE 6–16 A hydraulic cylinder converts pressure into back-and-forth movement at the cylinder rod clevis (A).

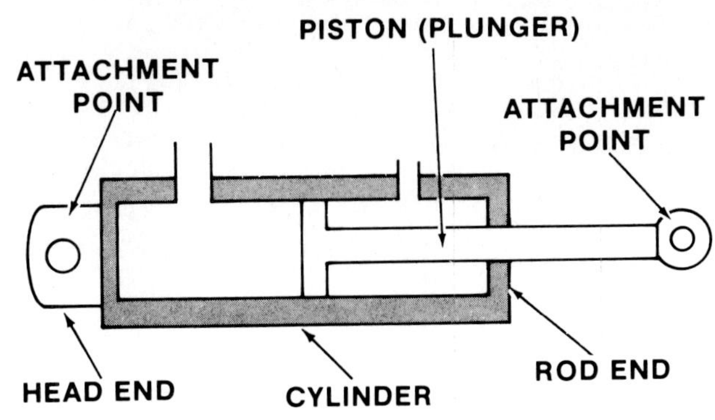

FIGURE 6–17 A double-acting cylinder. (Courtesy Deere & Company)

floorboard is pushed in, the hydraulic fluid in the master cylinder is pressurized. The pressure is equally transferred through the hose and pipes to the four brake cylinders that control the brake discs. When the brake pedal is pressed, the cylinder rods extend and push against the brake discs. These brake discs press against the brake lining and the automobile stops because of friction on the wheels. When the brakes are released, springs force the cylinder rods back into the four cylinders at the same time (Figure 6–18).

The hydraulic motor. Another point of work in a hydraulic system is the hydraulic motor. A hydraulic motor transfers hydraulic pressure into circular or rotary motion. This type of work is most common on hydrostatic-driven machines, such as lawn and garden tractors, that either do not need a mechanical drive or cannot have one. A hydraulic motor can be mounted to the axles of the rear wheels of a lawn tractor. As hydraulic pressure turns the drive shaft of the motor in a circular direction, the axle of the wheels will turn.

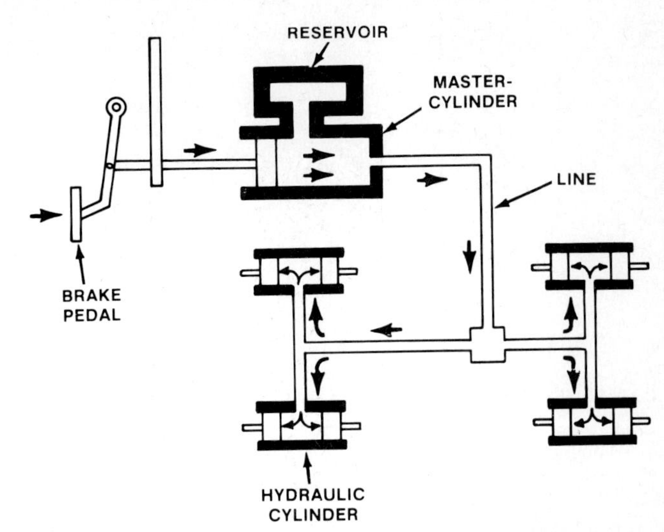

FIGURE 6–18 An automobile brake system is human-powered. Pressure at the brake is designed to control the cylinder on the brake pads at each wheel simultaneously.

Another example is a large off-the-road machine that is used in farming or construction. Often these machines must be articulated so that they can bend in the middle. This articulation eliminates the possibility of using a mechanical drivetrain made of a metal shaft. A hydraulic motor can be mounted at each wheel of the machine and a rubber hose can be used to carry hydraulic pressure across the articulated area (Figure 6–19).

The hydraulic motor. A hydraulic motor works exactly opposite of the hydraulic pump that is used to produce system pressure from the power of a drive shaft (Figure 6–20). The pump traps and forces hydraulic fluid into the system by the action of turning plates or pistons. The motor, on the other hand,

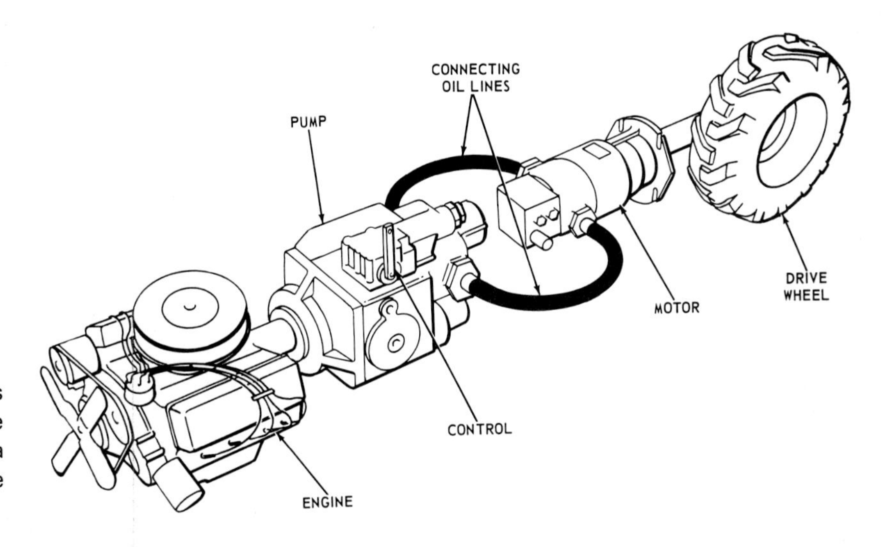

FIGURE 6–19 Rubber hose is used to carry hydraulic pressure from the pump to the motor in a hydraulic system. (Courtesy Deere & Company)

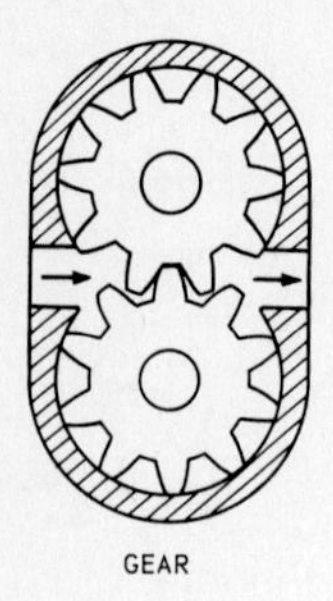

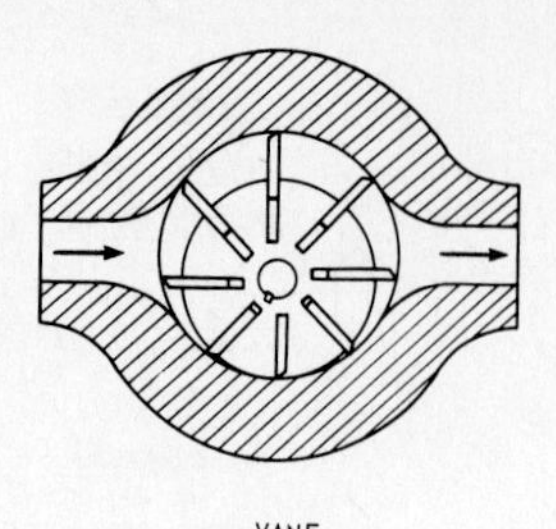

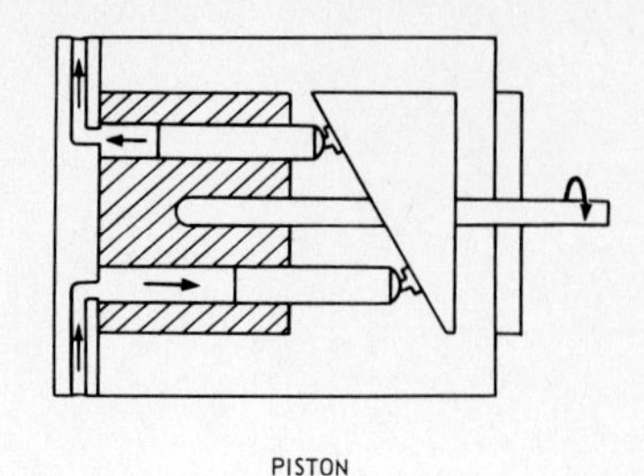

FIGURE 6–20 The three main types of motors work like pumps in reverse. (Courtesy Deere & Company)

converts this pressure into torque by reversing the action.

Figure 6–20 illustrates the three main types of motors. In the gear-type motor, the bottom gear is attached to the drive shaft. As hydraulic fluid is forced through the vanes, they rotate. As they rotate the drive shaft spins.

The vane-type pump uses spinning vanes. As they spin, they push outward against the wall of the motor. Hydraulic fluid pushes them in a rotating direction. The vanes in turn spin the drive shaft at the center of the vane mount. The piston motor is equipped with pistons that are pushed inward by hydraulic fluid force. As they are pushed inward they slide on a fixed swash plate that is set at an angle. The drive shaft is part of the piston casing and turns as the pistons slide along the swash plate.

Hydraulic pumps are designed like the motor. They, however, force the hydraulic fluid into the system with the power supplied by a power source.

Define the Control Points in a Hydraulic System

Without a control device, the hydraulic system cannot produce the correct amount and quality of work necessary for the safe and efficient operation of the machine. The three basic types of controls that you can look for in a hydraulic system are:

A method of applying varying amounts of presure or resistance.

A method of reducing hydraulic pressure or resistance at some point in the system.

A method of controlling the flow or path that the hydraulic fluid takes.

Pressure variation. In the automobile brake example, the hydraulic system pressure is varied by human power. The driver evaluates and adjusts pressure on the foot pedal in order to vary the stopping distance. Shorter stops take more pressure than longer ones. Foot pedals, hand levers, and other controls can be used in many different forms in order to vary the pressure in a hydraulic system.

Pressure reduction. A constant internal hydraulic pressure can also be adjusted by a reducing control that relieves hydraulic pressure from a certain part of the system. For example, a design engineer may want to limit the pressure in a certain section of a hydraulic system because the parts in that section can only withstand a certain amount of force. A reduction relief valve can be set to help keep the pressure within the necessary limitations.

Flow control. In a double-acting cylinder, movement of the rod is created by two streams of hydraulic fluid flow. One path is to the front of the plunger in order to retract the cylinder rod, and another is to the back of the plunger in order to extend the rod. Movement will not take place if fluid flow is allowed on both sides of the plunger at the same time. A "valve" that is controlled by the operator is used to close the fluid flow to one side of the cylinder and open the other. This action occurs because fluid takes the path of least resistance. In the cylinder example, fluid flow is either on or off. Some valves are extremely complex in design and operation. Yet all valves work on the same basic principle (see Figure 6–21).

A valve that lets the hydraulic fluid under pressure flow back to the reservoir when the valve is in "neutral" position creates an *open* hydraulic system. If the valve is in neutral position, and the hydraulic fluid under pressure is stopped dead at the valve, then the hydraulic sytem is called a *closed system.* Closed systems can create trapped hydraulic fluid under pressure and result in unsafe situations. Closed systems must be properly depressurized or drained before the operator shuts down the machine.

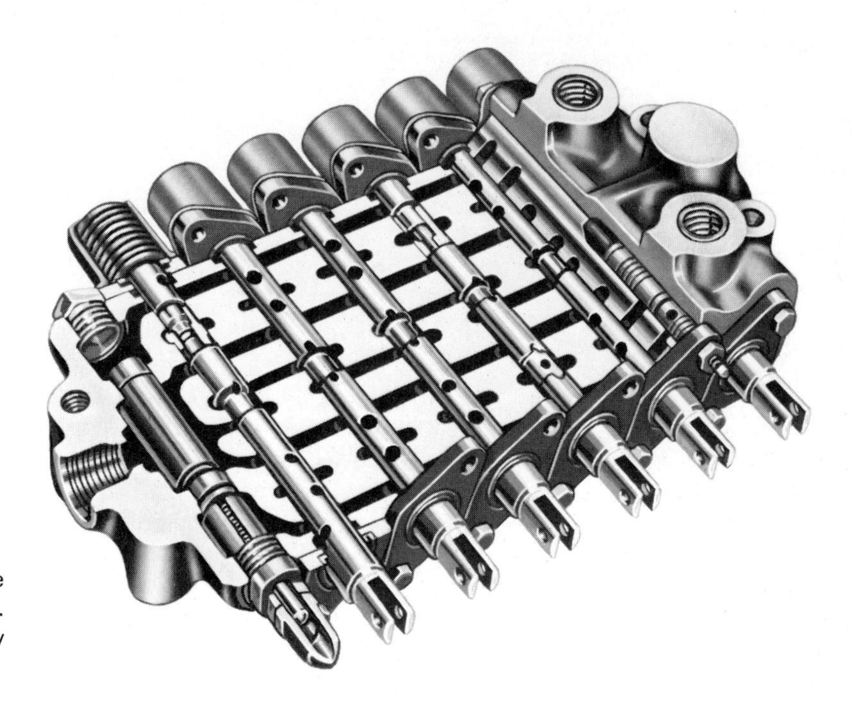

FIGURE 6–21 Valve stacks are many valves attached together. Each rod is one valve. (Courtesy Deere & Company)

On and off valves. Flow control also can be accomplished by simple on and off valves that are installed in the hydraulic lines and pipes. These valves work like the tap on your kitchen sink to turn on your home water supply. These on and off valves are especially useful for shutting down parts of the system for regular maintenance procedures.

Figure 6–22 is an example of a basic hydraulic machine power system. It contains the most common parts that you will find in a machine that uses hydraulic fluid under pressure to perform work. A brief explanation of the most common hydraulic system parts is given below. You can use this as a checklist when you begin your exploration of hydraulic machine systems.

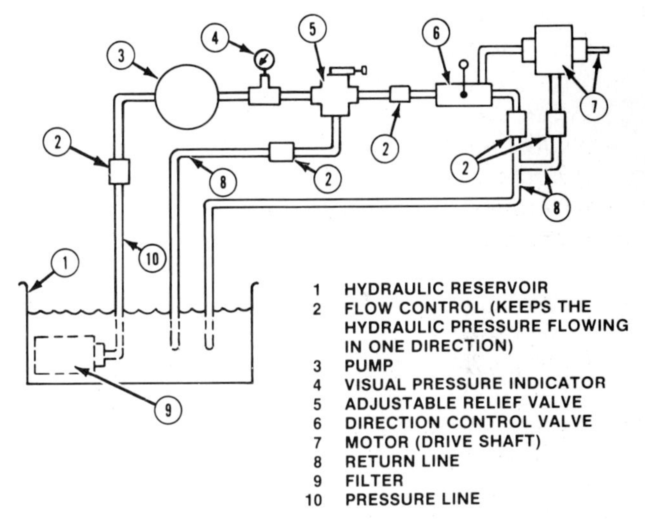

FIGURE 6–22 A basic hydraulic system.

A Basic Hydraulic System

1. *The hydraulic reservoir.* The reservoir is a holding tank for hydraulic fluid that is vented to the outside air. It must be vented to allow for the pressure that would otherwise be caused by hydraulic fluid flowing in and out of the reservoir in uneven quantity. Hydraulic reservoirs also help cool the fluid and keep it at a safe operating temperature. In some cases, a cooling mechanism may be installed outside of the reservoir to help control the fluid temperature. These heat exchangers help the reservoir cool the high temperatures caused by the friction of the system parts and the flow of hydraulic fluid. Depending on the machine, a reservoir may be the size of a bread box or the size of a small imported car. (See Figure 6–23.)
2. *Direction controls.* The lines that carry pressurized hydraulic fluid are equipped with controls that make sure the hydraulic fluid doesn't run backwards toward the power source. These direction controls allow the fluid to travel in only one direction.

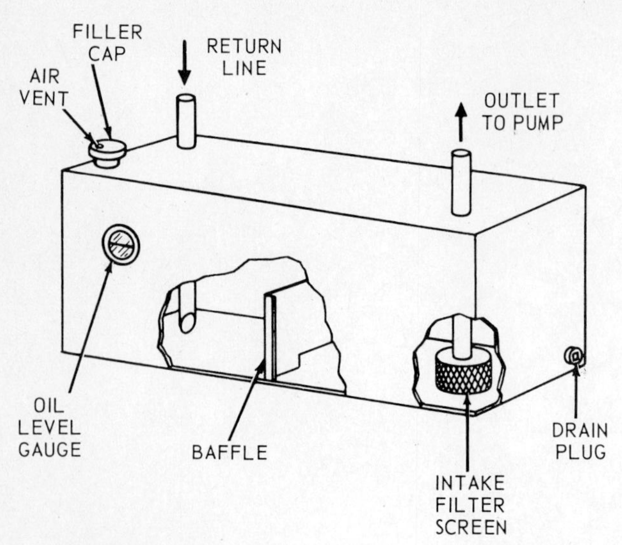

FIGURE 6-23 A typical hydraulic reservoir. (Courtesy Deere & Company)

3. *Pump.* A pump works like a motor except that a pump produces hydraulic pressure from the circular motion of a power system drive shaft. Pump output on a hydrostatic drive system such as lawn tractors is directly controlled by the operator to produce either more pressure or less pressure. This type is called *variable displacement* pump. On larger machines, as shown in the above example, the pump is *fixed* as to how much pressure it can produce. In some system designs an accumulator is added to store hydraulic fluid under pressure. This feature keeps the system from overworking when pressure variation is sporadic. Pumps are directly connected to prime movers, such as an electric motor or gasoline engine, by drive shafts, belts, and other linkage devices.
4. *Pressure gauge.* Pressure gauges are used as visual indicators. The operator of a hydraulic system can look at the gauge and find out whether or not system pressure is within the correct parameters. Gauges are often not used when a pilot line is designed into the system as an internal control. If a gauge is used, it usually means that there is a point of adjustment, such as a pressure relief valve, that must be made by the operator.
5. *Pressure relief valve.* Relief valves are a point of adjustment for the operator. They help control the internal pressure of the hydraulic system. When pressure is too high, the relief valve channels some hydraulic fluid back to the hydraulic reservoir. Relief valves are often preset at the manufacturing company. Sometimes they are designed as an integral part of the pump. Unlike control valves, the pressure relief valve is automatically actuated.
6. *Flow control valve.* Flow control valves are usually operated by hand, by pneumatic pressure, or by electricity. They direct the flow of pressurized hydraulic fluid to the point or points of work.
7. *Point of work.* The hydraulic cylinder and motor do the work in a hydraulic system.
8. *Return lines.* Because a hydraulic system is a closed system, return lines carry fluid back to the reservoir so that it can be reused.
9. *Filter.* The most destructive feature in a hydraulic system is dirty or contaminated hydraulic fluid. A filter is usually mounted somewhere in the hydraulic system to trap and eliminate any metal chips or other contaminates that might become suspended in the system. A common example of a hydraulic system filter is the lubrication filter that you change on your automobile engine block every 3,000 to 7,000 miles. (An interesting comparison can be made. An automobile lubrication subsystem is not a hydraulic system because work is not done by the lubricating oil. Lubricating oil only helps reduce engine friction and heat.)
10. *Lines.* The lines that connect each of the parts together represent metal pipe or special wound rubber hose. These reinforced pipes and hose must be able to withstand the highest pressure of the hydraulic system. Special metal and textile wound rubber hoses are mainly used in places where the hydraulic system must flex, such as the connection between the front and the rear half of an articulated off-the-road construction machine.

Understand the Difference Between a Hydraulic and a Hydrostatic System

A hydraulic system can produce movement and work by pushing and pulling a cylinder rod, by pushing a cylinder rod against a cam to produce mechanical motion, and by spinning a drive shaft. One of the most useful and most used types of hydraulic system is the *hydrostatic system.* The basic hydrostatic design begins at the input drive shaft that is usually powered by an engine. The pump is directly turned by the drive shaft. As the hydraulic fluid flow is created, the motor drive shaft is turned. In a

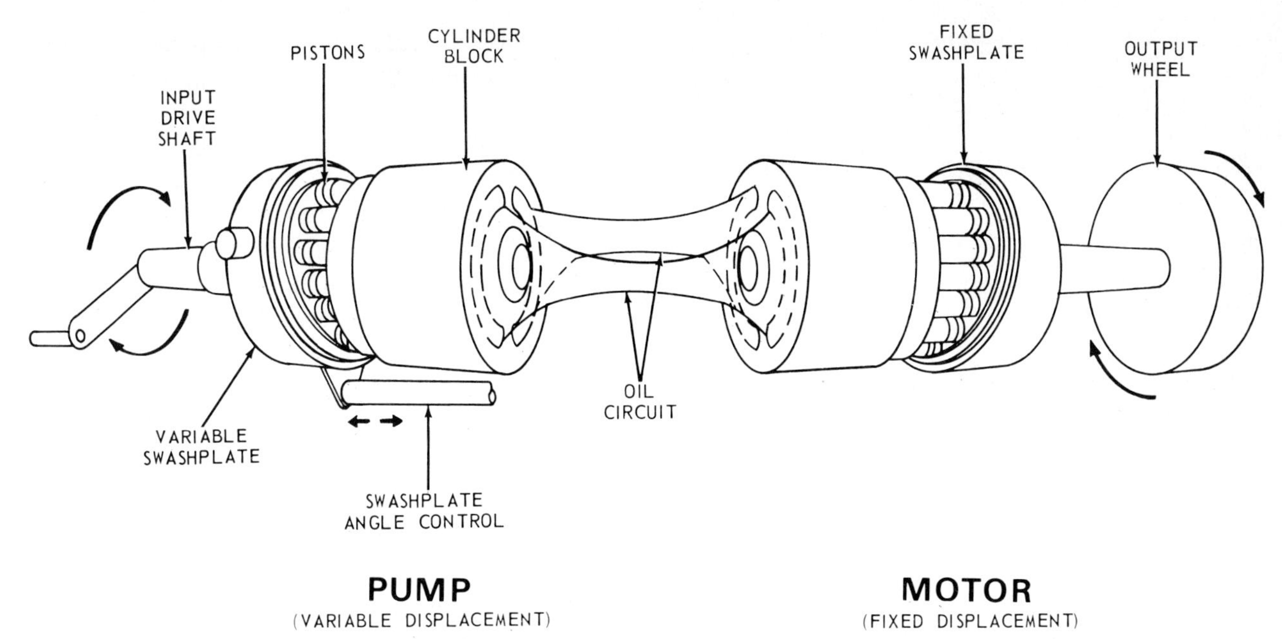

FIGURE 6–24 A typical hydrostatic system. (Courtesy Deere & Company)

hydrostatic system, the motor is attached directly to the output wheel. The hydrostatic system also is illustrated in Figure 6–19. (See Figure 6–24.)

As can be seen in Figure 6–24, a piston pump that can be adjusted at the swash plate angle control pushes oil into the system. A fixed swash plate on the motor ensures that the variation of the pump output will always react proportionally to the motor. Although hydrostatic systems are hydraulic systems, not all hydraulic systems are hydrostatic.

Learn the Role of the Operator in Maintaining a Hydraulic System

The first responsibility of the operator is to keep the hydraulic system as clean inside and out as possible. Because the hydraulic system is a closed but vented system, hydraulic fluid is constantly reused. Hydraulic fluid not only transfers power but it also internally lubricates and continually bathes the working parts of the system. For this reason, the operator must be given a schedule to monitor hydraulic fluid quality and fluid level.

Hydraulic fluid in small quantities can also be lost at places that are sealed, like the rod seal of a cylinder. Hydraulic fluid loss also can take place because of fluid deterioration. When the level becomes too low, hydraulic fluid will lose the ability to lubricate system parts properly.

Hydraulic fluid picks up and carries away small particles within the system as the parts internally deteriorate under operating conditions. Thus, hydraulic system filters must be replaced at regular intervals in order to keep the hydraulic fluid clean. Dirty hydraulic fluid creates undue wear on the internal parts of the system.

A property of hydraulic fluid is that it is practically incompressible when pressure is applied to it. One variable that will affect hydraulic fluid expansion and contraction is temperature variation. The heating and cooling of hydraulic fluid alters the fluid volume. Cooling apparatus of the system must be kept clean and in good working condition. The main cooling points for the hydraulic fluid in the basic system example illustrated above takes place at the reservoir and at the pipes and hose that transport the fluid. Depending upon the application, the operator may be responsible for keeping the hydraulic pipes and hose free of dirt, grease, oil, and often mud.

Cavitation. Hydraulic fittings and connections always must be kept tight. Hose and pipes must be kept sealed and in good condition. If there is a small crack in any part, then the system will *cavitate,* or suck in air. This sucking in of air changes the hydraulic fluid quality, and it can severely damage the system. It is essential, however, that the reservoir be allowed to breathe if it is designed to vent to the atmosphere. As hydraulic fluid flows in and out of the reservoir in many machines, the reservoir must be able to allow air in to compensate for the variation in hydraulic fluid volume.

The operator can enter into and control the basic hydraulic system illustrated above by performing certain actions:

- The on and off valves can be regulated.
- The pump can be shut off.
- The pump output can be varied.
- The relief valve can be changed to increase or decrease pressure.
- The flow control valve can be controlled in order to change the path of the pressurized hydraulic fluid.

ANALYZE PNEUMATIC SYSTEMS (AIR UNDER PRESSURE)

Pneumatic systems are not a common source of work in our everyday machines. Pneumatic, or air under pressure, systems have helped to automate our manufacturing factories. This type of system is mainly used in industrial machines such as:

- Robots.
- Newspaper and other publishing presses.
- Garage car lifts.
- Tire-inflation equipment.
- Factory production machines, such as metal presses, hand tools, and material-handling devices.
- Controls for other systems.
- High-pressure portable washers and sprayers. (See Figure 6–25.)

Basic Properties of Pneumatics

Pneumatics is the duty of air flow and how it can be controlled. Pneumatic system parts and the theory of operation of the pneumatic system are similar to hydraulics. Air, however, can be compressed to a greater degree than can hydraulic fluid (Figure 6–26). Compressed air tries to expand. Pressurized air, like all system power, seeks the path of least resistance. It can be transferred through an *unbroken* line of pipes or hose to the point of work.

FIGURE 6–25 Sprayers require portable pneumatic compressors to provide air under pressure.

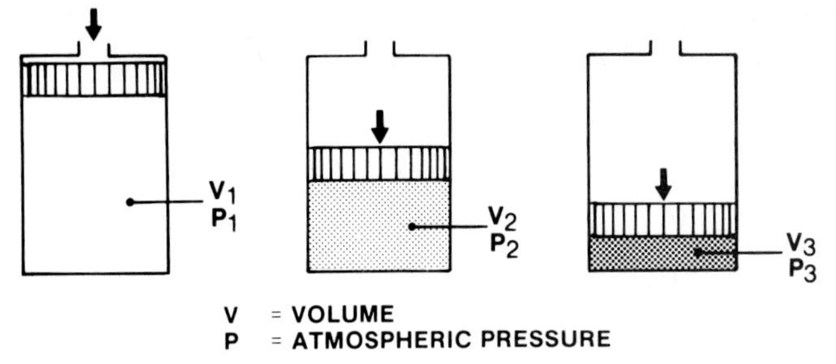

FIGURE 6-26 Air can be compressed more than liquids.

Pneumatic systems also differ from hydraulic systems in these important areas. Unlike hydraulic systems, pneumatic systems are used mainly in a stationary application. A pneumatic system, for example, can economically and conveniently be used to supply power to a complete manufacturing process in a factory. A single pneumatic power source can power the hand tools used by the workers, the movement of manufacturing robots, and the presses, lathes, and other production machines that cut, bend, and shape raw materials for the manufacturing process (Figure 6-27).

Air composition. Pneumatic systems are not able to reach the consistent reliability of hydraulic systems, especially for the higher pressure requirements. This is because air is a gas mixture. For that reason, it is less stable and less predictable than hydraulic fluid under pressure. Air contains approximately 78 percent nitrogen, 21 percent oxygen, and traces of hydrogen, carbon dioxide, neon, helium, xenon, and krypton. Also, as air becomes hotter, it absorbs more and more water vapor, which tends to increase the unpredictability of pneumatic system operation. Further, water vapor, if it is allowed to enter the system components, will rust the internal metal parts.

Air is readily available. Atmosphere air that we breathe is a readily available resource. Unlike hydraulic fluid, which must constantly meet a specification for quality and grade, air can be obtained at any time. Not all air, however, is equal. Some air may contain more suspended particles of dirt and other contaminates. Although it is readily available, some air must be treated and filtered more than other air.

Air is a safe transfer of power. Air can be transported through the pneumatic system safely because it is explosion-proof and noncombustible by itself. Because it is noncombustible, compressed air can be stored and used as needed. This storage of potential energy eliminates the need for the pneumatic system components to work continually. After compressed air is used, it usually can be vented back into the atmosphere because it is a nonpollutant and does not contaminate the environment, as does hydraulic fluid.

Compressed air is relatively expensive. Compressed air, however, is a relatively expensive way to transfer power to a point of work because of the high cost required to put it under pressure. Also, if it is not properly conditioned, contaminated air can cause expensive machine downtime due to parts failure. Air needs to be filtered in order to remove suspended dirt and other pollutants. Air also must be conditioned in order to remove water vapor. Dust, dirt, and water vapor can cause undue wear in pneumatic system parts. Finally, air must in some cases be properly treated with oil droplets in order to lu-

FIGURE 6-27 Curling lines that attach tools to the pneumatic system are common in factories. (Courtesy Deere & Company)

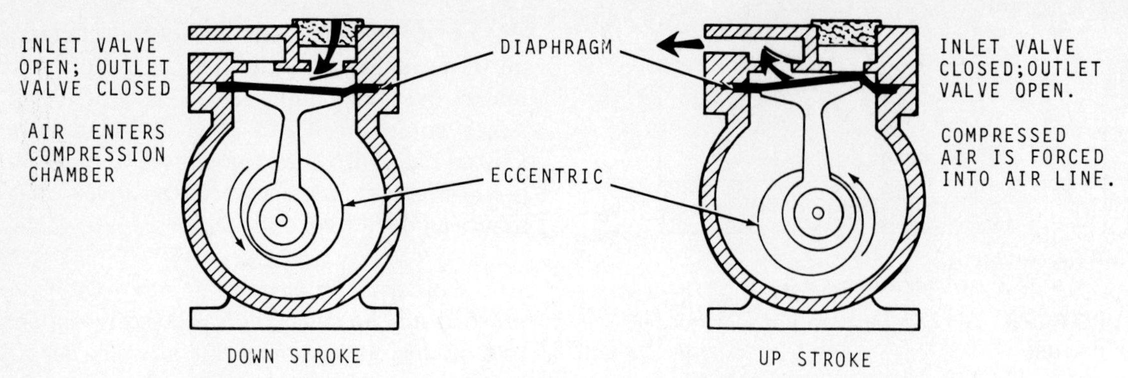

Figure 6–28 The pneumatic compressor pushes air into the pneumatic system.

bricate the internal working parts of the pneumatic system. The removal of this oil may add a cost to system operation if the system is vented.

Air transfers pressure equally in all directions. Air, like hydraulic fluid, is shapeless. It will transfer equally in every direction when it is put under pressure. If pressure is put on top of air that is sealed in a container, the pressure will be the same on all sides of the container. For this reason, air pressure or power can be transmitted to a remote point of work through hose and pipes. Like the hydraulic system, air that is pushed into the system becomes pressurized only when there is a resistance to the flow.

Isolate the Points of Work in a Pneumatic System

Because the pneumatic system design is similar to the hydraulic system, the points of work also are similar. Work is done by pushing or pulling and by creating torque or rotation on an axle. Because air is not a pollutant, air is often used as a means of control for other systems.

The pneumatic cylinder. The pneumatic cylinder is a hollow tube with a plunger that is moved back and forth by applying air on one side of the plunger at a time. The rod attached to the plunger extends or retracts as the plunger moves. Air cylinders are often returned to the rod retract position by a spring mounted on one side of the plunger. The cylinder concept is used frequently as a control device. Many controls, such as a hydraulic control valve, are on and off parts that must be shifted back and forth to control the hydraulic fluid flow. Air is an ideal control in these machines because it is not a contaminate and will not pollute other systems if a leak develops.

The pneumatic motor. Some pneumatic motors are similar to pneumatic compressors. The compressor is driven by a prime mover drive shaft and creates a pressurized air power flow in the system (see Figure 6–28). The pneumatic motor converts this pressurized air flow into mechanical power again at the motor drive shaft (Figure 6–29).

Learn the Control Points in a Pneumatic System

The three basic types of controls in a pneumatic system are:

A method of applying varying amounts of pressure.

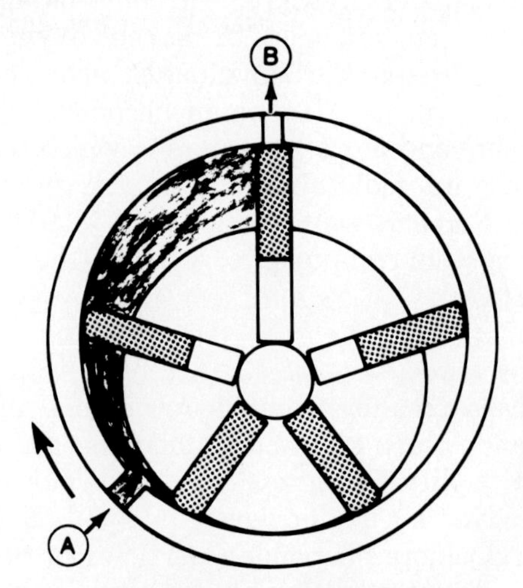

A PRESSURIZED AIR IN

B EXHAUST AIR INTO THE ATMOSPHERE

FIGURE 6–29 Pneumatic motors often resemble vane-type hydraulic pumps and motors.

A method of reducing air pressure at some point in the system.

A method of controlling the flow or path that the pressurized air takes.

Pressure variation. The speed of the pneumatic motor and the reaction of the pneumatic cylinder are both affected by the amount of pressure on the system. Pneumatic foot pedals, hand levers, and other controls can be used in many different forms in order to vary air pressure.

Pressure reduction. Because the pneumatic system is a pressure system, you can adjust the internal pressure by a reducing control that relieves hydraulic pressure from a certain part of the system. For example, a design engineer may want to limit the pressure in a pneumatic system in order to limit the work output. A relief valve (Figure 6-30) can be used to eliminate some of the pressure of the system.

A pressure reducing valve can also be used to adjust the pneumatic system pressure by hand. These valves open and close the system to increase or decrease the internal pressure of a system. A visual pressure readout gauge is usually provided with the valve.

Flow control. In order to use the power in a pneumatic cylinder, you must be able to vary the flow of the pressurized air. Valves that open and close lines, and on and off controls, can be used to change the direction of travel. The operator must be able to use these controls either manually or be able to set a timed sequence of valve movement in order to operate the system (Figure 6-31).

Figure 6-32 is an example of a basic pneumatic machine power system. It contains the most common parts that you will find in a machine that uses air under pressure to do work. What follows is a brief explanation of these most common parts. You can use this information as a checklist when you begin your exploration of pneumatic machine systems.

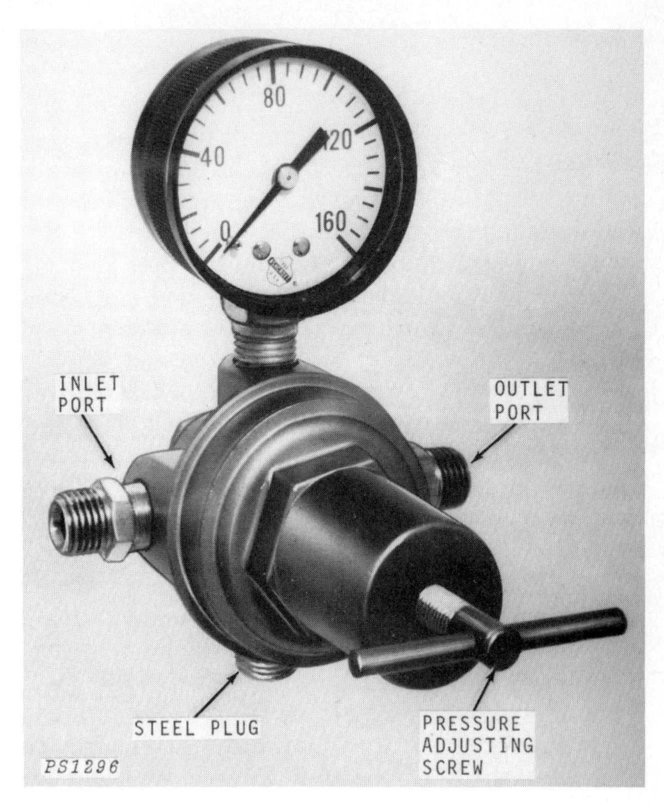

FIGURE 6-31 A pressure control valve is used to adjust internal system pressure.

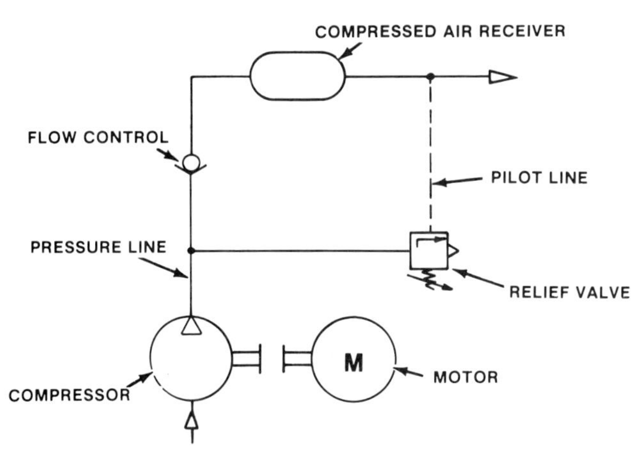

FIGURE 6-30 A flow adjustment can increase or decrease air pressure.

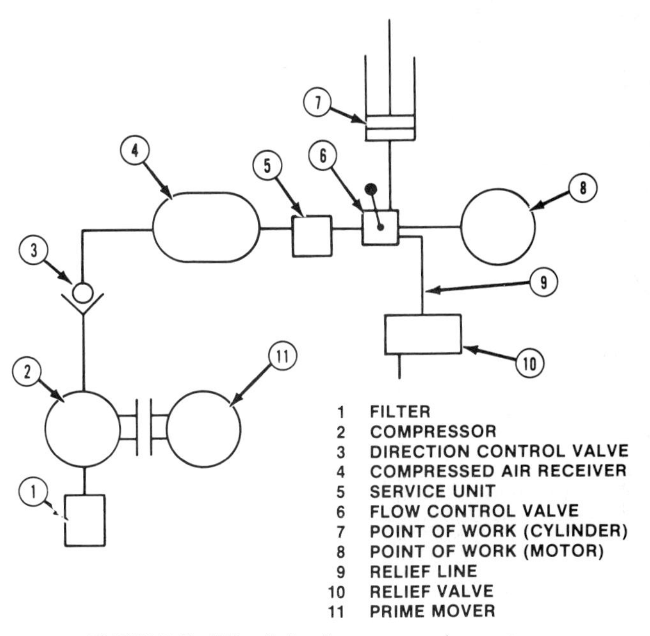

FIGURE 6-32 A basic pneumatic system.

A Basic Pneumatic System

1. *Filter.* Contaminates that float in the air are destructive if they get into a pneumatic system. Filters are mounted at the place where air is pulled into the system. This external filter is designed to remove all suspended particles. In harsh environments a primary filter can be used to remove larger particles and a secondary filter can be used to remove smaller particles.

   An exhaust muffler may look like a filter; however, the exhaust muffler is located at the point of work to quiet the sound of air under pressure escaping into the atmosphere.
2. *Compressor.* Air is pressurized two basic ways. The most common method is by a reciprocating or piston-type compressor. Air is pressurized by converting a prime mover drive shaft torque into an up and down motion. The second method is by a rotary compressor that turns with the drive shaft. The sliding vane rotary compressor is an example of this type. A blower-type compressor is another example. Two fan-type vanes push air into the pneumatic system.
3. *Direction control valve.* Like hydraulic fluid flow, air under pressure tends to flow in the direction of least resistance. Direction controls keep the air from flowing back to the compressor when the compressor is not running. Flow controls are a nonadjustable internal protection for the system.
4. *Compressed air receiver.* Air receivers come in all sizes. Often these receivers are big enough for a person to enter a depressurized receiver in order to clean it (Figure 6–33). Receivers have one or more of the following features:

   A pressure relief valve.

   A pressure gauge.

   A shutoff valve for incoming and outgoing air.

   A water drain.
5. *Service unit.* A service unit is one or more of the following parts. A compressed air filter eliminates the small suspended contaminates, such as dirt and water vapor, in the air by swirling the air around and through a trap in the filter chamber. These contaminates are removed at the bottom of the bowl. A compressed air regulator is often used to help control the internal pressure of the system. A gauge is usually mounted on the unit as a source of feedback information on internal pressure for the operator.

   After the water and solid contaminates are removed from the pressurized air, oil is often suspended in it. This suspended oil is picked up at a small venturi that restricts and speeds up the air flow. As the air speed increases, a vacuum is created in the oil tube. Oil flows through this tube, and droplets of oil are suspended in the air flow. This suspended oil helps internally lubricate pneumatic system parts (see Figure 6–34).
6. *Flow control valve.* The flow control valve opens or closes pneumatic air lines in order to channel the air to the correct point of work and to block it from doing work. These valves can be hand operated or controlled by any method that can slide the spool of the valve back and forth. These control valves are often similar to the valve stacks for hydraulic systems.
7. *Point of work.* A cylinder is a popular point of work in a pneumatic system. Cylinders can perform many functions. They can push or pull. They can be designed with cam and cam followers in order to produce a variety of movement.
8. *Point of work.* A pneumatic motor can be used to operate machines as large as a printing press or as small as a hand-held drill on an assembly line. In all applications, the motor transforms air pressure back into torque or circular movement.
9. *Relief line.* Relief lines can be connected to any part of the pneumatic system that may be

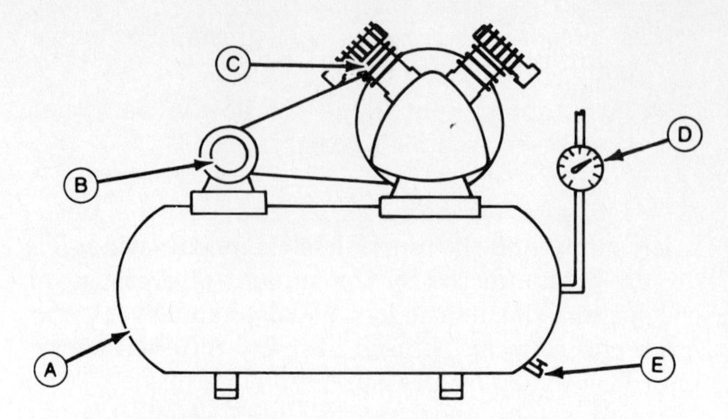

A TANK
B COMPRESSOR (CONNECTED TO THE TANK)
C PRIME MOVER
D VISUAL PRESSURE INDICATOR
E WATER DRAIN

FIGURE 6–33 A pneumatic receiver or "tank" stores air under pressure.

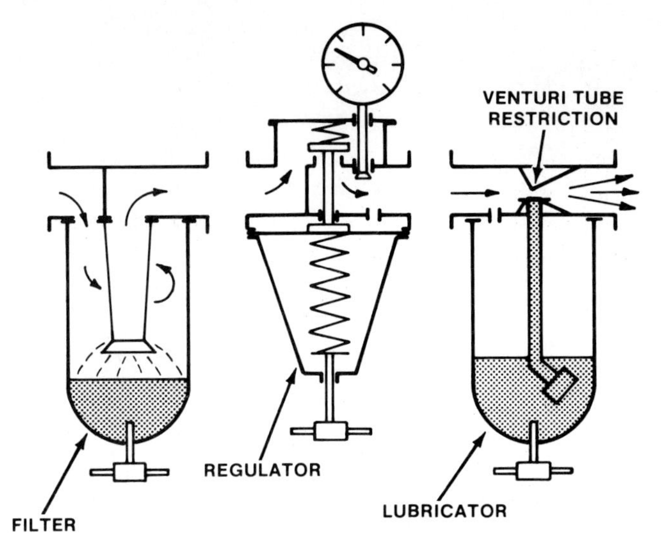

FIGURE 6–34 A pneumatic service unit further conditions the air.

affected by high pressure or pressure surges. Relief lines are connected to relief valves. Relief lines can be confused with pilot lines, which are direct connections between the pressurized part of the system and the prime mover or compressor. As air pressure reaches a set level, the air pressure is fed back to the source of pressure. Regulators on the prime mover or compressor start or stop the compression of the air, depending upon the pressure in the pilot line.

10. *Relief valve.* Like any valve, a relief valve is opened and closed at a desired time. In a relief valve, air pressure above a desired level forces the valve to open. Air is relieved into the atmosphere. Relief valves may have visual pressure gauges attached. This visual gauge usually states that the valve is adjustable by the operator of the machine.
11. *Prime mover.* The most popular prime mover for a pneumatic system is the electric motor. Electric motors, which can be shut off and turned on rapidly, reach their maximum output much quicker than a combustion-type prime mover.

Understand the Role of the Operator in Maintaining a Pneumatic System

The condition of the air is critical in a pneumatic system. The operator must be given basic schedules, cleaning, and maintenance instructions that explain how to ensure that the best possible air enters the pneumatic system. Filter, water removal, and air lubrication parts must be monitored. All parts must be kept tight to assure that pressurized air is not forced out.

The operator can enter into and control the basic pneumatic system at these points:

On and off valves.

Pressure regulators.

Air flow into the system.

The amount of air flow at some point in the system. An air restrictor can be adjusted to limit the amount of air through the system.

The direction of the air flow.

CONCLUSION

Every system must be powered by some prime mover. System parts transfer and help control the power generated by the prime mover and convert it into mechanical movement. The operator controls the output of the system at a control point. By evaluating the feedback, the operator can fine-tune the machine to do the quality and quantity of work required.

When you look at the systems concepts from the standpoint of a technical writer, you can see the value of technical operation instructions. When you use systems terminology, it is easy to explain why machine operation is a continuous evaluation and adjustment process (see Figure 6–35).

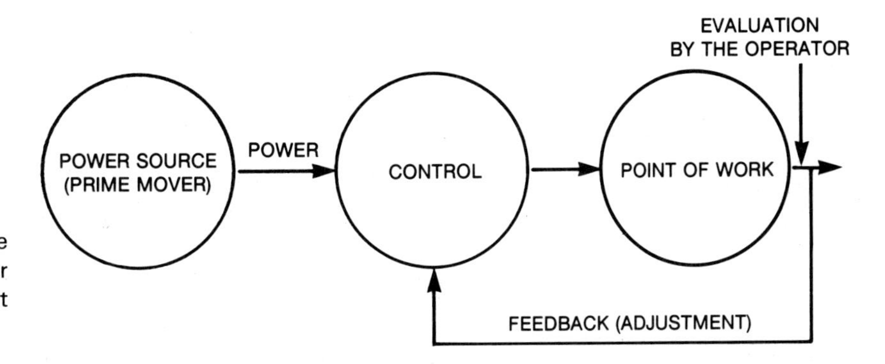

FIGURE 6–35 Basic machine power systems include a power source, control point, and a point of work.

This chapter covers only three basic systems. In Chapter 7 we will cover the electrical system, the most common source of power, and the mechanical system. Both systems work on the same principles as the three machine power systems that are covered in this chapter. Power, no matter what form it takes, flows to the path of least resistance.

After you cover the electrical and mechanical systems, the logic of the basic system design will be used to develop a working file system that you can use to collect your ideas and put like concepts together. Without a usable file system, you will find, unfortunately, that it is easy to miss important information when you prepare the technical operation instruction.

CHAPTER SUMMARY

- Systems are made of specially designed parts that are interconnected.
- Work is done in a machine power system when a resistance has been overcome. Work can be seen as a pushing, pulling, or turning operation.
- Machine-powered systems are like biological and ecological systems except for the work or output that they create.
- Machine-powered systems convert mechanical energy into power, transfer that power through the system parts, and convert it back into mechanical energy at the point of work.
- The operator of a machine system controls the system by evaluating feedback and by making adjustments.
- Prime mover power systems, such as the gasoline engine, convert fuel to circular power or torque. There are two basic types of fuel-burning power systems—rotary and reciprocating.
- Hydraulic systems convert power into fluid pressure. Work is usually done in hydraulic systems at cylinders and motors.
- A pneumatic system converts power into pressurized air.
- Pneumatic systems operate on the same basic principles as hydraulic systems.
- Hydraulic and pneumatic systems are similar in the types of parts, points of work, and adjustments that are used to control both systems.
- Pneumatic-powered machines have helped automate the manufacturing process.

EXERCISES FOR CHAPTER 6

Understand Machine Systems

1. How many hydraulic or pneumatic systems can you find in the machines in and around your home? The place where you work?
2. Sketch five simple systems. Use any type of system as your example. Concentrate on how the parts are interconnected.
3. Use your imagination to create a system sketch of a hydraulically operated robot. Begin by imagining a simple type of work that you want your robot to perform. Draw your idea in a diagram similar to those used in this chapter.
4. Write a 500-word report on how each of the following things work. Draw simple diagrams to support your ideas.
   a. A hydraulic cylinder.
   b. A hydraulic motor.
   c. A pneumatic compressor (pump).
5. Choose a machine that you are familiar with.
   a. How many points of control are there for this machine?
   b. What type of feedback must the operator interpret?
   c. Do the technical operation instructions that were attached to the machine adequately explain how to evaluate feedback from the machine output?

7

Use System Concepts to Outline Ideas

The information given in this chapter will enable you to:

- Explain the basic design of an electrical system.
- Describe the major control points of an electrical system.
- Explain the basic design of a mechanical system.
- Describe the major control points of a mechanical system.
- Define what is meant by a system part "rating."
- Explain how one system can be used to control a primary system.
- Trace the flow of power from the point of work to the power source.
- Use the concepts of machine systems to organize technical data collected in interviews and other types of research.

INTRODUCTION

A machine power system is an unbroken connection of parts designed to transmit, control, and convert power to useful work. Both the design engineer and the system engineer in a manufacturing company organize and synchronize systems so that they work together to make a complete machine. Every machine consists of one or more of the five power systems. Each power system is designed with basic parts. As you examine machines you will constantly see these basic parts. By breaking down the machine into simpler system concepts, even the most complex machines can be understood.

A *hydraulic* system is a specialized type of system that is usually found on machines that are impossible or difficult to design with a mechanical drivetrain. The hydraulic lines that transport power through the system can be bent to adapt to many situations that require movement and flexibility. Hydraulic systems are used in machines as big as an off-the-road compactor, which is used to press asphalt and compact it on an airport runway, and as small as a self-propelled lawn tractor.

However, no matter how big or small, a machine that is equipped with a hydraulic system will have "similar" parts. The same is true for the other systems. All machines with this type of system may require extensive care and operation instructions, because the general public may not be familiar with how they work.

Pneumatic systems are usually less understood than are hydraulic systems by the general public or homogeneous consumer audience. Pneumatic systems, which were not seriously applied to work situations until the late 1940s, were common power sources used to modernize and mechanize the factory shop floor. Today, this type of power system is often found on industrial robots and as a power supply for the manufacturing process in automated equipment. Much of the material-handling conveyor lines in manufacturing companies are controlled by pneumatics.

The *electrical* system, on the other hand, is the most widely known and applied power source in our society. Without relatively abundant and inexpensive electric power, modern life would be much different. Our communication industry, for example, would not be as advanced and reliable as it is. Our "convenient" existence would be much more labor-intensive without electricity to run clothes washers and dryers, vacuum cleaners, microwave ovens, electric lights, air conditioning, and garage door openers, to name but a few conveniences. Not only is electricity put to work in our daily life but it is the overwhelming source of power throughout the world.

This chapter will explain the basic electrical system. The technical writer must know the fundamental principles of how electrical power is both transfered and transformed into work in order to understand the many machines that make use of electric power. The major parts of the electrical system are listed and explained. With this fundamental look at an electrical system, you will also be able to understand the electronic circuits that operate in a similar manner but on a much smaller scale. Major system components such as internal protection devices and the system control points that the technical writer must isolate and explain to the consumer are also covered.

The *mechanical* system is another commonly used power source. Anyone who has disassembled an appliance, record changer, bicycle, spring-driven clock, or animated toy will recognize the gears, drive shafts, springs, drive belts, and clutches that comprise the drive train of the mechanical system such as a spring-driven clock or an automobile transmission. Two simplified examples of a mechanical system are illustrated in this chapter.

This mechanical transfer of power may be easiest to trace because the parts and components of the drive train are often visible from the beginning of the system to the point of work. Every part of the mechanical system must touch in order to close the system. Power in a mechanical system is not measured by unseen hydraulic or pneumatic pressure or by electrical voltage. Rather, it is measured by torque, or visible circular movement, at a drive shaft.

In order to complete the systems concept, the use of combined systems is covered. The machine designer may use parts of one system, such as an electrical system, to *control* a second system, such as a hydraulic system. To someone who is looking at a system controlled by another system, the design may seem extremely complex. By breaking down these systems into their components, they can be more easily understood.

The final topic in this chapter covers the practical use of systems concepts to group and pattern technical data into a file system where data is easily retrievable. Without a method of organization, the technical writer will find it difficult to write a comprehensive, accurate, and complete technical operation instruction. Without information that can be easily accessed, the accuracy of the data will be difficult to check and double-check. Logical and usable technical operation instructions are the product of well-organized technical writers.

KNOW THE BASICS OF ELECTRICAL SYSTEMS

We rely on electricity to power our televisions, telephones, computer data banks, and satellite communication systems. The modern manufacturing company office and shop floor would not be as efficient or productive as it is without the power of electricity. Electricity is one of the most abundant and available power sources known. We not only live in an "Age of Information" but we live in an "Age of Electricity" (see Figure 7–1).

Like every machine power system, the wires and parts of the electrical system must be connected in an unbroken chain from the prime mover to the point of work. If the wires that transport the electric

FIGURE 7–1 Without electricity, our computer revolution would not have taken place. (Courtesy Deere & Company)

current flow are broken, or if they touch each other before the point of work, a "short" will occur. This happens because electric current, like all forms of power, takes the path of least resistance.

Electron flow. Electricity is the movement of electrons from one atom to another. All matter is composed of atoms. The identified elements such as hydrogen, copper, aluminum, and silver are made up of atoms that are composed of a different number of neutrons and protons in their nucleus and in the orbits of their electrons. Atoms are arranged in a design similar to our own Solar System. Protons have a positive charge, and neutrons have a neutral charge. The electrons that orbit this nucleus have a negative charge. When the number of protons in the nucleus equal the number of electrons in orbit around the nucleus, then the atom is called a neutral ion (Figure 7–2).

Unlike the planets of our Solar System, electrons, however, can be forced out of their orbit. As one electron leaves an orbit it is replaced by one behind it. The action is much like that of a juggler who is forcing electrons from one hand, or from one atom, to another. This occurs much more readily in some elements than in others.

Conductors and insulators. Elements with fewer than four electrons in the outer shell have weak electron bonds. These elements are good *conductors.* They transmit electron flow with little resistance. Elements with more than four electrons in

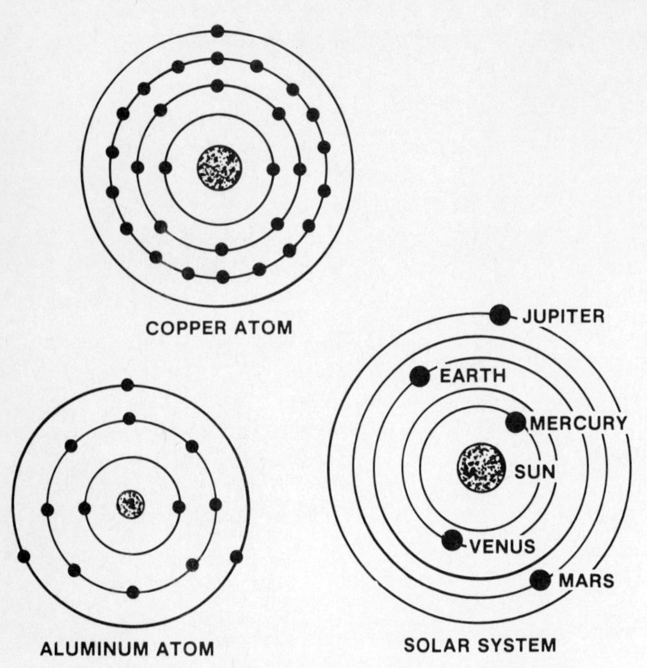

FIGURE 7–2 Atoms of elements are arranged like the Solar System.

the outer shell are good *insulators* and resist electron flow. Copper, aluminum, and silver are good conductors, whereas rubber is a good insulator. Some elements are neither good conductors nor insulators.

Semiconductors. Elements with four electrons in their outer shell are good *semiconductors.* The main types of semiconductors are *diodes,* which act like the direction flow controls in a hydraulic system. Diodes keep electric current flowing in one direction. Another type of semiconductor is the *transistor.* A transistor is useful because a small electric power flow can be used to control a much larger power flow to a point of work. Transistors can also be used to amplify or vary currents to such things as a stereo speaker.

Much of our advanced technology works because of two basic phenomenon related to electron flow.

Electricity can be *induced.*
Electron flow creates a *magnetic field.*

Understand the Methods Used to Induce Electricity

Electric current can be *induced,* or created. There are six primary ways to put the electrons in a conductive material into motion:

1. A conductive material such as a copper wire can be passed through the force fields of a magnet. Generators are the prime movers that create electron flow in this manner (Figure 7-3).
2. Dry-cell batteries like those used in a flashlight convert a chemical paste into electron flow. Wet cells, such as an automobile battery, can also create electron flow. An automobile battery, for example, creates an electric current by the exchange of electrons between a rod of zinc and a rod of copper submersed in a conductive solution of diluted sulfuric acid or electrolyte. The chemical breakdown of the metals creates the electron flow from the copper rod, which has a positive charge. Whereas generated electricity is used mainly on stationary machines, battery-operated machines are mobile (Figure 7-4).
3. Light from a source such as the sun is made up of small particles of energy called *photons.* When light strikes a conductive material, the photons cause a release of electrons. A "photo-voltaic cell" can be used to generate an electron flow. A "solar cell" stores this energy for a controlled output (Figure 7-5).
4. Electron flow can be made from pressure. When you listen to your stereo, or use a microphone, you are using the minute pressure variations that are sensed by "crystals." For example, a stereo cartridge needle rides in the groove of a record and vibrates according to the shape and depth of the groove cut. The needle puts pressure on the crystal material in the cartridge head. The electron flow is sent to an amplifier where it is controlled and then sent to speakers where sound or work is produced.

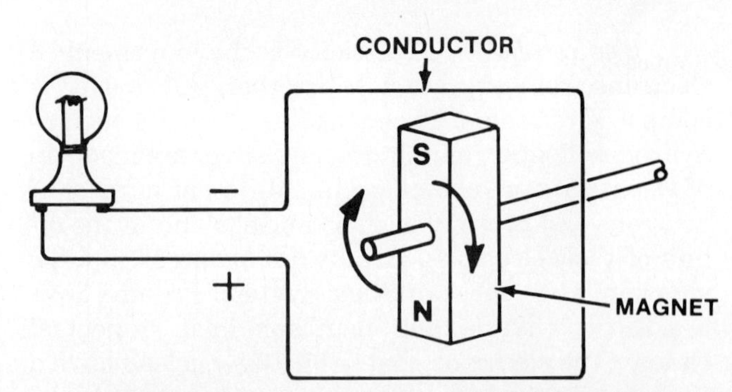

FIGURE 7–3 A magnet rotating inside or around a conductor can induce electron flow.

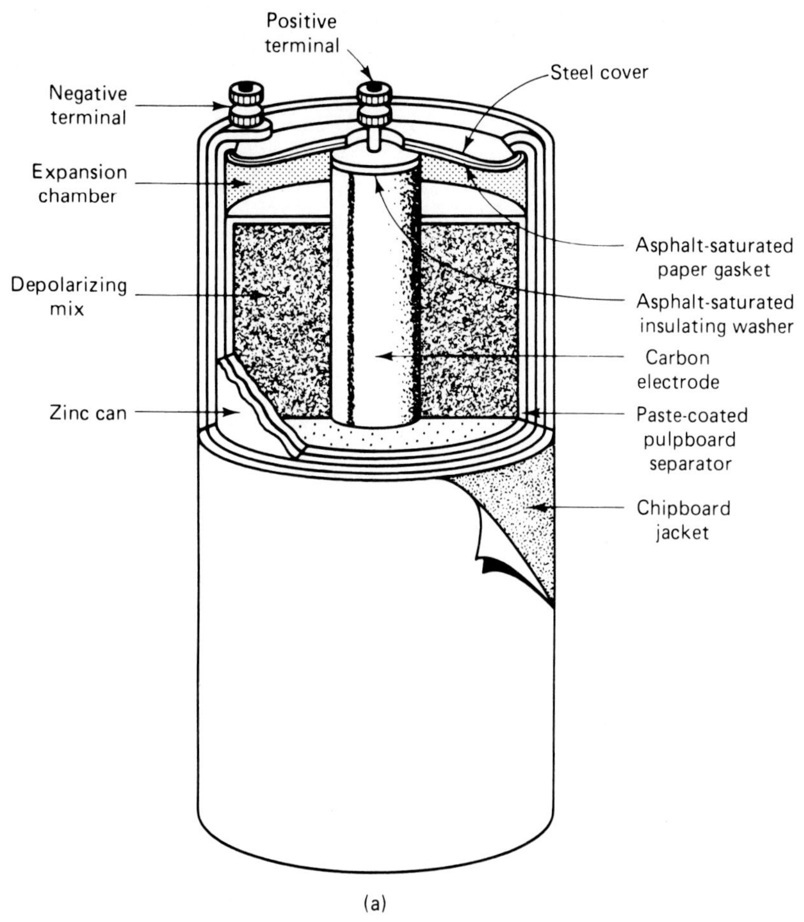

(a)

FIGURE 7–4(A) A zinc-carbon dry cell battery.

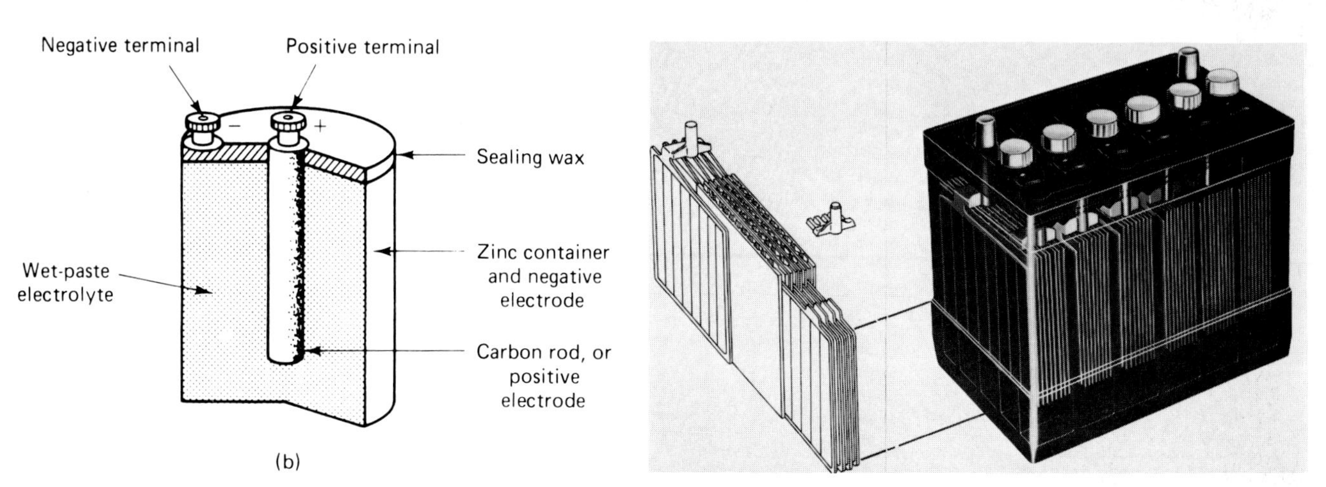

FIGURE 7–4(B) A wet cell battery with plates exposed. (Courtesy Deere & Company)

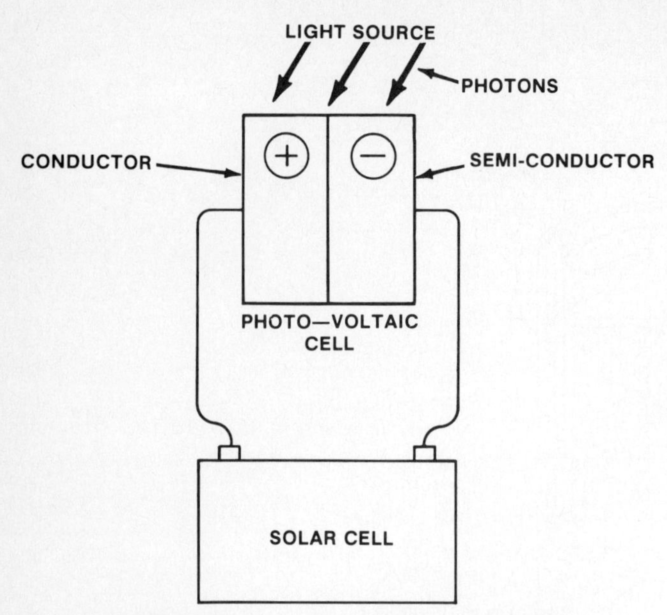

FIGURE 7–5 A voltaic cell creates electron flow from the rays of the sun.

5. Heat is another cause of electron flow. This basic concept is useful when sensing devices are designed. A "thermocouple," for example, is created by connecting two dissimilar metals. As they become hotter, the electron flow increases. This electron flow can be transformed into heat meters, which show how hot the internal part of a machine is running (Figure 7–6).

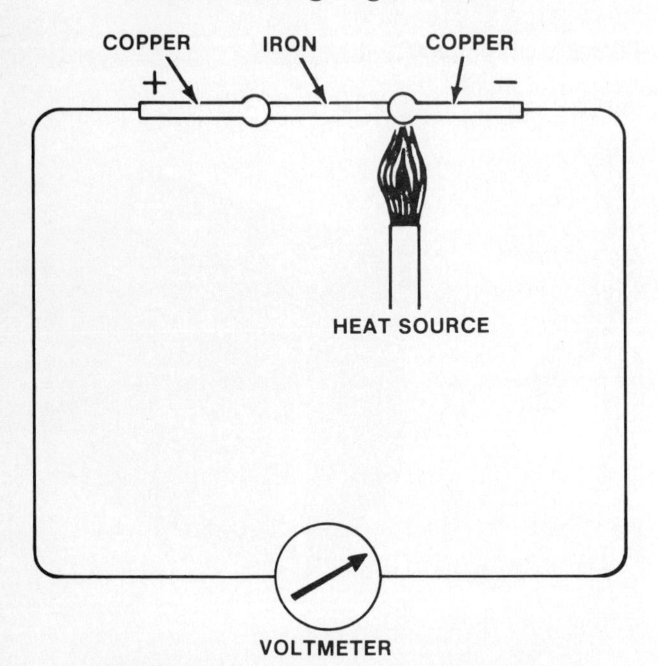

FIGURE 7–6 Electron flow is induced when two different metals that are connected are heated.

6. Friction is another source of electron flow. A good example of this is the static electricity that builds up when you walk across a rug on a cold, dry day. A practical application of static electricity doing work is in the office copying machine. When you feed an original document into the copy machine, the image is projected on a statically charged paper. The paper is then dusted with toner powder and fixed in place. The result is often a fast process that produces an exact duplicate of the original document.

Alternating and direct current. Generated-induced current is the source for electric current flow in our homes and offices. One characteristic of electron movement is that electrons are forced into a direction flow when a conductive material is moved across the force fields of a *magnet.* These invisible force fields knock electrons out of their orbit and create an electric current. Electron flow created in this manner is called *alternating current* or *AC.* Current or power flows backward and forward as the conductive material moves back and forth. Alternating current is used to light homes, schools, and offices and to operate machines (Figures 7–7 and 7–8).

The second type of electron flow is in one direction. It is called *direct current* or *DC.* In an automobile battery, for example, the electron flow is from one pole to the point of work to a second pole. One pole is negative (−), and one pole is positive (+). The negative pole is the "ground" and is connected to the earth. In an automobile, the ground is attached to the frame.

Note: Whenever you must show a DC current you must show the polarity or the positive and negative sides of the circuit.

Chemical reactions are not the only methods that are used to produce a DC electron flow. Direct current can also be generated when a copper wire or other conductive metal is moved in one direction across the fields of a magnet (see Figure 7–9).

Electromagnetism. Another property of electron flow that you must be familiar with is that it can create a temporary *electromagnet.* One of the most useful concepts that you as a technical writer can learn is the relationship between electric current flow and magnetism. Many of the controls of our machine systems are operated by electromagnets. Electromagnetism and magnets are usually considered an integral part of electrical theory.

FIGURE 7–7 The major source of electricity is from power generating stations like this fossil-fuel-burning installation. (Courtesy Commonwealth Edison)

FIGURE 7–8 A nuclear core being loaded at the Zion Nuclear Station. Nuclear fuel can be used to turn the generators at a power plant. (Courtesy Commonwealth Edison)

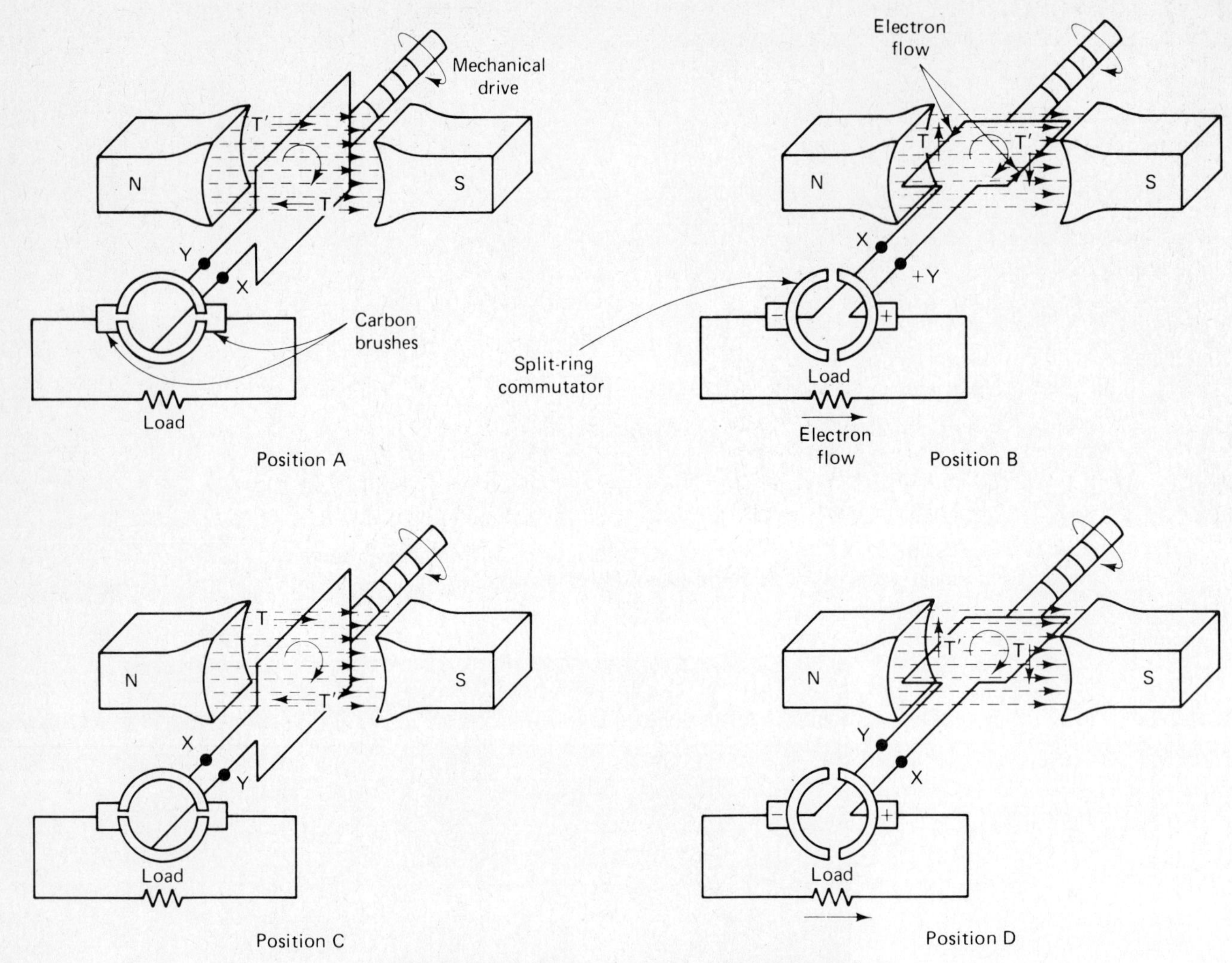

FIGURE 7–9 Direct current can be generated.

An electromagnet is created when you wrap a conductive element such as an iron rod with a conductive wire made out of copper. When an electric current flows through the wire, a magnetic field is created around the wire. The iron core intensifies and magnifies this current. When the electric current is turned off, the magnetic field is eliminated. This simple action can be used to move the inner openings of a solenoid valve back and forth and change the path of air or hydraulic fluid flow in another system. The force of the electromagnet is varied by changing the core conductor, electron flow, and the number of turns of wire on the core (Figure 7–10).

Learn the Definitions of Voltage, Current, and Resistance

The three main terms that you should be familiar with when researching the electrical system are:

Voltage.
Current.
Resistance.

Voltage. Voltage is a measurement of the "potential" of an electrical circuit. Voltage measures

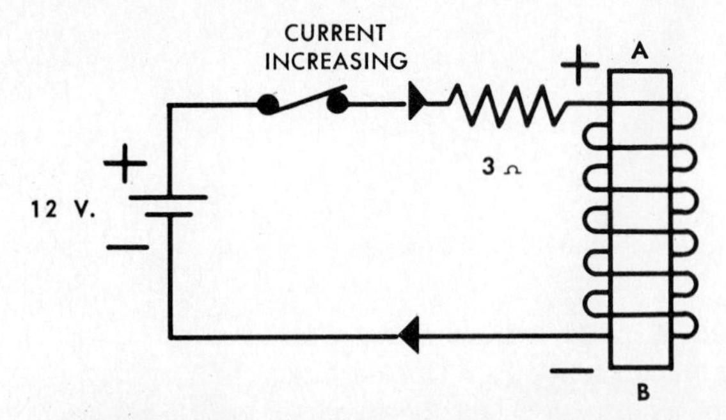

FIGURE 7–10 An electromagnet can be created by electron flow around an iron core. (Courtesy Deere & Company)

electromotive pressure or force. If the electrical circuit is compared to a water system in our home, voltage would be the water pressure. Voltage is continuously present in an electric circuit that has a positive and negative power supply. Electricity does not need to be converted to work before voltage can be measured. Voltage can exist without a current. An electric supply to a home is either 110 volts or 220 volts. This voltage is available as a potential no matter how much current is being used.

Current. In order to have a current, there must be voltage. Current is the rate of electron flow measured in *amperes.* Amperes measure electric flow just as gallons per minute is a measure of hydraulic fluid through a hydraulic system. The standard measurement for current flow is 1 ampere. One volt across 1 ohm of resistance causes a 1 ampere current flow. An ampere is 6.25×10^{18} electrons passing one point in the electrical circuit in one second.

Resistance. Resistance is present in all conductors of electricity. It also can be used to do work. For example, resistance can be used to produce light in an electric light bulb. It can be used to produce heat in heating coils, soldering guns, and clothes irons. It can be transformed into amplification to vibrate the speakers on a stereo unit. *Resistors* are control devices that can be either fixed or variable. You can turn an electric light on or off. You can replace the on-off switch with a variable resistor and adjust the electron flow and the amount of light that the bulb produces.

LEARN THE MAJOR POINTS OF WORK IN AN ELECTRICAL SYSTEM

The solenoid. In the example given of the electromagnet, you can see how electron flow can produce a north and south pole on a temporary magnet. These electromagnets often are used in solenoid valves. *Solenoids* are common control devices for machines of all sizes. An example of an electric-controlled solenoid valve is the hydraulic system flow control valve. The actuation of the magnet attracts the core of the valve. As the core slides, the paths for the hydraulic fluid flow are opened and closed. When the current is shut off, a spring or other return is usually used to push the core back to the original position.

Solenoids also are used in starter motors for engines. As the electrical circuit in an ignition system closes, the solenoid pushes the pinion gear to the engine flywheel. The electric starter motor then turns the flywheel until the ignition circuit is closed.

The electric motor. Electric motors are another example of the practical application of electromagnetism. Electric motors can either be direct-current or alternating-current types and can be made in all shapes and sizes. Estimates are that the average household has between 35 and 40 electric motors of various sizes. Electric motors are a common way that electric current is transferred into mechanical work. The basic operation of the motor is explained in similar terms as the solenoid valve or temporary electromagnet.

Electric motors work because of the electromagnetic phenomenon. Motors are designed on the principle that like charges in two electromagnets will repel each other. For example, a positive end of an iron core will repel the positive end of another electromagnet. In order to attract each other, two magnets must have a positive and a negative pole facing each other.

A simple universal electric motor that can work in an AC or DC system is constructed of two sets of wound coils of wire that conduct electricity. The stator coil is stationary. The rotor is attached to the drive shaft of the electric motor. The rotor is free to turn. The same electric current is sent through both coils of wire. Brushes allow the rotor to rotate while supplying the electric current. The magnetic fields of the stator and rotor oppose each other, and the repulsion of the rotor by the stationary stator turns the rotor and the motor drive shaft (Figure 7-11).

Electric motors are not considered to be prime movers. An electric motor must have a supply of

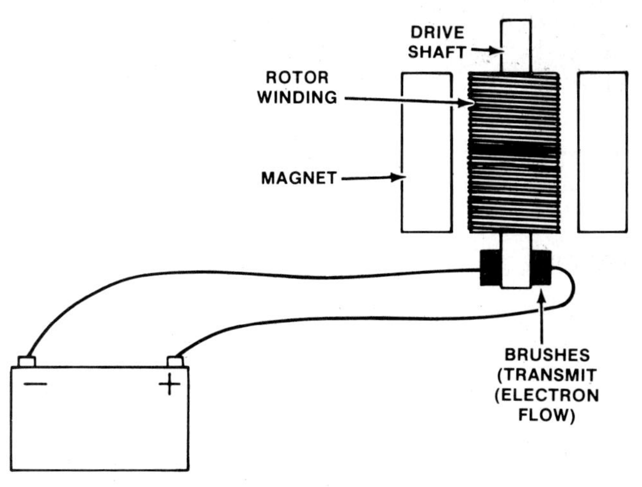

FIGURE 7-11 A simple electric motor.

electric current before torque can be generated on the drive shaft.

Resistance is a point of work. Other points of work are the creation of light and heat by forcing an electron flow through a resistor such as an electric light bulb. Electric current is transformed into work because heat and light are emitted when electric current is forced through a material that is not as conductive as the copper wire that transmits the current. Heating coils and other thermal devices may not produce light, but they produce a heat source that can be applied to a given job.

Learn the Main Control Points in an Electrical System

Electricity is a popular source of power because it is relatively easy and safe to control. Basically, the two ways to control an electrical system are to turn it on or off or vary the current flow. An example of a simple on-off control is the wall switch used to control a ceiling light. There are, however, many types of on-off controls. Some of them may not be as obvious as the switch on your wall.

On-off controls. A good example of how an on-off switch can be made into a complex device is the computer keyboard. The analog computer works through a series of on-off streams of electricity. The letter "A," for example, will have a different series of on-off patterns than will the letter "B." On-off controls also can be found in the form of pushbuttons, levers, or sensors that open or close the electrical system under certain conditions (Figure 7-12).

Timed on-off controls are popular ways to control machines. A common example of this type of control is a light timer that can be plugged into a wall socket. The operation of the light or appliance that is plugged into the timer is controlled manually at a time setting. Often machines are necessarily controlled through timers, gates, or regulators that open and close the electrical circuit.

Vary the current. By varying the current, electricity can be changed into voice patterns such as those that can be heard through stereo receiver speakers or over the telephone. Through an amplifier, current variations can be received and transformed into vibrations that create music and voice patterns. When they are properly set, amplifier vibrations create waves that can be transformed into voice communication and other sounds.

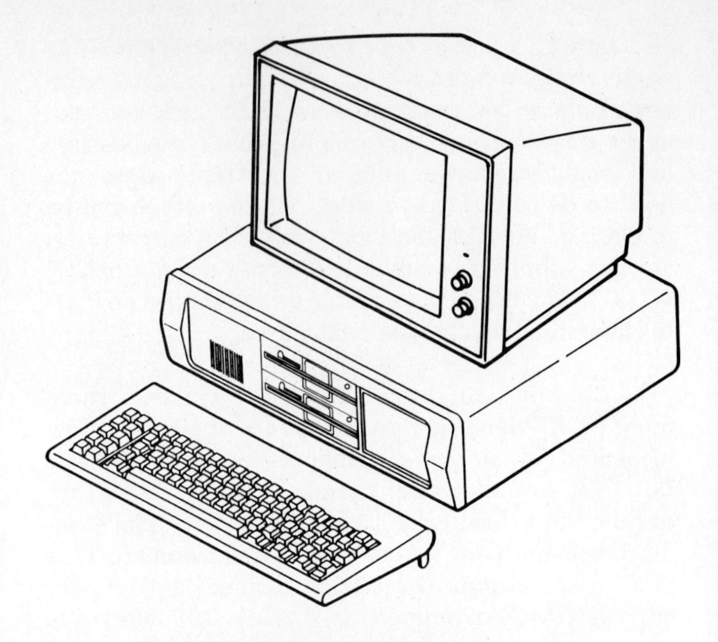

THE COMPUTER

FIGURE 7-12 The computer is an example of on and off switches and gates. (Courtesy of the Radio Shack division of Tandy Corporation)

Vary the resistance. When an electric current is directed through a material that is not a particularly good conductor, such as carbon particles, the flow is resisted. By varying the rating of the resistor, an electric current can be altered so that it can be made to do useful work at a transistor or memory chip. Common examples of how resistance is varied can be found in a rheostat used to dim or brighten a light at a control knob. Less resistance for the current flow means a brighter light because more power is reaching the point of work.

Like all systems that work efficiently, the electrical system must have a means of internal protection that helps prevent damage to the machine due to faulty internal components or an overload of power in the system. In an electrical system, these internal protection devices ususally are *fuses.* A fuse is *rated* as to how much power it can carry. By using a fuse, the designer of an electrical system can help prevent damage to the electrical system parts caused by spikes or overloads of electricity. The correct fuse must always be inserted into an electrical system for both the safety of the electrical machine system parts and for the safety of the operator. Fuses are made of a filament that will burn up and physically open the circuit which stops current flow.

A *breaker switch* can also be used. It is designed to open the circuit if a load or amount of electron flow becomes too excessive. A breaker switch is also rated. It can be reset instead of replaced after the cause of the overload has been discovered and rectified. A variation on the breaker switch is the surge protector, which can be used to protect electronic equipment from sudden increases in electrical supply.

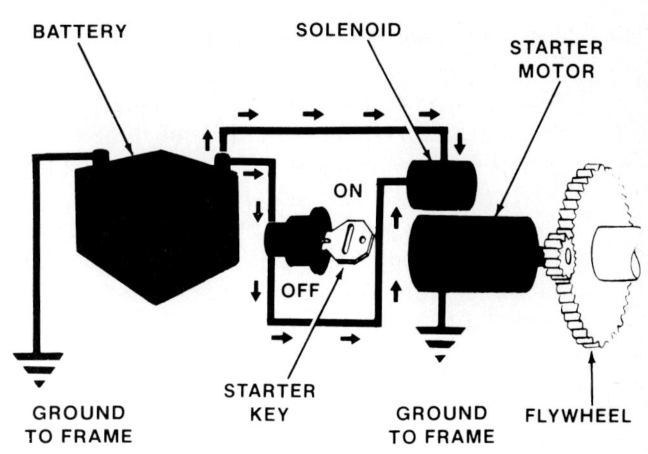

FIGURE 7–13 A typical ignition system. (Courtesy Deere & Company)

Design a Basic Electrical System

The common electrical system to most people is probably the wires, transistors, capacitors, motors, buttons, and buzzers that have been soldered together on a circuit board. The first computer designers, for example, used this technology to create their first memory banks, switches, and input apparatus. The first computers were extremely large, and had miles of copper wire and vacuum tubes. These early computers were the largest machines in offices during the late 1940s and into the 1950s. The first electronic computer called ENIAC weighed 30 tons and contained 18,000 vacuum tubes.

Today, electrical circuits can be built atomic layer by atomic layer. One small silicon chip can contain thousands of transistors, diodes, resistors, and other electrical system components. This miniaturization has produced pocket computers as powerful as the early giant computers. These circuits can only be traced by the technical writer from engineering diagrams. In a sense, these circuits, which carry relatively small currents, are called *electronic circuits.* In the remainder of this section, two circuits that *can be traced* are covered.

The basic system. The following example of an automobile direct-current starting system shows how the components of an electrical system are put together. The basic ignition system shown below can be used on any machine that requires a starter motor to turn over the cylinders in a fuel-powered engine. An engine such as a gas-combustion type is self-perpetuating. The pistons of the engine must first be put into motion. The starter motor turns the crankshaft and starts the pistons in motion (Figure 7–13).

1. *Battery.* The chemical battery has two poles. The negative pole is connected to the frame of the machine. The positive pole is the source of power for the system. Some batteries can be recharged up to a certain point by the recharge circuit of the machine.
2. *Low voltage flow.* A constant flow of electricity is wired to the ignition.
3. *Ignition switch.* When the ignition key is inserted and turned to the "on" position, the low current line is connected. The system is then "closed." The solenoid in the starter motor is actuated. The starter motor gear extends to the flywheel. The full charge of the electrical flow turns the starter motor. The solenoid is spring-loaded so that it quickly disconnects the circuit when the ignition key is released. However, if the switch remains open, the starter motor and engine flywheel can be damaged.
4. *Fuse.* Fuses of various load-carrying ability are put into electric systems to help protect the internal parts of the system. If the circuit is overloaded, then the fuse elements will break and the circuit will open. When the circuit opens, the electric current flow will stop. The fuse in Figure 7–13 is not illustrated. The location varies depending upon the design of the system.
5. *Solenoid.* A solenoid is a wound core. When the low-voltage electric flow magnetizes the core, it attracts a piece of metal in the solenoid. When this metal moves, the high-voltage line closes, and high-voltage electric flow goes to the starter motor. The drive shaft of the starter motor also extends.
6. *Electric starter motor.* The electric starter motor transfers electric power into work at the turning drive shaft. A gear on the drive shaft of the electric motor is meshed with the gear on the engine crankshaft. When the drive shaft

turns the gear, the gear turns the crankshaft. The engine is started.

7. *Ground.* A ground that is connected to the earth through the frame of the machine is necessary before the chemical reaction in the battery can take place. Alternating-current systems, such as your home lighting system, also must be grounded in some way to help eliminate shock that can develop when there is an overload or a short circuit.

Another basic example of an electrical system is the AC system in your home. Although it is not a machine power system, you can still find the basic parts of all electrical machine systems in the circuit. This basic system has two wires for electron flow and a separate ground wire. The electric current flow moves back and forth so quickly that we cannot perceive the shift. Usually, an AC circuit has a 60 cycles per second shift. That is, the back and forth electric current shift occurs 60 times every second. (see Figure 7-14.)

The electrical system is a popular way to transform power into work because electricity is relatively inexpensive, efficient, clean, and reliable. Electricity also is easy to control and transfer into work by the operator. With minimal care and adjustment, the electrical system usually can be controlled and maintained.

First, the operator of an electrical system must be sure that the circuit is protected from an overload by the proper fuse or circuit breaker. An overloaded circuit is a severe hazard to both the operator and the machine. Like all system parts, electric parts are rated as to how much power they can safely carry and transform into work.

Second, the operator must be responsible for replacing electric parts with correct replacements. All wires must be properly insulated to prevent a short circuit. All parts must be kept connected to assure that they work efficiently and properly. For most machines, wires are color-coded to ensure that they are connected properly. Operators must be told the proper *sequence* if they must connect or disconnect these color-coded lines or any electrical lines in a system.

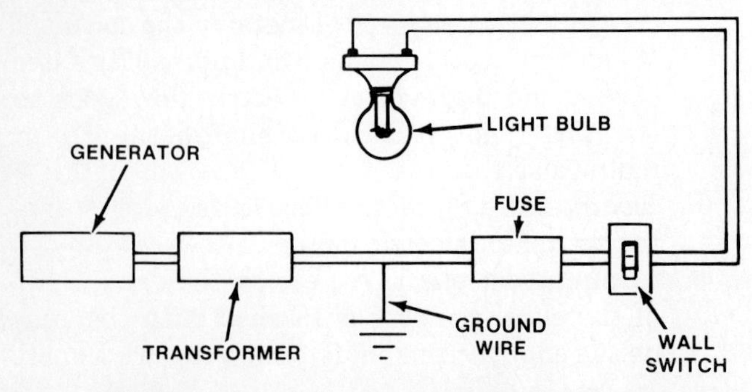

FIGURE 7-14 A basic home electric system.

An electrical system is controlled in a few basic ways. Operators can use many types of switches to open or close circuits. The operator also interacts with the electrical system at points where current is varied or resisted.

UNDERSTAND THE MECHANICAL SYSTEM

Mechanical systems, like electrical systems, are more common than hydraulic and pneumatic systems. Mechanical systems are found in many machines that are manufactured for the mass-consumer market. The gears, drive belts, and other system parts and components are usually more visible and easier to trace than in other systems that have been covered. Mechanical systems are used in products such as:

- Bicycles.
- Garage door openers.
- Motorcycles.
- Elevators.
- Drivetrains for rear-wheel-driven automobiles, tractors, road graders, motorcycles, and trucks.
- Transmissions for all types of vehicles and machines (Figure 7-15).
- Conveyor lines in factories.
- Gear movement clocks and watches.
- Household appliances.

A mechanical system can be connected to the point of work in another system, such as the drive shaft of an electric motor, or a gasoline engine, by means of gears, pulleys, belts, and chains. The mechanical system in an electric can opener begins at the drive shaft of the electric motor that turns the gears that eventually secure and cut the top of the can. Connection devices are used to transfer the mechanical power of power sources or other system points of work to the mechanical system.

Learn the Basic Mechanical System Concepts

The mechanical system is characterized by an unbroken transfer of power from a power source to a point of work by means of pulleys, levers, wheels and axles, inclined planes, wedges, and screws.

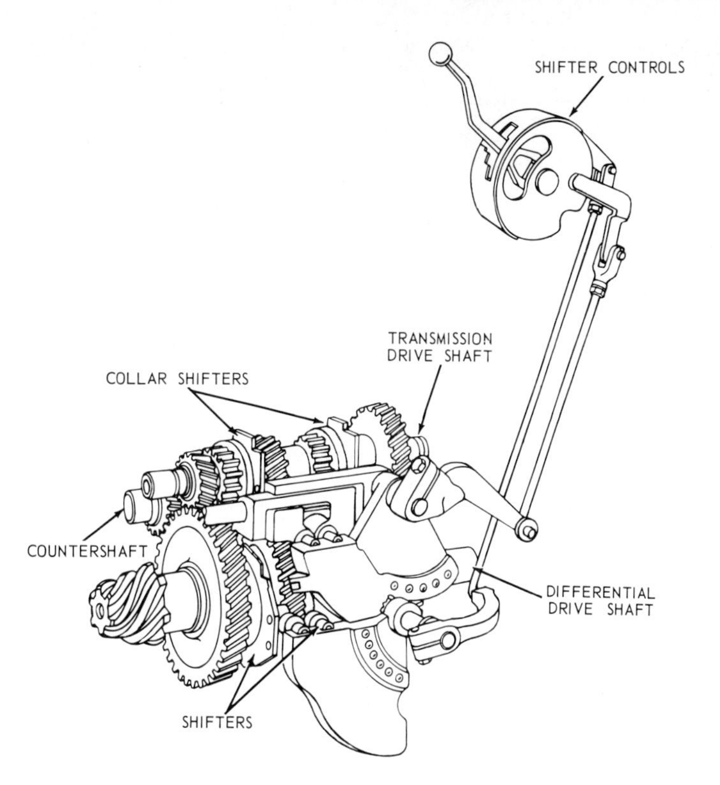

FIGURE 7–15 A mechanical transmission. (Courtesy Deere & Company)

These basic elements are used in principle to design all mechanical systems including the gears, axles, and transmissions that perform or control most of our mechanical work.

Care must be taken when following a mechanical system through a machine. All mechanical parts are in motion when they transfer power throughout the system. Unlike pipes, hose, and wire of the preceding systems, mechanical movement that transfers power creates points where parts meet that can crush, pinch, and sever anything that comes between them.

The gear. The basic carrier of power in a mechanical system is the gear. Gears are a combination of levers, a wheel, and an axle. It is constructed of a flat wheel that is mounted on an axle. Rods or levers extend from the edge of the wheel. These levers, or gear teeth, are meshed with another gear in order to transmit power from one gear axle to another. The power is turned into "torque," which is a circular motion of the axle (see Figure 7–16).

The amount of torque transferred from one gear to another is proportional to the distance from the center of the gear to the teeth. For example, a small gear driving a larger gear will create less speed but more torque, or pressure, on the driven axle. A large gear driving a small gear will create more speed and less torque (see Figure 7–17).

Other types of power transfers. Gears are not the only part that can be used to transmit power in a mechanical system. A drive shaft attached to the engine or motor can extend the torque produced by the engine beyond the original drive shaft. As we have seen, drive shafts can transmit power in a straight line to another part or point of work.

Belts and chains can be attached to a drive shaft or axle in a mechanical system in order to transmit torque from one shaft to another. Belts and chains come in many sizes, shapes, and styles. Probably the most common example of a chain-driven axle is the drivetrain on a bicycle. The pedals turn a drive gear or sprocket. The drive gear transfers power to the chain that links the front drive to the rear-driven gear and then the axle of the wheel. The wheels turn and the bicycle moves forward (see Figure 7–18).

Power can be transferred from the drive shaft of an engine or motor by coupling the drive shaft directly to the mechanical system. This is the most efficient method used to transfer power. A direct drive eliminates power transfers. Every time a power transfer is added to the mechanical system, added power must be put into the system in order to maintain the system output. Every contact between a mechanical part, such as a gear, produces friction that must be overcome.

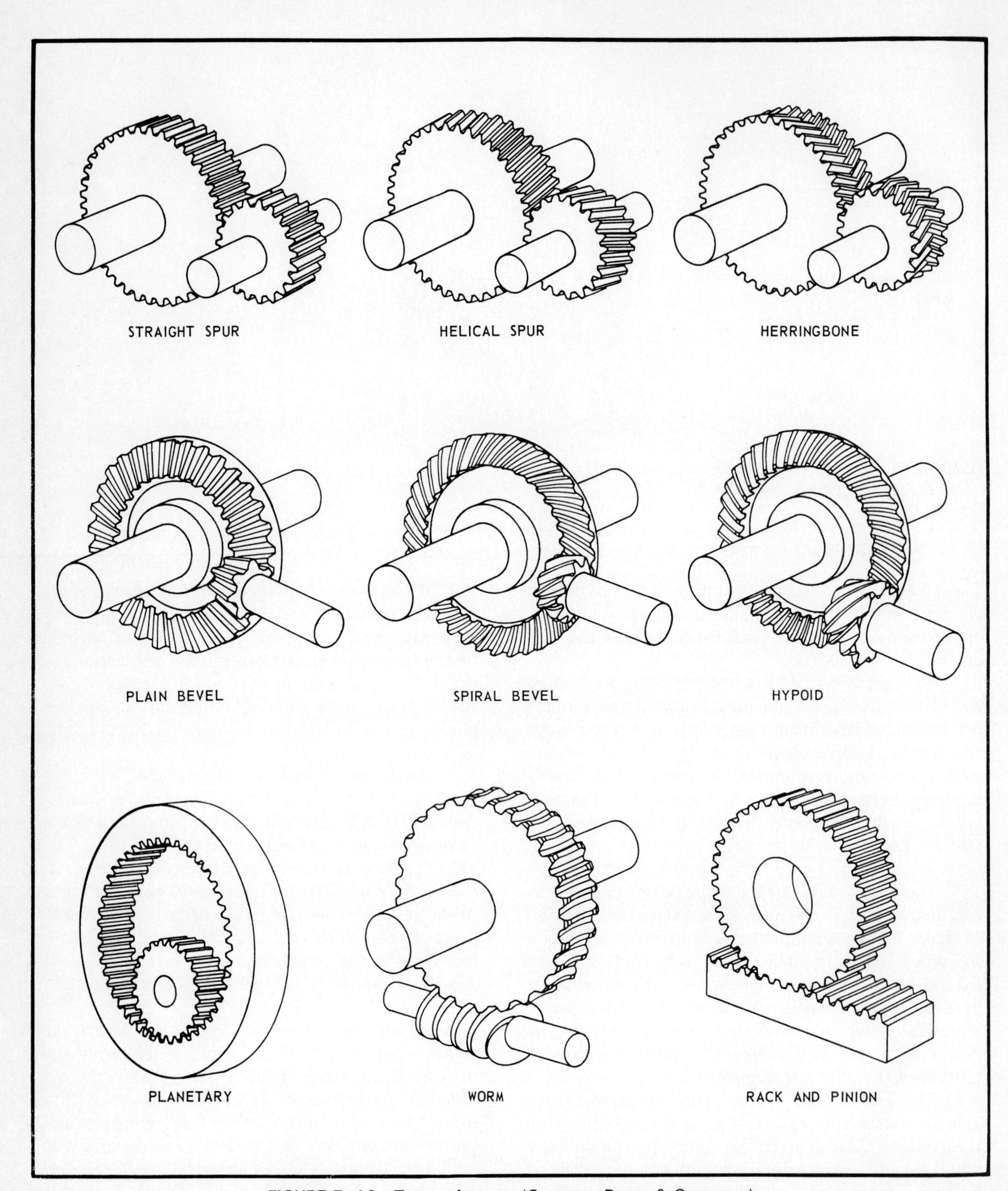

FIGURE 7–16 Types of gears. (Courtesy Deere & Company)

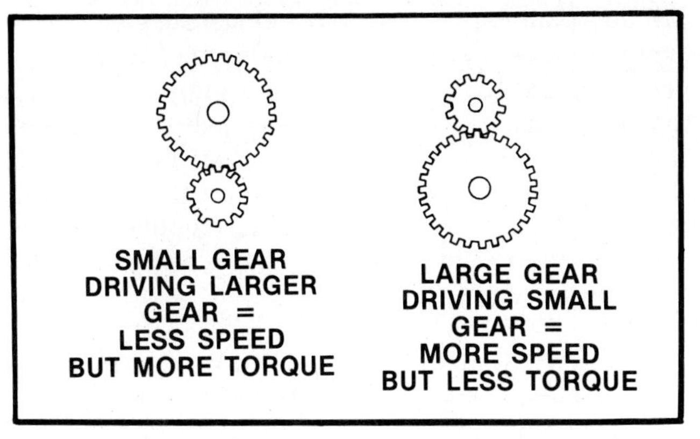

FIGURE 7-17 Gear ratios determine speed and torque. (Courtesy Deere & Company)

FIGURE 7-18 Belts connect drive shafts by riding in pulley or sheave grooves (B).

Friction. Although friction plays an important role in system dynamics, it is even more important in the basic mechanical system. Work is done when a force is overcome. Often friction is that resistance. Friction is a force that is acting upon every material thing, no matter whether it is at rest, rolling over something, or rubbing against something. The product of friction is *heat.* To the people who design them, heat is probably one of the most undesirable byproducts of a machine-powered system. Heat means that there is a loss of work through an inefficient transmission of power through the system. This is created mainly by misalignment of mechanical system parts, wear, or improper part replacement.

Properly controlled, friction is also an important form of work. Without friction, your bicycle or automobile would not move. Tires would not be able to develop force caused by friction between the tire and the pavement. By harnessing the properties of friction, the design engineer who creates the ma-

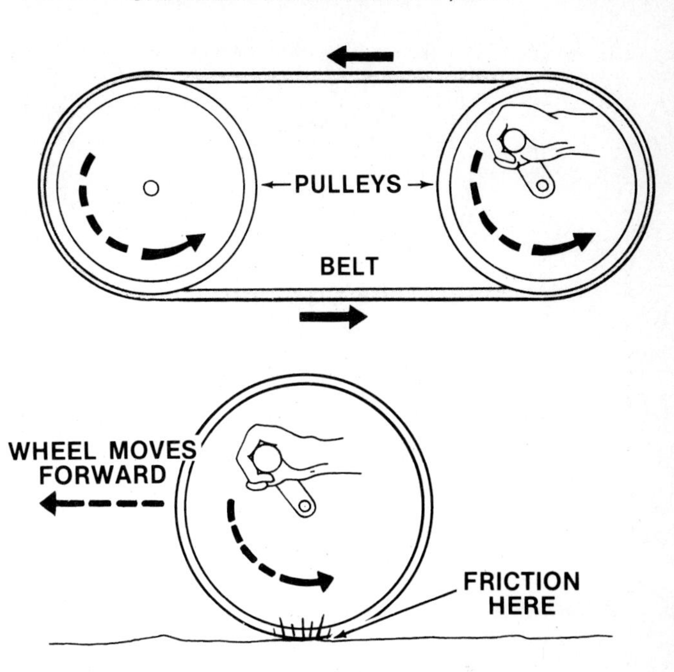

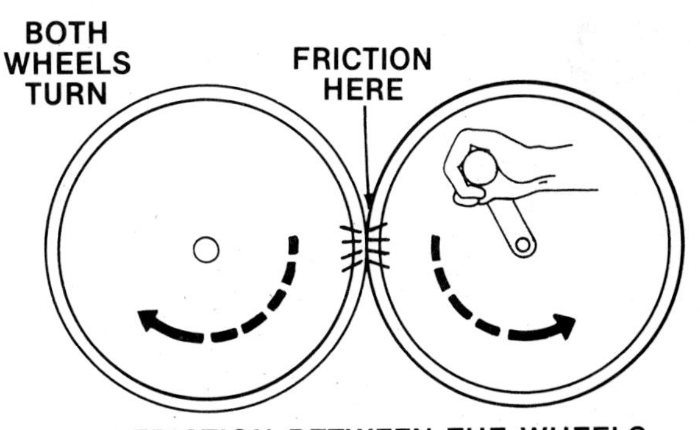

FIGURE 7-19 Not all friction is a waste of energy. Friction can be used to transmit power from belts to pulleys, and from wheels to surfaces. (Courtesy Deere & Company)

chine can use friction to transmit mechanical power or motion (see Figure 7-19).

Friction between two surfaces can take the place of gear interfacing. For example, friction between two flat plates serves as a clutch that closes the mechanical system. Friction is used to drive many cassette stereo units and turntables that play your favorite music (Figure 7-20).

If required, the most common way to overcome friction is to make the finishes on the metal faces that contact each other extremely smooth. Another way is to put a thin coat of lubrication, such as motor oil or machine oil, on the metal surfaces. For parts that are under high pressure and subject to heat, different grades of *grease* are packed around

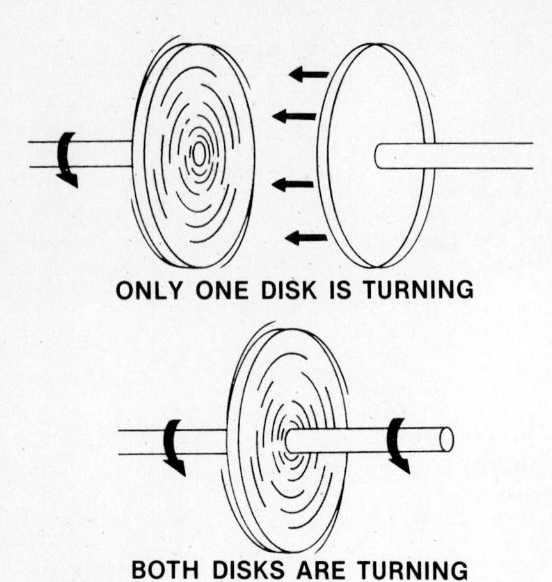

FIGURE 7–20 Friction can be used as a transmission link. (Courtesy Deere & Company)

the surfaces of two metal contact points. The type of contact, the heat or friction produced by the contact, and the sophistication of the lubrication supply determine the grade and quality of the oil that is used in a machine.

Engines are designed with features that continually bathe the mechanical parts of the engine in oil. If oil fails to reach the mechanical linkages and piston area for only a few combustion cycles, then the engine will "freeze up" and bind. The loss of lubrication in some situations is extremely critical. By using the wrong grade and type of oil, the same results can happen over a longer period of time.

Find the Points of Work in a Mechanical System

The mechanical system is designed to transfer power from a prime mover power source by means of the forces of the wedge, inclined plane, screw, lever, pulley, and wheel and axle. Not only do these simple concepts transfer power but they are often the point of work. An example of this is a gear-driven electric clock. The electric motor turns a gear that in sequence turns all of the gears and shafts in the clock. The point of work is the final gears and shaft at the clock hands. This is an example of simple circular movement at the three axles of the second, minute, and hour hands.

The point of work on a tire, for example, is the footprint or area where the tire meets the ground. Millions of dollars in research and testing go into the design of a tread for an automobile tire in order to get the right "footprint" or tread design. It is in this footprint that the friction is developed that enables the tire on the wheel to push forward. This point of work is extremely important because the footprint, or tread, determines how well an automobile tire corners and handles in different climates and under different road conditions.

Cams. Cams can be used to vary the simple circular movement of a mechanical system. A cam that is mounted on the axle shaft can vary movement in a variety of directions by controlling cam followers. The simplest example of this is the cam and cam follower that control the valves of an engine. In this design, the rotation of the cam produces a sequence of up-and-down movements on the cam follower, which is connected to the lifters that move the valves in and out of the engine cylinder combustion chamber (Figure 7–21).

Although the cam and other types of converters, such as linkages and actuators, are not the exact point of work, they do control the power and create up-and-down, in-and-out, and circular motions of other parts that do the actual work. Converters are a direct surface-to-surface contact between the mechanical system and the point of work.

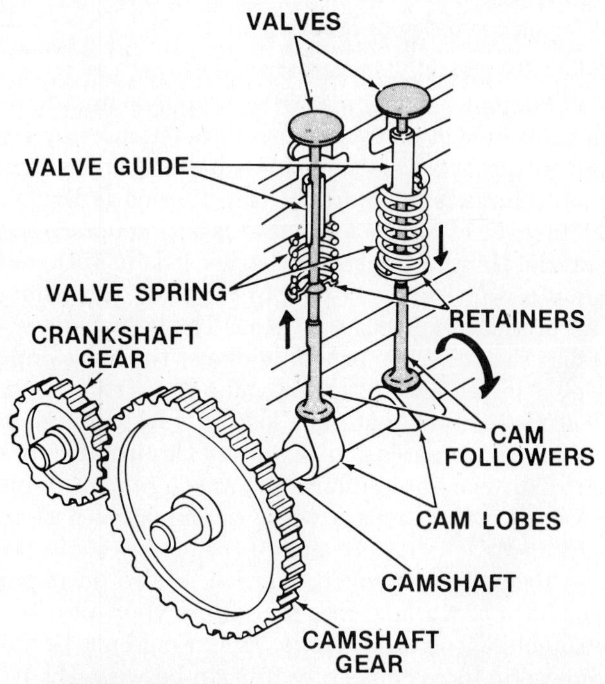

FIGURE 7–21 Cams help vary mechanical movement. (Courtesy Deere & Company)

FIND THE CONTROLS IN A MECHANICAL SYSTEM

Like the electrical system, there are both operator controls and internal controls in a mechanical system that can vary the resistance to the power as it travels through the various parts. Mechanical power can be turned on or off or altered by adjusting the power source. Adjustments also can be made at the following common points.

Gear ratios. As illustrated earlier, the size of the drive gear and the ratio of the driven gear will determine the speed and the amount of torque applied to the driven shaft. Often mechanical systems are designed with a control point that is accessible to the operator for the purpose of changing gears in some part of the drivetrain. Gear ratios also can be varied through control points such as a car shift lever.

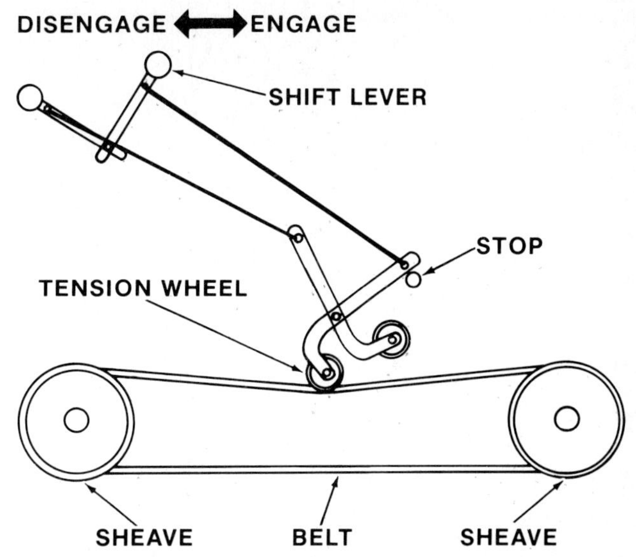

FIGURE 7-22 A tension wheel creates belt friction. (Courtesy Deere & Company)

Transmissions. Transmissions are the most convenient way either to change the gear-ratio arrangement or to break the flow of mechanical power. Transmissions are used to connect gears directly by meshing the gear teeth of the drive gear, the transmission gear, and the driven gear. Another way is to use a pressure-plate design. The drive plate and the driven plate are simply pressed together. Friction from the two surfaces transfers power. A third way to transmit power is by a tension wheel and a belt. By putting tension on a drive belt, power can be transmitted from one pulley or sheave to another. When tension is not put on the belt, the belt does not generate enough friction to be turned by the drive sheave (see Figure 7-22).

Transmissions are usually the point where the mechanical system is turned on or off. Transmissions are used when it is inefficient to shut down the power source whenever the power flow must be kept from the point of work. An "automatic" automobile transmission is an example of two of the above types of control. In neutral, the transmission breaks the power flow from the engine. In the "drive" position, the operator selects the best gear ratio for speed. In drive 1 and drive 2, the transmission progressively is set to create less speed and more torque or turning power at the wheels.

Internal protection. Internal parts, such as gears and drive shafts, are usually protected from overload or freeze-up by means of shear pins and other load-rated connectors. A shear pin, for example, can be placed along the side of a drive shaft and a driven shaft in order to connect them. If one of the shafts freezes up or stops because of overload, the shear pin will break and set the two shafts free from each other. The mechanical system is internally protected because only the shear pin needs to be replaced after the problem is solved. If a shear pin is not used, the shafts and other parts may be damaged from the overload (Figure 7-23).

Useful overload. An overload can also be useful. An example of this is the automobile brake. As the brakes close around the wheel axles, a force is created through friction between the axle and the brake that the drivetrain cannot overcome. The automobile stops. Some pressure-plate types of transmissions and belt-driven sheaves are set to slip at a certain resistance in the system. In this example, the adjustment of pressure on the transmission

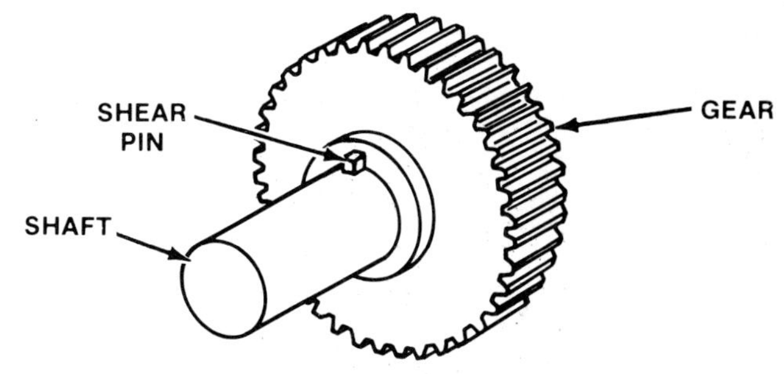

FIGURE 7-23 A shear pin is a type of internal protection.

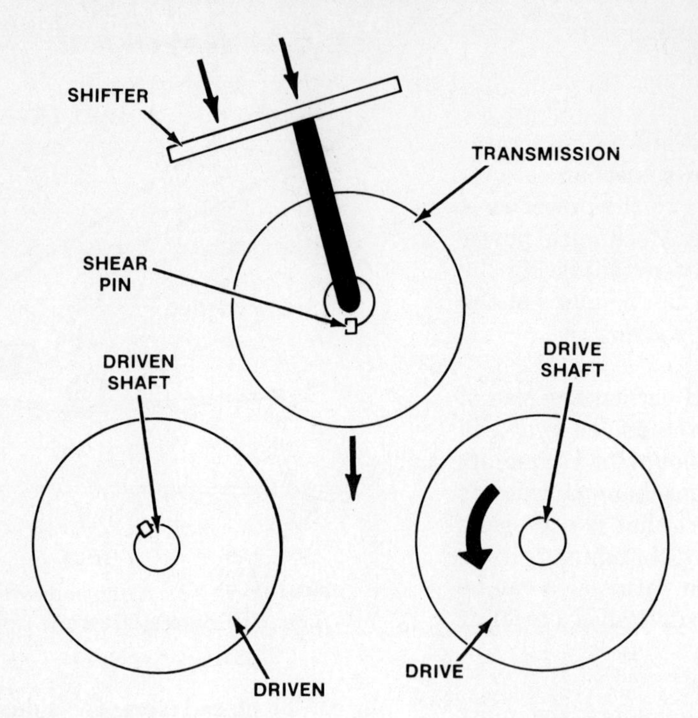

FIGURE 7–24 A basic transmission.

plate or the belt tension may be an adjustment point for the operator.

Design a Basic Mechanical System

A simple mechanical system can be illustrated by showing a basic transmission. The transmission is designed with a drive wheel that is powered by a drive shaft connected to a prime mover or motor. As the transmission is lowered into place, it connects drive to wheel. The transmission wheel is the part that closes the mechanical system (Figure 7–24).

1. *Drive Shaft.* The drive shaft of a prime mover is connected mechanically to the system.
2. *Drive wheel.* The drive wheel transmits power that will be converted to work.
3. *Transmission wheel.* The connecting wheel opens and closes the mechanical circuit. The mechanisms used to push and pull this wheel can become extremely complicated; however, all transmissions work by fitting into the "gap" in the mechanical system that must be closed in order to complete the system.
4. *Shifter.* A shift controls the transmission wheel.
5. *Shear pin.* A shear pin is an internal protection for the system. A shear pin is designed to take a certain amount of pressure. It connects the wheel or a moving part to a drive shaft. If one of the parts on either side of the transmission wheel becomes frozen or too difficult to turn, then the shear pin will break and release the wheels from the drive shaft. The remaining parts of the mechanical system then will not self-destruct. This type of internal protection is not as critical to the friction-driven transmission as the gear- and chain-driven one. Friction will burn up the two interfacing parts before the entire system can be damaged.
6. *Driven wheel.* The work is done at the driven wheel, which is the axle of the driven gear. Whether the driven shaft is connected to a bicycle wheel or to the gears of a Ferris wheel, the movement of the driven shaft is the point of work for the mechanical system. The drive wheel can be mechanically connected to a number of movement-producing points of work.

A second example of a mechanical system is the simple chain-driven mechanical linkage. This type of mechanical system is found on ten-speed bicycles, motorcycles, and other types of machines propelled by a power source that is transferred to the rear wheel (Figure 7–25).

1. *The driving sprocket.* The driving sprocket, or gear, ratio determines how much torqued speed is produced at the driven sprocket. This config-

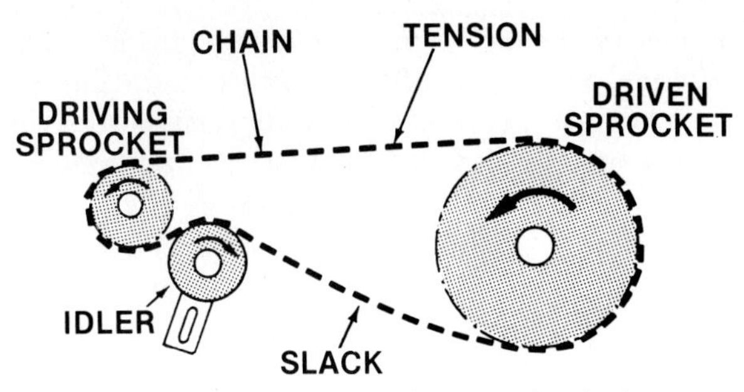

FIGURE 7-25 A gear-driven mechanical system. (Courtesy Deere & Company)

uration produces less speed but more torque. If you ride a multispeed bicycle, you use this gear ratio on a smooth road surface to give you increased torque and a good start.

2. *The chain.* Chains come in many types and styles to fit the application. Chains must interface the parts of the system as efficiently and dependably as possible. The chain in the illustrated system also acts as the internal protection device. When system parts bind or freeze, the chain will withstand only the amount that the weakest link can carry. This is usually the connecting link pin. This pin strength can be varied depending upon the strength rating of the system parts.
3. *The driven sprocket.* The driven sprocket on a bicycle, for example, is connected to the rear-wheel axle. Reciprocating or pumping action of the bicycle rider is transferred to the rear axle as torque. The torque then is transferred into friction at the tire footprint.
4. *The idler.* In a mechanical system like the ten-speed bicycle, an idler is used to take up chain slack. As the drive gear is increased, the excess chain slack is decreased.

In the next illustration, a normally closed transmission is shown. The clutch, in other words, must be pushed inward in order to "open" the mechanical system. As an exercise, follow the transmission to the place where it controls the mechanical system. There are four major movements in the sequence that are required to disengage the clutch (see Figure 7-26).

1. The operator must depress the clutch pedal.

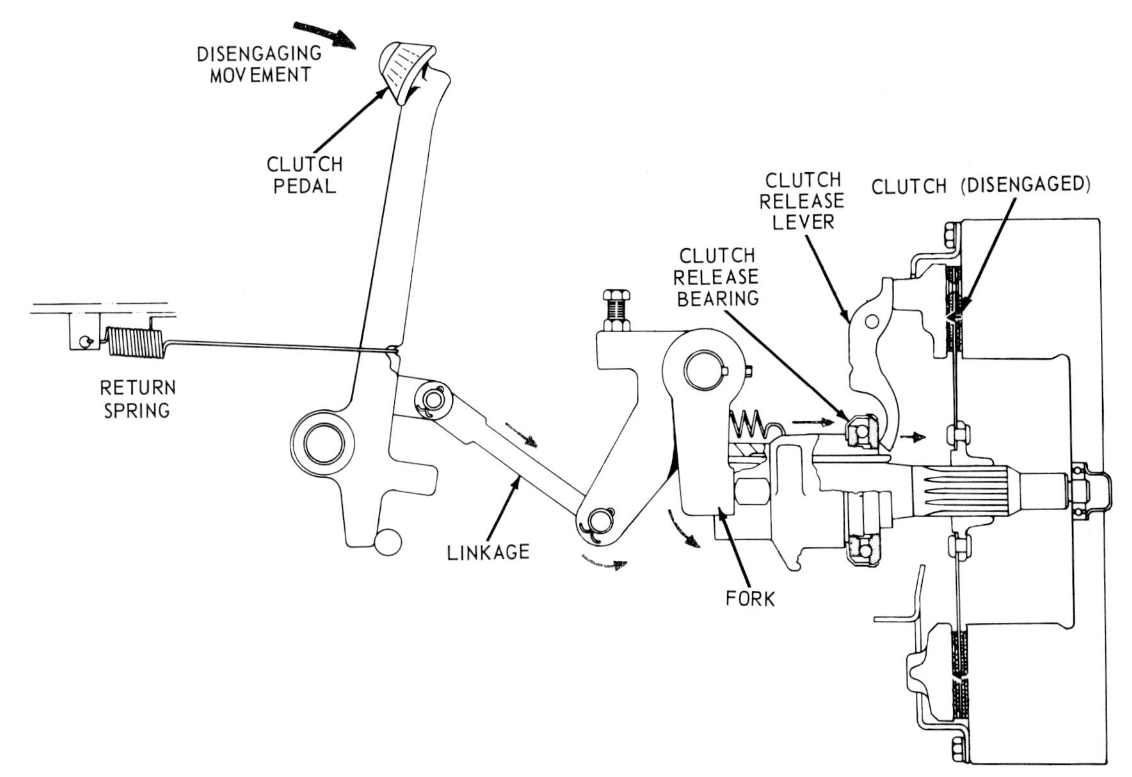

FIGURE 7-26 A normally actuated transmission. (Courtesy Deere & Company)

2. The linkage is pushed forward.
3. The clutch-release bearings push against the clutch-release levers.
4. The clutch is pushed away and opens the system.

Clutches are valuable parts of some mechanical drivetrain systems. When they are released, the gear ratios can be changed to produce more speed or torque from the output of the engine or motor.

Understand the Role of the Operator in a Mechanical System

The mechanical system is characterized by a direct contact of all system parts. All parts are usually metal to metal, plastic to plastic, or plastic to metal connections. Friction and misalignment can be the worst enemies of a gear-driven mechanical system. Friction can be reduced by lubrication. For example, oil must be applied to the bearings that support the pedals and the wheels of a bicycle or undue wear will occur. A wet transmission, like an automobile transmission, must be kept filled with lubrication to assure that it operates correctly. Improper lubricant and lack of lubrication are the chief culprits of most mechanical system failures.

On the other hand, oil must not be allowed to contact dry drives such as dry transmissions that rely on friction to transmit power. Oil must not be applied to drive belts that use friction to turn sheaves and pulleys. The operator of a mechanical system must be given the lubrication points as well as checkpoints that must be kept free from lubrication and lubrication leaks. The points that must be kept free from grease and oil are as important to the machine operator as are the lubrication points.

The operator must be given enough information to make simple adjustments as necessary. Mechanical systems, like all systems, will wear during use. Operators must be told where the *quick-wear* parts in the mechanical system are, and what they must do when these parts become worn or misaligned. A good example is the drive sheaves on a lawn and garden tractor. If they become loose from misalignment or part failure, the operator must be told what to do or what to replace in order to keep the machine running on a daily basis.

Belts, such as automobile fan belts, and chains, such as bicycle chains, must be constantly checked. As they wear, they have a tendency to stretch. A timing belt may keep an accurate tension of about 93 percent to 95 percent. A belt, like the ones that drive the waterpump and air conditioner on your automobile, may lose 15 percent of its tension under normal use after only a few weeks of operation. Operators must be given guidelines or parameters that tell them exactly how much stretch is allowed in a belt before adjustment is necessary.

BECOME AWARE OF RATINGS FOR ALL SYSTEMS

Machines can be likened to the human anatomy. All machine parts are connected together on a machine frame or skeleton to form systems. Like the human body, machines are designed to push, pull, or overcome limited amounts of resistance. If these limits are exceeded, both the skeleton and the components of the system may become damaged. All machine parts are rated for the amount of work that they can perform. Rated parts include hose, pipes, wires, gears, shear pins, and other load-carrying system members.

Every part in a system is made to carry a certain amount of force. If parts that are too weak to carry the system force are used, then that part will become a safety hazard to the machine, to the operator, or to both. The overloaded part will become the weakest link of the system. A system is only as strong as the weakest member. You must inform operators of the rating of all replaceable parts that they are capable of maintaining. An example is that the operator must be told the rating of the fuse in the electrical system. A fuse that has too great a load-carrying ability can damage all of the system parts because it will allow too great of a charge through the system. A fuse that is too small to carry the charge will constantly blow out.

The parts ratings that you give the consumer are called *parts specifications.* These specifications tell consumers exactly what they should use in order to replace the fast-wearing parts that they can change *without the aid of a professional service technician.* Parts specifications are usually printed with the general product specifications in many operator's manuals and directions (see Figure 7–27).

COMPOUNDED SYSTEMS

Often you will find that complex machines have *compounded systems.* This situation is usually found when one system is used to *control* a primary work system. The most popular compounded systems are "air over oil," or a hydraulic system that is controlled by a pneumatic system; an electrical sys-

LUBRICANT SPECIFICATIONS 177

LUBRICANT SPECIFICATIONS

THE TRANSAXLE AND STEERING SYSTEMS IN YOUR VEHICLE ARE FILLED AT THE FACTORY WITH HIGH-QUALITY, LONG-LASTING LUBRICANTS OR FLUIDS THAT DO NOT REQUIRE PERIODIC DRAINING OR REFILLING EXCEPT UNDER SEVERE DUTY CONDITIONS. HOWEVER, THE LUBRICANT OR FLUID SHOULD BE CHECKED PERIODICALLY AND REFILLED WITH THE PROPER LUBRICANT OR FLUID, MEETING FORD TECHNICAL SPECIFICATIONS.

Item	Ford Part No.	Part Name	Ford Specification
Brake Master Cylinder	C6AZ-19542-A, C6AZ-19542-B	Ford Heavy Duty Brake Fluid	ESA-M6C25-A
Wheel Bearings, Rear	C1AZ-19590-B	Multi-Purpose Long Life Lubricant	ESA-M1C75-B
Hinges, Door Latch, Hood Latch, Auxiliary Catch, Trunk and Liftgate Latches	D7AZ-19584-A	Polyethylene Grease	ESR-M1C159-A
Lock Cylinders	D8AZ-19587-A	Ford Lock Lubricant	ESB-M2C20-A
Transaxle, Automatic	XT-4-H	Motorcraft Type H Automatic Transmission Fluid	ESP-M2C166-H

FUSE PANEL 185

FUSES

FUSE PANEL DESCRIPTION

Fuse Panel Cavity Number	Fuse Rating	Color	Protected Component
1	15 Amp	Light Blue	Hi-Mount Stoplamp, Brake Lamps, Front and Rear Turn Signals and I/P Turn Indicator Lamps, EEC Module, Speed Control Function
2	8.25 Amp Circuit Breaker		Windshield Wiper Motor, Interval Wiper Module, Windshield Washer Motor
3	Spare		(Not Used)
4	15 Amp	Light Blue	Front Park, Side Marker and Tail Lamps, "Headlamps-On" Indicator Chime, Front "Laser" Lamp
5	15 Amp	Light Blue	Electronic Cluster, Heated Rear Window Switch, Electronic Flasher, Backup Lamps, Heated EGO, Illuminated/Keyless Entry Module, Cluster Indicator Lamps
6	15 Amp	Light Blue	Station Wagon Rear Window Wiper and Washer Motors, Diagnostic Lamp Module, Warning Chime, Headlamp Switch Illumination, Clock Illumination, Radio Illumination, Cluster Illumination, EATC Control Illumination, Power Window Relay
7	Spare		(Not Used)

NOTE: The turn signal flasher is located on the front side of the fuse panel. The hazard warning flasher is located on the rear side of the fuse panel approximately behind the turn signal flasher.

FIGURE 7–27 Two charts of specification information for the consumer. (Copyright 1986 by the Ford Motor Company, *Mercury Sable 1987 Owner's Guide,* FPS-12059-87A. Used with permission)

tem that controls a hydraulic or pneumatic system; and an electric system that controls the mechanical transmission in a drivetrain.

An electrical control for a system is popular with machine designers because of the wide application of a solenoid or electromagnet. Flow-control valves can be operated on this principle. Solenoids also can be used to push transmission parts together. In a machine such as a farm tractor, or an off-the-road grader or hauler, an electrical system might be subdivided into many separate circuits such as lighting, ignition, and charging that are powered by one prime mover.

It may be difficult to understand a machine that has many compound systems. You must remember that even the most complex systems are designed one part and one connection at a time. If the machine has been designed with system logic, then

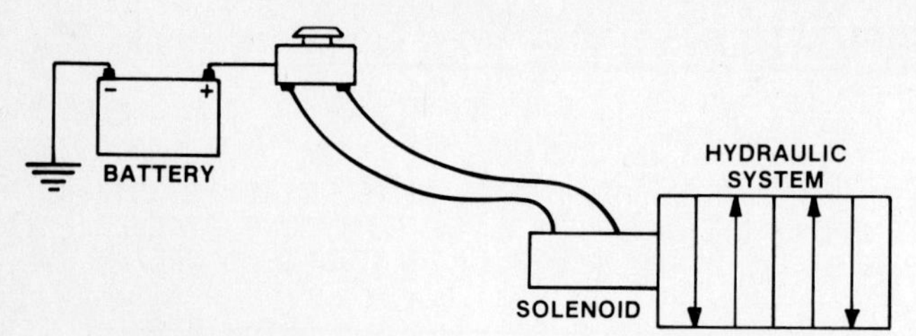

FIGURE 7–28 Systems are used to control other systems.

you can use system logic to understand the machine (Figure 7–28).

TRACE THE FLOW OF POWER IN A MACHINE SYSTEM

In order to give structure to your research and to help give you a concrete mental picture of the machine, trace the systems of the machine in rough drawings. Trace the flow of power in a system *from the point of work to the power source* indicated in the engineering department drawings and diagrams and informal interviews. (If you find that the electrical system is too complex to follow in this way, begin at the power source.) After the systems have been successfully discovered on paper, you then have the option of tracing the systems either on the prototype or the machine. This hands-on tracing of how system parts are *connected* will help you *visualize* the machine, and it will give you a basis for organizing your technical operation instruction in visual concepts.

Warning: Before you physically attempt to trace a machine system on a machine, trace it on paper with the documents of the manufacturing process. Interview the machine experts. Know what to expect from machine systems. Before you trace a system on a machine, read the next chapter on machine safety.

By tracing a system from the point or points of work, you will not be as confused by compounded systems, which may make your research more difficult. You should trace the machine systems for two reasons. First, this road map that you can sketch from your information will become your first attempt logically to find out how the machine works and is worked on. Second, the sketch of the machine system will enable you to create a foundation for the hanging file or similar method of storage and retrieval that you need to write the technical operation instruction.

A rough sketch of the system diagram from the point of work to the power source is an extremely helpful tool that you can use later in the writing process. This sketch can be roughly drawn without details other than component connections and controls from a compounded system design. A drawing such as this is especially useful for the next step in your writing process. It is a tool that you can use to connect the machine parts together mentally and isolate the important points.

OPRGANIZE YOUR TECHNICAL DATA IN A SYSTEM FILE

The most important step in any writing exercise is the time when the writer organizes and groups data into patterns of like concepts. Organization of data saves time and helps you write more accurate and up-to-date technical operation instructions. The most practical way to organize technical data is by using the outline of the machine systems.

You can further refine your rough system-connection drawings and make the information usable for others by translating them into a system profile. A *system profile* is a short character sketch that explains how each system functions in a machine to do work. You can customize the profile to fit your writing situation. A sample of a machine profile form can be put together with the following information:

SYSTEM PROFILE—WORKSHEET

System ______________________
Machine name ______________________
Machine model number ______________________
Point of work ______________________
Internal protection points ______________________
____
Adjustment points (controls) for the operator __________
____
____
Maintenance points ______________________
____
____
Power source ______________________
Safety points ______________________
____
____
Special operation notes ______________________
____
List of parts in order from the point of work
______________ ______________
______________ ______________
______________ ______________
______________ ______________
______________ ______________
Last date of compilation ______________
____
____

Staple the system rough sketch to the system profile. Use all of the rough drawings and profiles to develop a hanging file or similar information storage and retrieval system. An organized grouping of data in a storage and retrieval system such as a hanging file is essential. Every piece of data may be important to the overall technical operation instruction. If one piece of data is missing, then the entire validity of the instruction can be endangered. Files help you keep together the hundreds of pieces of paper gathered from your research and interviews.

A file system. A file system is also useful if you or a member of your company is involved in a legal court case. Your professional survival as a technical writer may depend on how well you can prove the accuracy of what you have written. In some liability cases, the design engineer, or other machine expert from your company, may be called to answer questions formally presented in a court of law or testify under oath. This testimony may include the reasons why safety information was or was not given in your technical operation instruction. The expert may be closely questioned and asked to examine the safety information word by word. With a permanent file of data built upon machine systems logic, you will have easy access to your records in the months and years to come.

As a responsible technical writer, you should keep all safety-related material at all times. Here is a basic list that outlines some of the main safety records that you may collect during your research of the machine:

Test results on safety.
Industry standards for safety warning requirements.
Safety reports in industry literature that may affect your machine and instruction.
Reviews by insurance companies (usually these are kept in the engineering department files).
Government regulations, such as the "Occupational Safety and Health Act" standards that were followed when the safety message was written.
Customer complaints and other customer feedback that you have received and acted upon.

Freedom of information. Remember, however, that your files may be made available to attorneys representing the person who is bringing a court action against your company. Do not file false or exaggerated claims. Do not write memos or other documents that say such things as:

This machine is totally unsafe in the area of. . . .
An accident is sure to happen if. . . .

You must leave the documented comments about your machine to the safety experts who are supposed to generate documents. Your comments or documented evaluations may complicate the court process.

Develop a Hanging File Storage and Retrieval System

A hanging file is created with a file cabinet, large hanging folders, and small folders that can be inserted in major larger folders. The following hanging file outline can be divided into two parts. (See below.) Files I through V must be completed before you can create VI through XII. You can use files VI through XII as a means to fine-tune your ideas and double-check the accuracy of the general data. The logic of the example of how a file system is outlined is that information is divided into basic types of information for quick reference. Second, it is cross-referenced into specially grouped information that also will help you verify whether or not you have completed the necessary research.

A Basic Machine Hanging File Topic Outline:

Note: Each file, such as FILE I, is a single large hanging folder. Each small lettered topic, such as "a. Standard operation sequence," is one manila folder inserted into the larger hanging folder.

FILE I GENERAL INFORMATION.
a. Standard operation sequence.
b. Basic machine parameters.
c. Test results.

FILE II HYDRAULIC SYSTEM (IF THERE IS A HYDRAULIC SYSTEM).
a. System profile.
b. System-connection drawing.
c. Interviews on specific parts or components.
d. Drawings, diagrams, and bills of material.
e. Collected artwork.
f. Basic maintenance and care instructions.

g. Safety documents and important safety information for support of your safety statements.

FILE III PNEUMATIC SYSTEM (IF THERE IS A PNEUMATIC SYSTEM).
a. System profile.
b. System-connection drawing.
c. Interviews on specific parts or components.
d. Drawings, diagrams, and bills of material.
e. Collected artwork.
f. Basic maintenance and care instructions.
g. Safety documents and important safety information for support of your safety statements.

FILE IV ELECTRICAL SYSTEM (IF THERE IS AN ELECTRICAL SYSTEM).
a. System profile.
b. System-connection drawing.
c. Interviews on specific parts or components.
d. Drawings, diagrams, and bills of material.
e. Collected artwork.
f. Basic maintenance and care instructions.
g. Safety documents and important safety information for support of your safety statements.

FILE V MECHANICAL SYSTEM (IF THERE IS A MECHANICAL SYSTEM).
a. System profile.
b. System-connection drawing.
c. Interviews on specific parts or components.
d. Drawings, diagrams, and bills of material.
e. Collected artwork.
f. Basic maintenance and care instructions.
g. Safety documents and important safety information for support of your safety statements.

Develop a cross-reference file on information dealing with these major topics:

FILE VI POINTS OF WORK FOR ALL SYSTEMS.

FILE VII POINTS OF CONTROL FOR ALL SYSTEMS.

FILE VIII SYSTEM PROTECTION INTERNALLY FOR ALL SYSTEMS.

FILE IX OPERATOR ADJUSTMENTS FOR ALL SYSTEMS.

FILE X SAFETY INFORMATION AND WARNINGS FOR ALL SYSTEMS.

FILE XI MAINTENANCE INFORMATION FOR ALL SYSTEMS.

FILE XII POWER SUPPLY FOR ALL SYSTEMS.

FILE XIII SPECIFICATIONS FOR ALL SYSTEMS.

FILE XIV ASSEMBLY INFORMATION IF NECESSARY.

FILE XV STARTUP INFORMATION FOR ALL SYSTEMS.

FILE XVI SHUTDOWN INFORMATION FOR ALL SYSTEMS.

When the hanging file or similar storage and retrieval system is developed, you will have the most complete and up-to-date picture that is available of the machine. You will have more information at your fingertips than any one machine expert will have in the entire manufacturing company. Your file will be the combined knowledge and work of people in every department of your factory.

CHAPTER SUMMARY

- A machine system is an unbroken line of parts that transfer power to a point of work.
- The electrical system is the most common and useful power system in the world.
- Electricity is the movement of electrons from one atom to the next through the path of least resistance.
- Atoms are structured like our own Solar System. A nucleus containing neutrons and protons is orbited by shells of electrons.
- Conductors are elements with fewer than four electrons in the outer shell. Electrons move easier from atom to atom when an electric charge is sent through these elements

than they do in elements with more than four electrons.

- Insulators are elements with more than four electrons in the outer shell. Electron movement is more difficult than the electron movement in a conductor.
- A semiconductor is an element with four electrons in the outer shell of the atom. Semiconductors can act like conductors or insulators.
- Electric current can be induced from these major prime movers: a generator, a battery, photons of light, crystals under pressure, heat, and friction.
- Current flowing through a conductor creates a magnetic field around the conductor. A conductor such as a copper wire wound around an iron core can produce a temporary electromagnet. A solenoid is a good example of the application of this principle.
- Alternating current or AC flows forward and backward through an electrical circuit as a conductor moves forward and backward through a magnetic field.
- Direct current or DC flow is in one direction.
- Voltage is a measure of potential current. Current is the rate of electron flow. Resistance is a blockage to electron flow.
- A motor works because similar electrical charges repel.
- A fuse, surge protector, or breaker switch are common parts that help protect the electrical system internally.
- An electrical circuit can be created with wires, transistors, or other parts, or they can be formed on small silicon chips.
- The most common controls for an electrical system are on-and-off switches and controls that vary current flow.
- A mechanical system is an unbroken connection of belts, chains, pulleys, gears, levers, wheels and axles, inclined planes, and wedges.
- Friction plays both a harmful and beneficial role in some mechanical systems.
- The usual point of work in a mechanical system is torque on an axle. A cam or similar device can change this torque to up-and-down or circular motion.
- The control of a mechanical system is usually at a transmission or a point where gear ratios can be changed.
- Internal protection in a mechanical system is provided by shear pins, keys, or other parts that break under excessive pressure such as a binding or freezing of parts. By breaking, they help prevent further damage to other parts of the mechanical system.
- Each part of a system is rated for the amount of work that it can do or the power that it can carry. Ratings are usually given in the specifications section of an instruction or manual.
- Compound systems are systems that are controlled by other systems. An example is a hydraulic system controlled by electrical on-off switches or pushbuttons.
- A hanging file or similar file-and-retrieval methods of organization must be used to group like bits of information. The organization of information is the first step in the actual writing process.

EXERCISES FOR CHAPTER 7

Organize Your Information

1. How many electrical systems can you find in the machines in your home? The place where you work?
2. Sketch five simple systems.
3. Use your imagination and create a system sketch of an imaginary hydraulically operated robot. Begin by imagining a simple type of work that you want your robot to perform. Draw your idea in a diagram similar to those used in this chapter.
4. Write a 500-word report on how each of the following things work. Draw simple diagrams to support your ideas.
   a. A chemical battery.

b. A solenoid valve.
c. A hydraulic cylinder.
d. A hydraulic motor.
e. A variable resistor in an electrical system.
f. A silicon chip from a computer.
g. An automobile transmission.
h. An automobile differential gear box.

5. Choose a machine that you are familiar with.
   a. How many points of control are there for this machine?
   b. What type of feedback must the operator interpret?
   c. How many systems are there in the machine?

8

Learn the Languages of Safety, Warranty, Measurement, and Specifications

The information given in this chapter will enable you to:

- Give a basic definition of product liability.
- Explain the role of the technical writer in the company safety program.
- Develop usable safety messages for the technical operation instruction.
- Define "point of use" for safety messages.
- Develop the levels of safety messages to increase their impact on the consumer.
- Increase your machine safety I.Q.
- Define warranty.
- Explain the difference between warranty and instruction information.
- Use measurements to create specifications for the consumer.

INTRODUCTION

Mass production and automation of the manufacturing process have had greater effect on the lifestyle of the "average" consumer than any social revolution or scientific discovery. More power is available to the consumer today than at any time in history. This availability of power often places the consumer in an awkward position. With the increase in power, our machines have become more complex and more difficult to operate. They are also more difficult to maintain and more costly to service than comparable machines in use only a few years ago. Even the simplest machines can be the result and embodiment of state-of-the-art technology. Information that correctly explains how to operate and take care of current machine technology has become a necessity.

Before the advent of mass-produced power equipment and other technologically advanced machines, the manufacturer could make and sell a product directly to the consumer. Operation of these basic machines usually could be explained in a few simple verbal operation instructions. All that was necessary to complete the sale of the machine was a handshake and payment. The machine manufacturer often was not legally or morally responsible for the reliability of the machine.

There was also no "truth in advertising" law that helped assure the machine worked as it was depicted in sales pitches and advertisements. The marketplace then can be symbolized today by the picture of the "medicine show" pitchman who not only sold an amazing medicine that could treat all of the consumer's ills but also a machine that could be used in a thousand ways to make life easier and more productive.

It was a time of *caveat emptor* or "let the buyer beware." In that uncontrolled and deregulated era, it was the responsibility of the consumer to inspect the machine thoroughly and to see that it functioned properly *before* the sale or trade was made. If the machine failed to perform or failed to do the work that the manufacturer promised, the consumer had no recourse but to either buy another machine or have the broken parts fixed or replaced. If the consumer purchased a machine that created hazards to the operator, the consumer or operator suffered the consequences (see Figure 8-1).

Today, the concept of "let the buyer beware" has been replaced by *caveat venditor* or "let the seller beware." The major shift in popular opinion to *caveat venditor* gained ground in the late 1950s and early 1960s with the growth of the *consumer awareness movement*. Leading this movement were books such as Rachel Carson's *Silent Spring* and Ralph Nader's *Unsafe at Any Speed*. The attitude of the American consumer began to reverse gears. Through court rulings and precedent-setting cases, manufacturers were legally required to accept greater responsibility for the design, fabrication, and assembly of their products.

At times, the machine manufacturer is placed into seemingly impossible situations. The legal climate that governs our marketplaces has dictated that the manufacturer and seller of a product must assume a greater responsibility to:

Provide the safest machine possible.

Back up all claims of machine performance made before and after the sale of the machine in the marketplace.

Warn of potential dangers, and instruct the operator how to use the machine properly.

In a sense, the manufacturing company that designs and makes a machine must somehow continue to extend into the workplace. The design must be the safest one possible in that it must meet or surpass accepted industry and legal standards for safety. The manufacturer also must assure that the advertising claims of the company are conservative estimates of what the machine can do. The best way found to help ensure the machine is operated correctly is to package or attach a technical operation

FIGURE 8-1 *Caveat emptor* was the attitude of the marketplace until the early 1960s.

instruction to the machine before it goes out of the back door of the manufacturing company.

The first part of this chapter covers some of the legal aspects of doing business in our marketplace, factors that affect the job of the technical writer. Providing on-site instructions and comprehensive training to every consumer at times is impossible in our global marketplace. Thus, the technical writer in a worldwide marketplace must be aware of the extended legal responsibility of the manufacturing company to instruct the consumer no matter where the machine is sold or used. The information that the consumer needs to know, or must know, goes beyond what the step-by-step instruction and simple factual sentence can express.

Technical writers must be aware of the problems created if they use the technical operation instruction to "resell" the machine to the consumer. Technical writers must realize that the instruction is a post- or after-sale document. The technical operation instruction is a specialized form of communication that interfaces the machine to either the consumer or the machine operator. Misinterpreted or incorrect claims about the ability or design of the machine that are written in the instruction can create an implied warranty. This cannot only be hazardous to the consumer but also costly to the warranty program of the company manufacturing the machine.

As the *collective voice* of the manufacturing company, the technical writer must be aware that the technical operation instruction must not be written without the collaboration of safety and legal experts in all of the areas of the manufacturing process. The following sections cover the basic concepts of product liability and warranty, especially as they apply to the machine and machine systems. Liability and warranty concepts require a language beyond the visual instruction style and word choice that have been explained in previous chapters.

The second half of this chapter covers the remaining specialized languages of *measurement* and *specifications*. These "quantity" languages have become a necessary part of technical operation instructions. They are used to inform consumers of vital statistics and facts about the machine that they must know in order to keep the machine working on a daily basis. Since measurement is the science of *metrology*, the technical writer must strive to put the terms and techniques of measurement on a scientific level that the audience can understand and use.

Measurement differs from specifications in that the overall dimensions of the machine are required by consumers in many cases to tell them exactly how large the machine is. Measurements are also given as guidelines that the consumer or machine operator can use to judge the output of the work that the machine is producing. Without measurements, the consumer cannot make the proper adjustments that ready the machine for day-to-day use.

Specifications, on the other hand, inform consumers of the capacity of the machine to do work. Specifications list the ratings for quick-wear and replacement parts that the consumer can personally replace in the machine as they wear out. For most machines, the most important specifications given to consumers are the power requirements and limits that must be provided to the machine for a safe and efficient operation.

UNDERSTAND THE ROLE OF THE TECHNICAL WRITER IN A COMPANY SAFETY PROGRAM

Ideally, every machine should be made so that it is absolutely fail-safe. This means that machines would never create a hazardous situation to the consumer or operator. This is a challenging task for someone who is designing a machine. For example, how do you make lawnmower blades safe and still do the job of cutting grass? Often the machine designer is not able to look into the future and imagine the ways that the machine will be used and misused. On some complex equipment, only warranty and liability cases may point out all of the ways that a machine can be misused. It is impossible, for example, to imagine the numerous ways that a simple tool such as a screwdriver or hammer can be used for jobs other than those they were designed to do.

Dix W. Noel and Jerry J. Phillips address this problem.[1] Although they are expressing an interpretation of the responsibility of the manufacturer of a particular chemical, their explanations can be used to help illuminate the machine technical operation instruction. They divide the technical operation instruction into two parts. First, the instruction is intended to tell the consumer how to use the machine efficiently. The second part is how to use it safely.

> Directions and warnings are intended to serve different purposes. The former are designed to assure an *effective* use of the product; a warn-

[1] Dix W. Noel and Jerry J. Phillips, *Products Liability Cases and Materials*, 2nd ed. (St. Paul, Minn: West Publishing Co., 1982).

ing, on the other hand, is intended to assure a *safe* use.[2]

> Instructions which fail to mention necessary safety precautions, or which are themselves hazardous to follow, fall in the same category as failure to warn. Furthermore, they must be exceptionally clear when intended for use by minors or by illiterate or even inexperienced adults.[3]

Everyone who uses the machine must be furnished with the easy-to-use information on how to use the product efficiently and safely. In a sense, the company safety program can be divided into the technical operation instructions that you develop to tell the consumer how to use the machine efficiently, the safety messages that you must use in the instruction to help ensure the safety of the consumer or machine operator, and the safety warnings that are part of the product.

Safety messages often are used to warn against misapplication. Development of safety information by the safety and legal experts of a company is mainly an attempt to determine the consequences of misuse of the machine. Safety warnings must alert the machine operator of the consequences of using the machine in a manner in which it was not designed to work. In a sense, the writing of instructions on how to operate the machine is an "offensive" or positive type of writing. The creation of safety warning information is a "defensive" type of writing.

Realize the Importance of Safety

The first seven chapters of this book concentrated on helping you develop a clear and concise technical writing style that is both visual and active. You have been given the words and definitions that you can use to build a solid writing foundation. You can use this basic information to develop your personal writing style. A writing style for the technical writer must include a comprehensive program of safety-data collection and evaluation.

With all of the words and definitions that have been given in the preceding chapters, the most important word in your technical writing vocabulary must be "safety." You must use safety warnings in your instruction to make the operator of a machine stop and think about his or her actions before a hazard can develop. Two main reasons exist why safety is such an important concept in a technical operation instruction. They are:

Consumer well-being.
Company liability.

First, as a technical writer, you must not overlook the safety needs of your audience. If you are writing technical operation instructions for a homogeneous audience, or an audience that you are able to analyze completely, you may not need to state the obvious. However, if you are writing to a heterogeneous audience that you are unable to analyze, you may need to add safety messages that otherwise would not be necessary.

Accidents can be reduced. Accidents can happen because of four major reasons. First, the machine that the consumer buys may be poorly designed or incorrectly manufactured. The technical writer has no control of this situation. Second, the consumer can make a mental error in judgment in the operation of the machine. With usable technical instructions, this type of accident can be reduced. Third, the consumer/machine operator can ignore the instruction. Often this cannot be controlled by the technical writer unless warning signs are added to the machine and the cover of the instruction that caution the consumer to read all tasks and warnings before operating the machine. Fourth, the consumer can be put into an unsafe position because he or she was not properly warned of the hazards of the machine. A manufacturing company has a duty *to warn* the consumer of the unsafe situations created during the standard operation cycle of the machine.

Your job as the company technical writer is to *help* eliminate as many *preventable* machine accidents as possible. You will never be able to make an unsafe machine safe simply by writing an instruction replete with warnings and cautions. You can, however, work toward the goal of making the consumer aware of the consequences that may result from machine misapplication and misuse. You must "warn" the consumer of any unsafe situations created by the particular machine design and do it in such a manner that the consumer stops and considers the safety message before completing the task.

You must make it clear to both the homogeneous audience and the heterogeneous audience using new technology that even the most innocent-looking machines must not be taken for granted. No

[2]Ibid., p. 485.

[3]Ibid., p. 473.

"powered" machine is *absolutely* safe if it is incorrectly used and maintained. Every powered machine must be handled and controlled as safely as possible. Many deaths a year are caused by the simplest mistake or misapplication of a machine. For instance, electric appliances can become extremely hazardous when they or the operator come into contact with water. A person standing in a puddle of water in the bathroom is in a hazardous situation when operating a hand-held hair dryer. A plugged-in appliance on the edge of a wash tub or basin of water is a potential hazard.

Safety I.Q. is not the same as operation I.Q. Do not take the safety I.Q. of your audience for granted. For example, everyone who passes a state-run road test can legally drive an automobile. Automobile operators are not tested on safety and maintenance. Consumers in a heterogeneous audience may not be able to list the basic safety hazards created by misuse of the family automobile. Use the following safety checklist to test a sample audience. Ask the audience to list the safety concepts associated with the *underlined* word. Be sure to tell them of all the safety concepts that they fail to identify.

> *Tires.* Tire sidewalls can blow out during tire inflation. The consumer must not stand parallel to the sidewall of the tire during inflation. Badly worn tires create a problem in that they can blow out and can lose traction in turning and stopping.
>
> *Radiators.* Radiators can spray hot coolant into the eyes and under the skin of someone who removes the radiator cap of a hot engine.
>
> *Battery.* Battery acid is harmful to eyes, skin, and clothing. Incorrectly handled or connected batteries can explode. Batteries that are jump-started or replaced must be properly connected *in sequence* according to the battery manufacturer's instructions.
>
> *Moving engine parts.* Belts and radiator fans are extremely dangerous if the consumer opens the engine hood when the engine is running. Whenever movement is present, heat often builds up. Hot metal can cause serious burns to skin.
>
> *Gasoline.* Gasoline is extremely dangerous if improperly mishandled. Pinhole leaks or breaks in fuel lines can spray gasoline into a fine mist that can explode.
>
> *Seat belts.* Seat belts are such an important safety apparatus that some states mandate their use by law.

Your instructions must explain safety information as clearly and accurately as possible for every machine. The consumer must not be frustrated by unusable and overly complex information in a safety message. You must help warn against consumer overfamiliarity and complacency. These can create unnecessary safety hazards on a power machine. Safety messages must grab the attention of the consumer and machine operator.

As a technical writer, you must first think about the welfare of the people who are buying and using the machine and reading your technical operation instruction. This includes the consumer who buys the machine and others who may operate it. The operator may not be the person who purchased the machine. Not only do you have an obligation to the consumer but your company has a legal responsibility to ensure that purchasers use the machine as safely and efficiently as possible. A technical writer must assure that complete safety and operation information for a machine is included in the technical operation instruction sent into the marketplace with the machine. The technical writer must try to force consumers to be conscious of safety procedures at all times when operating their machine. This is a difficult job. Machine "accidents" can happen to the most safety-conscious consumer.

Liability costs. Your second duty is to your company. There are many individuals who can become personally involved in a court case involving consumer safety. In order to do a complete job of collecting safety information, you must realize that there are many people in a manufacturing company who are legally responsible for machine safety. These include:

> The company attorney and/or safety expert, who may end up defending the safety of the machine in a court of law. A wrong word or the absence of a word can sometimes mean the difference between a successful defense of a machine in a liability case and loss of the case.
>
> The engineering experts who designed and developed the machine concept. Engineers have been forced to defend the safety of their machine design in a court of law. They are the most vulnerable to lawsuits when either the machine or the consumer fails to work safely.
>
> The production experts who are responsible for the methods of production. Production tech-

niques can have as much effect on machine safety as the original conceptual design developed by the engineer.

The quality assurance experts, who must maintain the company standards for machine production throughout the manufactured life of the machine.

The advertising specialists, who must not advertise or make claims for something that the company cannot produce. For example, an advertisement cannot say that a machine never needs adjusting if adjustments are necessary. If the machine malfunctions and injures someone because the necessary adjustment was not made, then the advertisement can be used against the manufacturing company in a court of law.

The after-sales service experts, those who are involved with safety problems. Often safety problems will not be obvious in a machine until it has been sold to and used by the consumer. The after-sale experts must monitor any post-sale problems and add their expertise to the development of safety information in the technical operation instruction. The development of complete safety information often is an ongoing process.

A duty of the technical writer in a manufacturing company should be to review the entire safety attitude of all of the machine experts. With most of the departments involved to some degree with machine safety, there may develop inconsistent safety messages and safety literature. Because you, the technical writer, will be the only machine expert in a position to see the entire company communication process, you must review the safety needs of all of the departments and machine experts. If an inconsistent attitude is present, you must bring this to the attention of the experts involved with the safety program. Do not attempt personally to mediate these inconsistencies in the technical operation instruction.

Define Liability

Many types of responsibilities that must be assumed by the machine manufacturer have been identified through court cases. The most common is the concept of "strict liability." A simple definition of strict liability is that a consumer or machine operator who is injured by a machine through the fault of the manufacturing company has the legal right to seek damages from the machine manufacturer. A machine can be deemed unreasonably dangerous because of many things. Among them is the machine manufacturer's *failure to warn or instruct the consumer.* Because of this right to seek a legal settlement by the injured operator or user, the machine manufacturer must:

Warn the consumer about unsafe or potentially unsafe situations that develop as the machine is cycled to do work.

Instruct the consumer about how to operate the machine safely.

Make sure the machine is manufactured to the highest standards possible.

Ensure the design of the machine is as safe as possible.

Ensure there is adequate warning *on the machine* in the form of warning signs and decals.

Liability can be a confusing and costly subject for both the machine manufacturing company and for those consumers who buy the machine and are injured by it during operation. Liability laws have developed on the state level in the United States. Each state without the benefit of federal guidance has created a body of liability law somewhat independently of the other states. This means that machines sold in interstate commerce, or sold across state boundaries, may be the subject of liability cases that differ greatly from market to market. At this time there is no national law that standardizes the concepts of liability. For the past several years, lawmakers have been debating federal legislation that will define the limitations and content of liability on a nationwide level.

Liability cases brought against a company can be extremely costly to the company, which must develop and retain legal council to handle liability claims. Settlements to a consumer or machine operator may be in the millions of dollars. Large settlements mean that the company's liability insurance premiums will increase. Costs for liability cases and insurance premiums must be passed along to the consumer who buys the machine. This added cost can price some machines out of the marketplace. A machine with added cost on the price tag may not be able to compete successfully with both domestic and imported machines produced in an environment where liability costs are not as great.

The liability revolution of the early 1970s has turned into a product liability crisis. The number of successful liability claims is rising, and liability insurance has become unaffordable or unattainable for some manufacturers. News reports in the mass media and in business publications cite more and more problems in the area of obtaining insurance for any company or individual dealing with a product or providing service to consumers.

An example of this is reported by *Products Liability Reports* in a recent issue. The James Hunter Machine Company, Inc., of Massachusetts, filed for a Chapter 11 reorganization under the U.S. Bankruptcy Code. According to a company spokesperson this action was taken

> ... not so much because business conditions are poor, but because of a little understood, yet all too familiar concern of American manufacturers—product liability.[4]

This is not an isolated case. There have been other companies who have gone out of business or severely limited their machine line because of the effects of products liability. Many companies refuse to make or design machines that are in demand and would sell because they are subject to liability suits. Liability cases have severely affected entire industries and limited new machine design and production.

The prospect of dealing with actual and perceived liability claims against a company may put more stress on the manufacturing process than it can withstand. Small- and medium-sized manufacturing companies may not be able to absorb the cost of liability cases. It is your job as the technical writer in a company to ensure your technical operation instructions are not the cause of expensive court proceedings. As the safety voice of the machine experts, you must:

Document all safety messages. You must be able to tell why the message appears in your technical operation instruction.

You must *report* and not create safety information such as warnings and cautions.

Be aware of discrepancies that might exist between the machine experts as to what safety messages are needed in the instruction and *on the machine* and how they are worded. Bring all discrepancies to the attention of the machine experts.

Remember that a single word can cause an accident if it is improperly used or left out of a safety warning or caution.

[4]*Products Liability Reports* (Chicago: Commerce Clearing House, June 8, 1984), No. 456, p. 9.

The first step in a liability case may be a request by attorneys for your technical operation instruction. They may begin constructing their case based on your technical operation instruction. For this reason, you must constantly strive to explain to your audience as completely and accurately as possible how to operate the machine in the proper manner.

Correctly Display Safety Messages

There are two main avenues for safety messages. First, safety messages that are developed by the company machine experts are put directly on the machine in the form of warning decals and signs. A lawnmower, for example, may have one warning sign that tells the operator where to stand when pulling the starter rope, and another sign that warns the operator that there is a rotating blade under the mower deck. Although it is not a machine, a common stepladder may have five or more warning signs and decals in strategic locations to warn the consumer of misapplication.

A guideline that is often followed is that these machine warning and other safety signs on the machine are also put in the technical operation instruction along with other safety information that the consumer must know. Safety signs may not appear verbatim in the instruction. Often they are paraphrased or more fully explained in the document as the information is needed (see Figures 8-2 and 8-3). This depends upon the safety itself, legal department, and industry standards.

These machine warnings are often directly and indirectly paralleled in the directions or manual. Safety information is often put in two locations in the publication. First, it may be placed in a safety section at the beginning of an instruction. These safety pages generally appear in bound manuals that are larger than a few pages. This isolated safety warning section lists all of the safety warnings that the manufacturing company must tell the consumer. These grouped warnings are intended to alert operators of a machine *before* they read the instruction (Figure 8-4).

These isolated sections are often included in a technical operation instruction or manual because of industry standards and requirements. These warnings and safety information usually provide only general information. They normally do not cover specific applications of the machine.

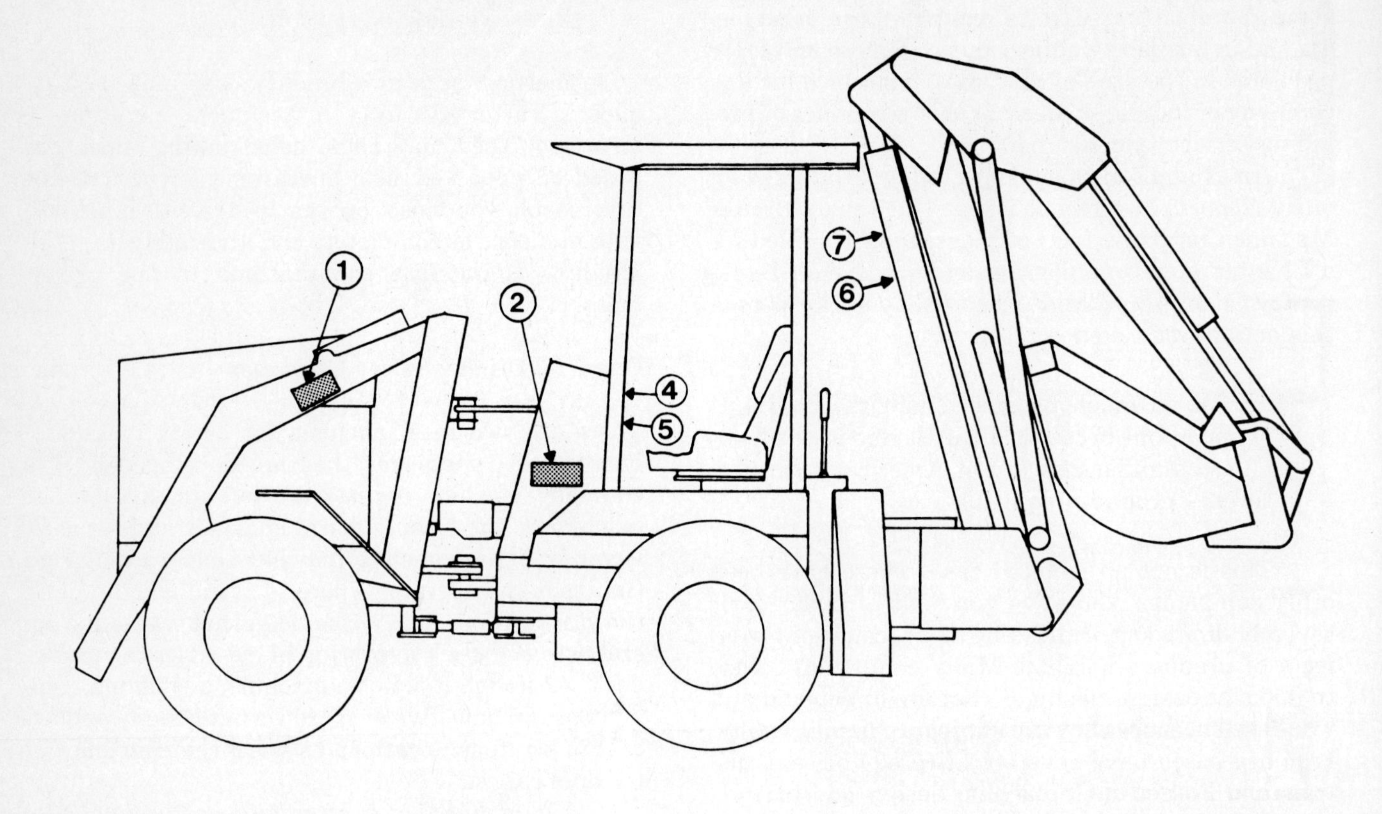

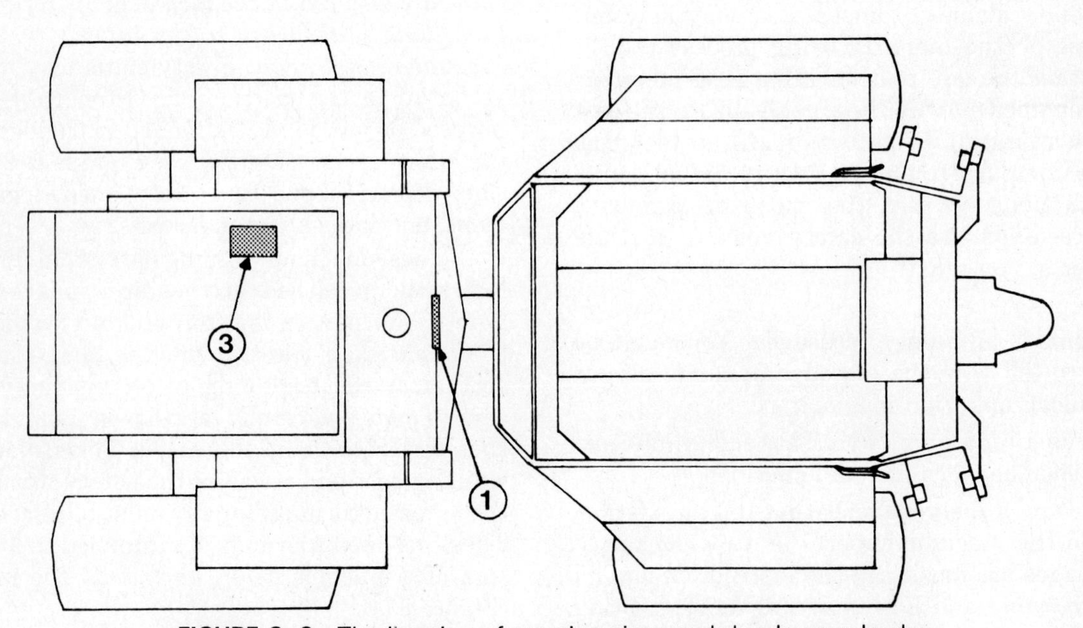

FIGURE 8–2 The location of warning signs and decals on a loader backhoe. (Copyright 1986 by Ford New Holland, Inc., *Operator's Manual Loader Backhoe General LB-620,* Form No.0(LB-620)-2-2M-486P. Used with permission)

DANGER

BOOM MAY FALL AND CRUSH YOU WITHOUT WARNING.

BOOM LOCKS WILL KEEP BOOM UP. ALWAYS MOVE BOTH BOOM LOCKS FORWARD INTO LOCK POSITIONS AND REST BOOM ON LOCKS BEFORE WORKING UNDER RAISED BOOM.

FAILURE TO DO SO MAY RESULT IN SERIOUS INJURY OR DEATH.

794840

(2)

WARNING

ARTICULATING JOINT CREATES A PINCH POINT

DO NOT USE STEERING WHEEL FOR HANDLE WHEN MOUNTING OR DISMOUNTING

TURNING OF STEERING WHEEL MAY CAUSE YOU TO GET CRUSHED

(3)

DANGER

DO NOT USE STARTING FLUIDS.
THEY MAY CAUSE AN EXPLOSION.

SEE OPERATOR'S MANUAL

FAILURE TO FOLLOW INSTRUCTIONS MAY RESULT IN SERIOUS PHYSICAL INJURY OR DEATH.

680289

WARNING

THIS MACHINE CAN BE DANGEROUS IF NOT OPERATED IN A SAFE MANNER. KNOW THE SAFETY AND OPERATING INSTRUCTIONS GIVEN IN THE OPERATOR'S MANUAL BEFORE OPERATING THIS MACHINE. IF YOU DO NOT HAVE AN OPERATOR'S MANUAL, CONTACT YOUR SPERRY NEW HOLLAND DEALER OR PHONE THE CONSUMER SERVICES DEPARTMENT OF SPERRY NEW HOLLAND IN NEW HOLLAND, PENNSYLVANIA AT 717-354-1545. A FEW RECOMMENDATIONS ARE LISTED BELOW

ALWAYS KEEP SEAT BELT PROPERLY ADJUSTED AND FASTENED. DON'T GET THROWN AROUND IN OR OFF OF THE MACHINE

NEVER ALLOW RIDERS ON THE MACHINE OR IN THE CAB

WARN BYSTANDERS BEFORE STARTING ENGINE

OPERATE ONLY WITH APPROVED ATTACHMENTS AND ACCESSORIES

CARRY LOAD AS LOW AS POSSIBLE. AVOID STEEP SLOPES AND ABRUPT STARTS AND STOPS AND SLOW DOWN ON ROUGH TERRAIN OR WHILE TURNING TO AVOID TIPPING OR ROLLING OF MACHINE

ALWAYS ENGAGE BRAKE, LOWER BOOM ONTO GROUND OR ENGAGE BOOMLOCKS AND STOP ENGINE BEFORE LEAVING SEAT

NEVER SERVICE MACHINE WITH BOOM RAISED UNLESS BOOMLOCKS ARE ENGAGED

KEEP ALL SHIELDS IN PLACE

FAILURE TO OBEY WARNINGS MAY CAUSE SEVERE PERSONAL INJURY

790770

WARNING

UNDER CERTAIN OPERATING CONDITIONS, OPERATOR MAY EXPERIENCE SUDDEN VIOLENT MOVEMENT.

ALWAYS WEAR SEAT BELT.

FAILURE TO DO SO MAY RESULT IN SERIOUS BODILY INJURY OR DEATH.

LOADER MAY TIP OR ROLL OVER IF NOT OPERATED IN A CAREFUL MANNER. ALWAYS CARRY LOAD AS LOW AS POSSIBLE. OPERATE AT SPEEDS CONSISTANT WITH CONDITIONS. AVOID STEEP INCLINES, HIGH SPEED TURNS AND NEVER CARRY MORE THAN THE SAE LOAD LIMIT OF:

TIPPING OR ROLLING MAY CAUSE SERIOUS INJURY

790603

WARNING

THIS BACKHOE CAN BE DANGEROUS IF NOT OPERATED IN A SAFE MANNER

ALWAYS LATCH BACKHOE BOOM IN TRANSPORT POSITION BEFORE MOVING MACHINE

OPERATE UNIT FROM SEAT ONLY

BE ALERT FOR BYSTANDERS AND KEEP THEM AWAY FROM BACKHOE WORKING AREA DURING OPERATION

LOWER BUCKET TO GROUND OR LATCH IN TRANSPORT POSITION AND STOP ENGINE PRIOR TO LEAVING MACHINE UNATTENDED

LEARN EXACT LOCATIONS OF ALL UNDERGROUND UTILITIES BEFORE STARTING TO DIG

FAILURE TO FOLLOW THESE GUIDELINES COULD RESULT IN SERIOUS INJURY TO OPERATOR OR BYSTANDERS

790740

(7)

AVOID METAL CONTACT WITH ELECTRICAL POWER LINES

CONTACT MAY RESULT IN SERIOUS INJURY OR DEATH BY ELECTROCUTION

238521

FIGURE 8–3 The signs that are designated in Figure 8–2. (Copyright 1986 by Ford New Holland, Inc., *Operator's Manual Loader Backhoe General LB-620,* Form No.0(LB-620-2-2M-486P. Used with permission)

HANDLE FUEL SAFELY—AVOID FIRES

Handle gasoline with care. It is highly flammable. Use an approved gasoline container.

Fill the fuel tank outdoors.

Do not fill fuel tank to top. Allow room for fuel to expand.

Do not smoke while you fill fuel tank or service fuel system.

Do not remove fuel cap or add fuel to tank if engine is hot or running.

Clean up spilled fuel.

Move chain saw at least 3 m (10 ft) from refueling area before attempting to start engine.

Let engine cool before you store chain saw in a building.

Do not store chain saw where fuel fumes could reach an open flame or spark.

Drain fuel before transporting chain saw.

1A3;TY1294 8 Y05;46SA G 200386

PROTECT AGAINST NOISE

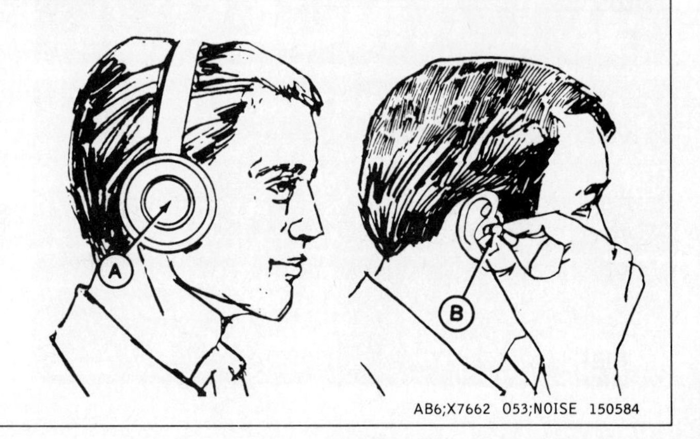

Prolonged exposure to loud noise can cause impairment or loss of hearing.

Wear a suitable hearing protective device such as earmuffs (A) or earplugs (B) to protect against objectionable uncomfortable loud noises.

AB6;X7662 053;NOISE 150584

WEAR PERSONAL PROTECTIVE EQUIPMENT

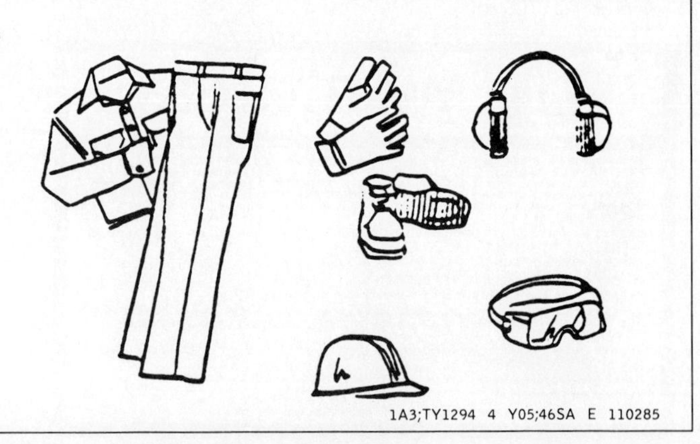

Safety glasses or goggles.
Ear plugs or muffs.
Hard hat.
Heavy work gloves.
Safety shoes with slip-resistant soles.
Close fitting work clothes.

1A3;TY1294 4 Y05;46SA E 110285

FIGURE 8–4 A general safety page. (Copyright 1986 Deere and Company, *Operator's Manual 450V and 800V Chain Saws,* OM-TY20829-F6. Used with permission)

Point of use. A second location for the use of safety messages is the point of use. *Point of use* is the location in the instruction where the safety message affects the task. Safety warnings and cautions are usually written above the task that they apply to. For example, the standard point-of-use warning tells the consumer about a potential hazard before the task is given. These warnings usually are more specific than the ones found in safety sections at the front of manuals. They explain exactly what to do and the consequence if it is not done. (Figures 8-5 and 8-6.)

Safety signs on a machine and safety messages in the instruction can also be used to help protect the machine. A machine that is misused can become a safety hazard to the machine operator. This information is used to help reduce the misapplication of the machine by the consumer. Misapplication is simply the use of the machine for work that it was not designed to do. For example, a misapplication of your automobile would be the hauling of 10,000 pounds of building material to a construction site. This is a job for a heavy-duty off- the-road truck, and not for the suspension and frame of an over-the-road automobile. Automobiles often are equipped with a sign on the door panel that tells the consumer the maximum weight that the vehicle will haul.

Develop Levels of Safety Messages

For maximum impact and influence upon the reader of your technical operation instruction, you must develop your safety messages in "urgency" levels. This use of levels must be put into your technical writing style sheet so that every instruction that your company develops has the same method of instructing the machine user about safety. For example, every safety message that warns the operator about a potential hazard to oneself should be preceded by such statements as: WARNING!, STOP!, HALT!, OR POISON!. Urgency-level signs require one major usage rule. You must be consistent whenever you use these safety message headings.

If you develop your company style sheet, you must use the industry standards created to provide your company with manufacturing guidelines. You also must look carefully at government standards and guidelines for your industry and for the creation of technical operation instructions. One organization that you must be familiar with is:

The American National Standards Institute (ANSI)
1430 Broadway
New York, N.Y. 10018
[General information phone: (212)354-3300]

ANSI. ANSI is a nonprofit organization devoted to helping develop industrial standards for several areas of manufacturing. Industry representatives work together in committees to create the standards. ANSI then approves and adopts the standards. These accepted procedures are put into publications that are available to both the general public and industry personnel. The *ANSI Catalog of American National Standards* lists the standards

COPIER MAINTENANCE
CLEANING THE DOCUMENT GLASS AND DOCUMENT PAD

Keeping the Document Glass and Document Pad clean will help ensure clean copies and, if your copier has the optional Automatic Document Feeder it will help prevent misfeeds. Xerox suggests that you clean the Document Glass and Pad at the start of each day, and during the day as needed.

- On copiers equipped with the optional Automatic Document Feeder, refer to **Cleaning the ADF Belt**, page 65.

1. Turn the copier off.

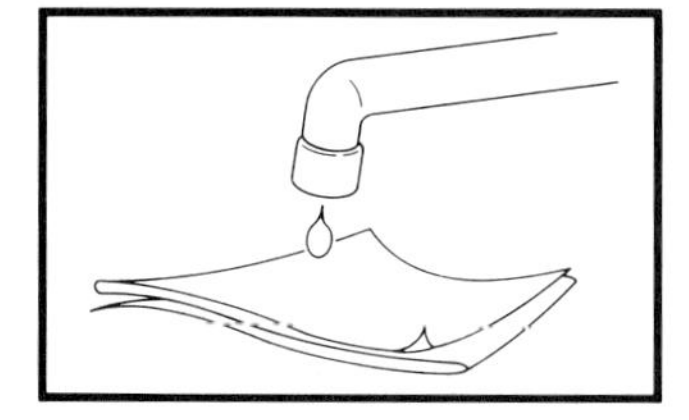

2. Slightly dampen a soft lint-free cloth or paper towel with water.

CAUTION
To avoid damaging the copier, DO NOT pour or spray water directly onto any part of the copier - always apply to cloth first.

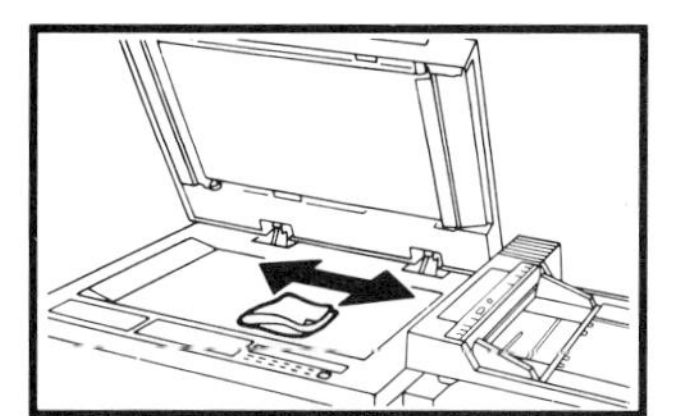

CAUTION
Use only WATER to clean the glass. Soap or other cleaners will remove protective coating from the glass.

3. Lift the Document Cover.

4. Wipe the surface of the Document Glass until it is clean and dry.

5. Wipe the surface of the Document Pad until it is clean and dry.

6. Close the Document Cover.

7. Turn the copier on.

FIGURE 8-5 Point of use "caution" messages. (Copyright 1985 by the Xerox Corporation, *The Xerox 1038 Reduction/Enlargement Copier User Guide,* 600P88316. Used with permission)

Installation

STEP 7 *Install Lights and Lenses*

PROCEDURE: Install a 75 watt maximum light bulb in each socket as shown. The lights will turn on and remain lit for 4-1/2 minutes when power is connected. Lights will REMAIN ON when Light Switch on Wall Control is **ON**.

If light bulbs burn out prematurely due to vibration, replace with "rough service" bulbs.

INSTALLING LENSES: Make sure antenna is straight down before installing lenses. Slide each lens into the lens guides as shown. Snap bottom tabs into lens slots. (The force adjustment controls are located on the back panel behind the lens.)

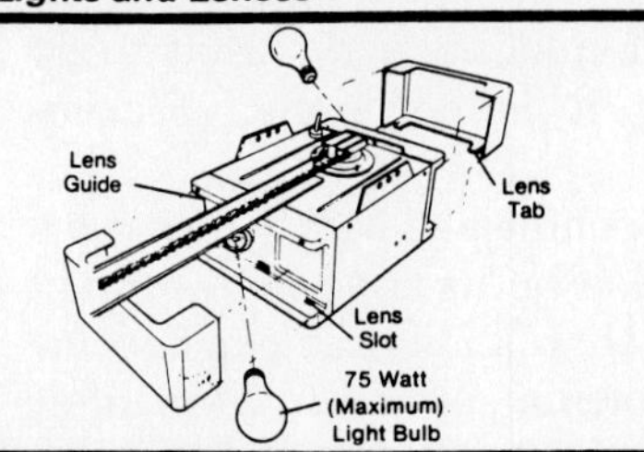

STEP 8 *Connect Electric Power*

TO AVOID SERIOUS PERSONAL INJURY FROM ENTANGLEMENT, REMOVE ALL ROPES CONNECTED TO THE GARAGE DOOR BEFORE OPERATING OPENER.

TO AVOID DAMAGE TO GARAGE DOOR AND OPENER, MAKE DOOR LOCKS INOPERATIVE BEFORE CONNECTING ELECTRIC POWER. USE A WOOD SCREW OR NAIL TO HOLD THE LOCKS IN "OPEN" (UNLOCKED) POSITION.

INSTALLATION AND WIRING MUST BE IN COMPLIANCE WITH LOCAL ELECTRICAL AND BUILDING CODES.

OPERATION AT OTHER THAN 120V 60Hz WILL CAUSE OPENER MALFUNCTION AND DAMAGE.

Opener MUST be permanently wired or plugged into a grounded 3-prong receptacle wired according to local electrical codes. **DO NOT** use a 2-wire adapter. **DO NOT** use an extension cord.

IF LOCAL CODES REQUIRE PERMANENT WIRING:

DISCONNECT POWER AT FUSE BOX BEFORE PROCEEDING.

PROCEDURE: Refer to illustration. Make connection through the 7/8 inch diameter hole in top of opener chassis.

1. Remove opener chassis cover by removing the cover screws.
2. Remove attached 3-prong cord.
3. Connect the black (line) wire to the black wire on terminal block; white (neutral) wire to the white terminal wire; the green (ground) wire to green ground screw.

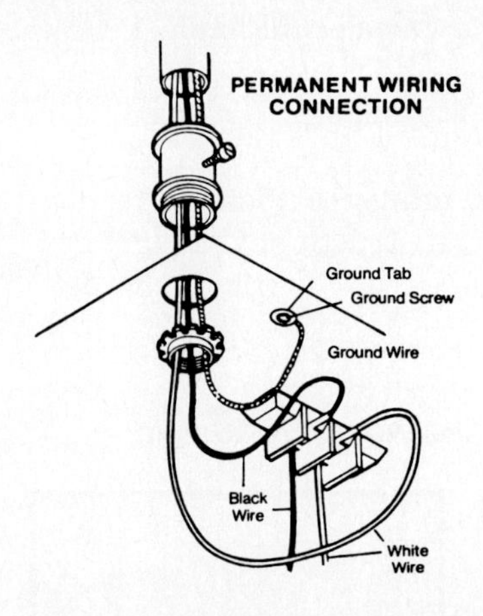

CAUTION: BE SURE THAT UNIT IS GROUNDED ACCORDING TO LOCAL CODE.

IMPORTANT NOTE: TO AVOID INSTALLATION DIFFICULTIES, DO NOT RUN OPENER NOW.

FIGURE 8–6 Point of use "caution," "important," and "STOP" messages. (Copyright 1987 by the Chamberlain Group, Inc., *Owners Manual Chamberlain Garage Door Opener System 550-6 & 350-6,* 114A1003. Used with permission)

that are available. A biweekly periodical entitled *Standards Action* is also available to help industry personnel keep up with standards that are being approved and newly published ones that are not in the annual catalogue.

ISO. For machines and information that are being developed for the international marketplace, a worldwide standards organization has been developed to help guide manufacturing companies. It is the:

International Organization for Standards (ISO)
Central Secretariat
1, rue de Varembe
CH-1211 Geneva 20
Switzerland

You can contact the American National Standards Institute (ANSI) for information concerning the ISO agency.

Industry standards. Some companies may belong to many organizations that generate industry standards. Membership usually depends upon the types of products that are produced by the manufacturing company. As a technical writer, your job can become easier if you know and are able to communicate with these people on a regular basis. New standards and requirements often mean that your technical operation instruction must be "revised" in order to keep current with industry standards.

Industry standards can tell you how safety signs should be worded, where they should be used, and often how they should be printed on your instruction. Follow these industry standards if they are available. If you use standards, then your technical operation instruction will be written in compliance with the requirements of your particular industry. Because standards are constantly changing or being revised, you must also obtain guidelines and recommendations from the safety and legal experts in your company who keep abreast of industry requirements.

Construct a Usable Visual Safety Statement

Safety information is given to the consumer or machine operator in three basic layouts. There are variations on these layouts depending upon the amount and type of information that must be presented to the consumer. They are:

Safety decals and signs *on* the machine.

Safety *panels* in a technical operation instruction that are usually placed in a general safety section at the front of a manual.

Point-of-use safety information placed in the numbered steps of an instruction.

First, both the safety decal and sign on the machine are good sources of safety data for you, but you must be able to interpret these signs. Their purpose is to visually attract the attention of the consumer. They are generally printed in red and white, or yellow and black. Other colors, such as orange, also may be used. These decals and signs are attached to the machine at the place where an unsafe situation may arise.

The basic layout of these signs includes a *picture box* where a symbol that is easy to identify is used to introduce the safety message. The safety message is usually forwarded by a *signal word* such as WARNING! or POISON!. Next is the *safety message box*, which explains the hazard in simple and easy-to-understand words. Again, the design of these safety messages will depend upon industry or government standards if they are available. Safety signs and decals are created by the machine experts. Safety information *on the machine* is an integral part of the overall machine design (see Figure 8–7).

Some technical writers reprint the entire safety sign or decal that appears on the machine as a safety panel in the technical operation instruction. Other safety panels can be developed that express the same basic information if the safety decals and signs are not usable in the instruction. These panels may be designed with a similar picture box, signal word, or safety message as the machine decal or sign. Often the signal word is eliminated and the safety message is expanded to contain further explanatory information. The method of displaying this information must be the responsibility of those who will defend the information in a court of law (Figure 8–8).

Signal word. The layout of a point-of-use safety warning in most technical operation instructions begins with the signal word. You can divide the use of signal words into two main levels of urgency. The first and most important level is an immediate or potential physical danger to the operator if a certain situation occurs. Signal words that are often used in technical operation instructions to warn the operator of a danger are:

WARNING! (see Figure 8–9)
DANGER!

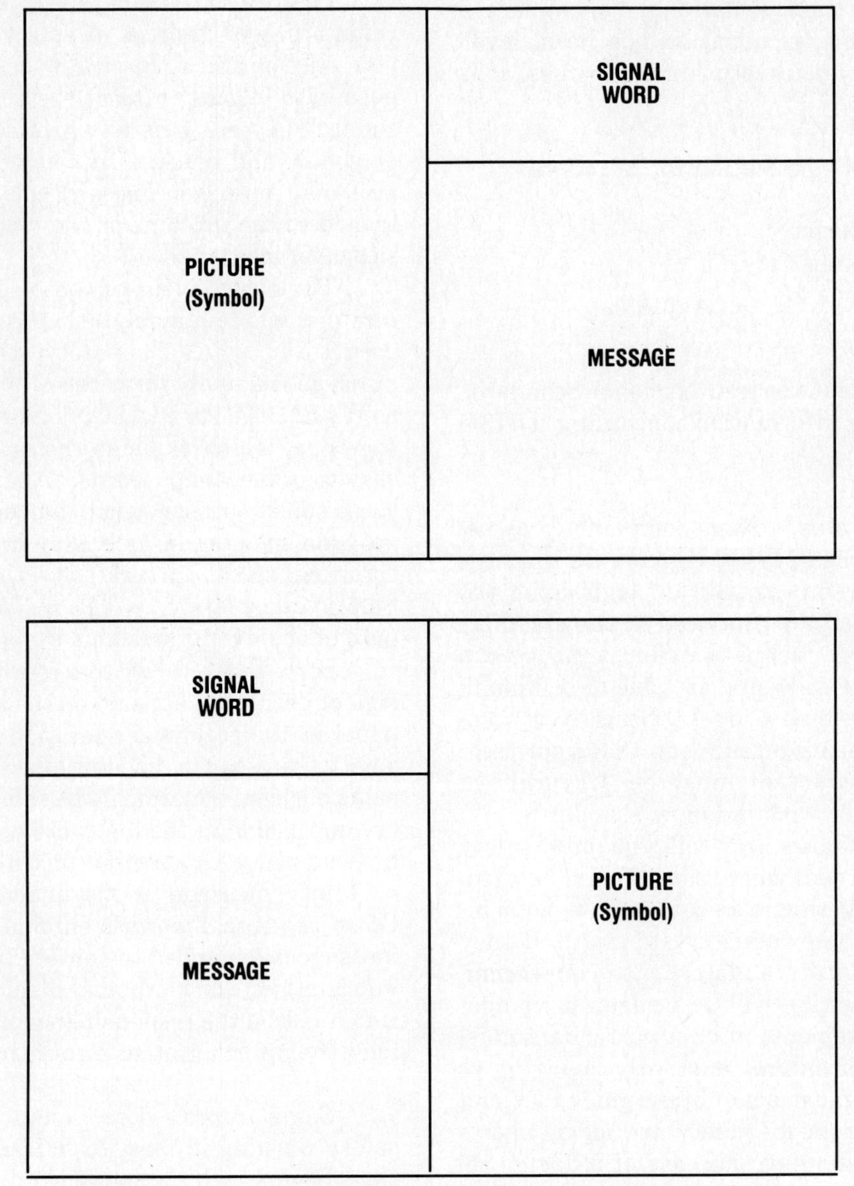

FIGURE 8–7 A basic safety sign layout includes three blocks of information.

20. Always read and follow manufacturer's instructions on packages of laundry and cleaning aids. Heed all warnings or precautions. To help prevent poisoning or chemical burns, keep them out of reach of children at all times (preferably in a locked cabinet).

21. Under certain conditions, hydrogen gas may be produced in a hot water system that has not been used for two weeks or more. HYDROGEN GAS IS EXPLOSIVE. If the hot water system has not been used for such a period, before using a washing machine or combination washer-dryer, turn on all hot water faucets in your residence and let the water flow from each for several minutes. This will release any accumulated hydrogen gas. As the gas is flammable, do not smoke or use an open flame during this time.

22. Always follow the fabric care instructions supplied by the garment manufacturer.

23. Refer to the INSTALLATION INSTRUCTIONS for the proper grounding of the washer.

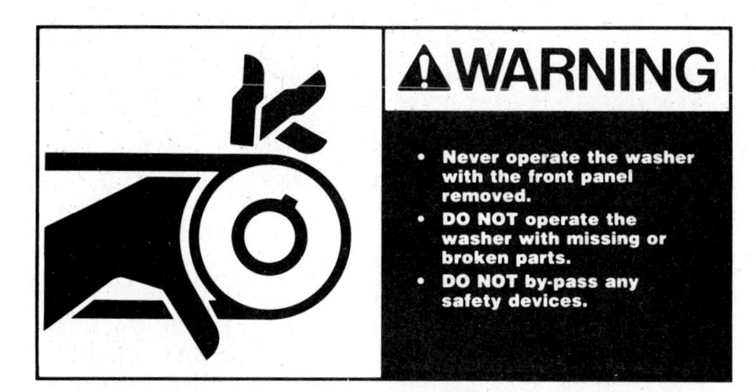

FIGURE 8–8 A safety sign can be printed in a manual. (Copyright 1987, Speed Queen, A Raytheon Company, *Operating instructions for Home Laundry Automatic Washer Model NA 3311-3059,* Part No. 31575 4/87. Used with permission)

POISON!
HALT!
STOP!
CAUTION!

The second level of urgency is to protect the machine from damage through misuse or misapplication, or to protect the work that one is performing. An example of this is the use of a note to add important information at a point in a computer sequence or operation. These signal words tend to be less dramatic. They might include:

MAINTENANCE POINT:
IMPORTANT:
NOTE: (see Figure 8–10)
STOP:
READ:

In addition to the use of a signal word, other techniques can be used to get the attention of the consumer. These include:

Safety information can be indented or set off from the regular text of the instruction by indenting it, or printing it in boldface or larger letters.

A few safety messages may be highlighted in colors (usually yellow or red).

Symbols can be placed in front of the message (Figure 8–11).

WARNING
RISK OF ELECTRIC SHOCK
DO NOT OPEN.

CAUTION: TO REDUCE THE RISK OF ELECTRIC SHOCK, DO NOT REMOVE COVER (OR BACK). NO USER-SERVICEABLE PARTS INSIDE. REFER SERVICING TO QUALIFIED SERVICE PERSONNEL.

The lightning flash with arrowhead symbol, within an equilateral triangle, is intended to alert the user to the presence of uninsulated "dangerous voltage" within the product's enclosure that may be of sufficient magnitude to constitute the risk of electric shock to persons.

The exclamation point within an equilateral triangle is intended to alert the user to the presence of important operating and maintenance (servicing) instructions in the literature accompanying the appliance.

WARNING
TO PREVENT FIRE OR SHOCK HAZARD, DO NOT EXPOSE THE UNIT TO RAIN OR MOISTURE.

NOTE TO CATV SYSTEM INSTALLER:
This reminder is provided to call the CATV system installer's attention to Article 820-22 of the NEC that provides guidelines for proper grounding and, in particular, specifies that the cable ground shall be connected to the grounding system of the building, as close to the point of cable entry as practical.

FIGURE 8–9 A "Warning" safety message. From Zenith Electronics Corporation, *Operating Guide and Warranty Zenith System 3 Color TV,* EP-EDCB 206-01228. Used with permission)

7. Improperly filed.

8. Improperly filed.

9. Unevenly filed.

NOTE: If chain has to be replaced, use chain which meets the low-kickback performance requirements established by the ANSI B175.1 (safety requirements for gasoline-powered chain saws) when tested on the representative sample of chain saws below 3.8 C.I.D. specified in ANSI B175.1.

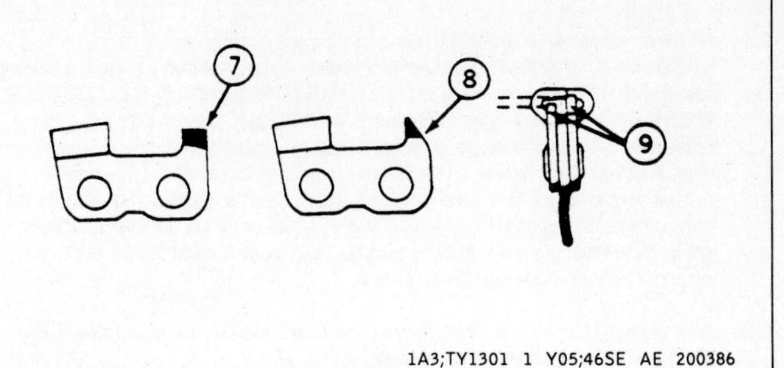

FIGURE 8–10 "Notes" also can signal important information necessary to complete a task. (Copyright 1986 by Deere & Company, *450V and 800V Chain Saws,* OM-TY20829-F6. Used with permission)

Remember, whenever you write technical operation instructions be sure the machine experts and the legal and safety departments of your company develop the layout and content of safety messages. Safety messages must fulfill the needs of the consumer. A general guideline to use to measure the usability of a safety message is that the message must *alert* the consumer to the danger. Second, safety messages must explain what the danger is and the consequences that can ensue if the message is ignored. Third, safety messages must tell the consumer how to avoid the danger.

Warnings halt the rhythm of the consumer or machine operator. What follows are a few examples of how safety messages can be written for a technical operation instruction to meet the general need of the consumer. If you are writing an instruction other than one for a classroom exercise, do not use these examples word for word. The wording of a safety warning is extremely important and must be developed by your safety experts for a *particular* machine design. In the examples below, note the use of heavier type and indentation. These devices break up the rhythm of the consumer as he or she reads the instruction or operates the machine.

WARNING! *ACETYLENE GAS IS HAZARDOUS. TO PREVENT AN EXPLOSION,* TURN ON THE GAS VALVE BEFORE YOU TURN ON THE OXYGEN VALVE.

STOP! Prevent accidents. Read all safety messages carefully before you operate the machine.

POISON! **BATTERIES CAN BLOW UP IF MISHANDLED.** ***BATTERY ACID CONTAINS SULFURIC ACID THAT CAUSES SEVERE BURNS. AVOID CONTACT WITH YOUR SKIN AND EYES. WEAR PROTECTIVE CLOTHING.***

DANGER! GASOLINE IS HIGHLY FLAMMABLE. STORE IN PROPER CONTAINERS. HANDLE WITH CARE.

CAUTION! **Do not remove shields or protective guards while the machine is ON. DO not operate the machine with guards and shields removed. Failure to do so can result in bodily injury.**

HALT! DO NOT CHARGE A FROZEN BATTERY. IT WILL EXPLODE.

IMPORTANT: USE ONLY 10W-40 WEIGHT OIL WITH AN SAE CLASSIFICATION. INCORRECT OIL GRADE CAN DAMAGE THIS HIGH-PERFORMANCE ENGINE.

NOTE: Keep the machine clean. Remove dirt and oil to keep the machine running efficiently.

MAINTENANCE POINT: Keep all fittings at 50 lbs. of torque. Improper torque will damage delicate machine parts.

A "Ten Commandment" effect can be created if every safety message begins with "Do not. . . ." This phrase actually may be difficult to translate into certain languages other than English. Vary the statement, but keep it as simple and descriptive as possible. The effect of safety messages should be that they make consumers hesitate in their normal action of operating the machine and to read and think about the safety information that you have presented.

In general, point-of-use safety messages are used to warn consumers that it is important to follow the tasks exactly as they are written. If the tasks are not followed, then an unsafe condition will exist that may harm the operator or the machine.

FUEL

CAUTION: Handle fuel carefully. If the engine is running, do not fill the fuel tank. If engine is hot, let engine cool several minutes before you add fuel. Do not smoke while you fill the fuel tank or service the fuel system. Fill fuel tank only to bottom of filler neck.

IMPORTANT: DO NOT mix oil with gasoline.

Unleaded fuel is recommended. Regular leaded gasoline with an anti-knock index of 87 or higher may be used. Avoid switching from unleaded to regular gasoline to prevent engine damage.

Use of gasohol is acceptable as long as the ethyl alcohol blend does not exceed 10 per cent. Unleaded gasohol is preferred over leaded gasohol.

Fuel tank capacity is 2-1/2 gal (9.5 L).

Lift seat. Fill fuel tank at end of each day's operation.

A98;TS185 M21;;FLS A 040985

FIGURE 8–11 Safety messages can be highlighted with symbols and type variation. (Copyright 1985 by Deere & Company, *130, 160, 165, 180, and 185 Lawn Tractors,* OM-M89684-K5. Used with permission)

The most important concern of the technical writer is the safety of the operator. For that reason, safety messages must be clearly written in short, active sentences that are easy to read and understand by the operator. They must be laid out or designed *consistently* throughout the text of the technical operation instruction.

As a technical writer, you must remember that operator safety is the responsibility of many people in the manufacturing process. The technical writer must report these safety facts to the operator. Be sure that your safety and legal experts approve the exact wording of each message. Every word is extremely important when it appears in a safety message. One wrong or poorly chosen word in one safety message can trigger an accident and can result in a costly liability judgment against your company.

Use Artwork to Depict Safe Operation

The safety message that you include in your technical operation instruction is not complete if you do not depict the consumer or operator of the machine in safe situations throughout the tasks. Artwork is as important in your attempt to get your audience to follow safety messages as are the words that you use. All artwork must be scrutinized to ensure it "illustrates" safety at all times. Always show the operator of a machine using the correct protective clothing such as ear, eye, face, and body wear. Make sure that "specialized" safety clothing is up to date and properly worn. When you show someone working with a chainsaw, for example, show the person wearing all of the necessary ear, eye, and hand protective clothing that is necessary to keep the user as safe as possible (Figure 8-12).

Another related point is that the operator of a machine must always be shown in the correct spatially logical position. The individual depicted in a task must always be shown in the safest position possible. The safety message that you may write to comply to an industry standard or a company legal department requirement can be undermined by artwork that shows unsafe practices.

Many machines are designed to be operated only with protective guarding in place. Photographs and line drawings that show the operator controlling a machine must have these guards in place. Always show the person in your pictures doing the job safely and correctly.

Much debate focuses on the use of cartoons to show the results of improper actions. For example, a negative cartoon would show an operator in a painful situation. The hoped-for reaction would be that the operator would be humorously educated about safety procedures. Companies who use cartoons claim that this is an effective way to get the consumer's attention. Companies that do not use cartoons say that cartoons show a negative action that does not positively reinforce the consumer or machine operator. Both arguments may have merit. A suggested guideline would be to eliminate safety cartoons from all technical instructions.

Develop Your Safety I.Q.

If you are reviewing safety messages you must be aware of the major safety procedures to be followed when someone works with a machine power system. *Every machine design is unique. Each machine requires that the manufacturing company develop an individual set of safety messages for the consumer.*

PROTECT PEOPLE AND PETS

Do not allow other people near the chain saw when starting or cutting.

Keep people and pets out of work area.

Do not let children or untrained persons operate the chain saw.

Never leave saw unattended.

Never hand the saw to another person while the engine is running.

FIGURE 8-12 Always show the consumer or operator wearing the correct clothing and in the correct position. (Copyright 1986 by Deere & Company, 450V and 800V Chain Saws, OM-TY20829-F6. Used with permission)

The following lists of safety facts are presented as general information. As you review machine safety messages to make sure they are complete, you will find that not all of the following safety information may apply to your machine. You may also find that your machine requires additional safety information beyond that contained in the following lists. In any case, you must be aware of these safety guidelines when you operate a machine or trace machine systems.

GENERAL SAFETY INFORMATION

Wear tight-fitting clothing if there is a danger of getting clothing entangled in machine parts.

When working near machines, do not wear ties or scarves that are tied around the neck.

Remove all jewelry when working near machines.

Car tires and other machine parts under pressure are hazardous if improperly handled.

Wear safety goggles and other eye protection if there is a danger of impact and flying material.

Wear a hardhat or other protection if there is a danger of falling objects.

Wear gloves that protect hands against sharp objects, abrasive material, or other destructive agents.

Wear steel-toed shoes if there is a possibility of objects falling on feet.

Wear a respirator, mask, or other protection when working around chemicals or harmful machine emissions.

Keep long hair tucked up under a hat or secured in some way when working near a machine, especially one with a mechanical power system.

Keep the work area free of clutter, spilled oil and water, and other objects that can cause falls.

Use proper lifts, hoists, and slings to lift heavy loads. Heavy parts such as liquid-filled large tires can crush.

Use the correct tools for the job. Use metric tools for metric hardware and standard tools for standard hardware. Do not use hand tools improperly, such as a screwdriver for a chisel.

Before operating the machine, always read as much as possible about the potentially dangerous situations that can be created on a machine.

Properly block and secure a machine before entering it for maintenance or to trace machine systems. Prevent accidental moving or shifting of the machine. For example, block an automobile properly before removing a tire.

Keep all protective guarding and covers on a machine when it is being cycled or work is being done. Guarding and covers that are removed can create an unsafe situation.

All machine power systems must be properly shut off and locked before entering the machine for any reason.

All machine decals and signs that give warnings or information must be clean and visible at all times. If a decal or sign is damaged or cannot be read, make sure that it is properly replaced before working on or with a machine.

All system internal protection, such as electrical system fuses and circuit breakers, must be properly rated to handle system power correctly.

Know all machine control points and operator controls. Always be in total control of the machine at all times.

BASIC HYDRAULIC AND PNEUMATIC SYSTEM SAFETY INFORMATION

Before working on a hydraulic or pneumatic system, be sure to drain it of all pressure. Sudden and unexpected machine movement can occur if pressure is not drained and someone breaks into the system by removing a coupling or similar item.

Certain types of hydraulic fluid or lubricated air under pressure can cause gangrene if it enters the skin and is not removed by a surgeon familiar with the procedure. Never feel a hydraulic or pneumatic line for leaks. Pinhole leaks can be invisible. Keep all hose and pipe in good condition. Keep all couplings properly connected.

Keep all hose and pipes clean. Usually they are a major point of heat dissipation for the machine.

Hydraulic fluid can become extremely hot.

BASIC ELECTRICAL SYSTEM SAFETY INFORMATION

Capacitors in such electrical machines as a television can cause a severe shock long after the electrical power is turned off. Do not attempt to work on an electrical circuit unless you are familiar with maintenance procedures.

Electric current becomes greater in water. Never use an electrical machine such as a radio or hair dryer near water.

Bare wires and short circuits are dangerous. Visually check for them before working on an electrical system. Never attempt to short-circuit an electrical system.

Batteries are hazardous and can explode if the gas they give off is ignited.

BASIC MECHANICAL SYSTEM SAFETY INFORMATION

Watch out for and avoid mechanical pinch-points and crush-points.

Mechanical systems may also have rotating blades and other parts that can sever limbs.

Do not loosen a part of a mechanical system that is spring-loaded and has tension in it. Parts may fly outward and cause severe injury.

Moving chains, belts, fans, and rotating shafts can cause injury if they are not properly guarded.

UNDERSTAND THE LANGUAGE OF WARRANTY

When a machine is sold by a salesperson in a retail store, a receipt is given to, and often signed by, the customer who buys the machine. The salesperson, store, or dealership is acting as a sales representative for the company that manufactured the machine. The sales receipt, which contains the price of the machine and the date sold, becomes a contract. This contract gives the consumer certain legal rights that vary from state to state and product to product.

At the time of sale, the machine manufacturer must also include or make available a printed warranty. This document is made available to the consumer in several ways. A warranty is often packaged as a separate document with the machine. Or, it can be a printed part of the sales slip or receipt. Sometimes it is printed as a document in the technical operation instruction (see Figures 8–13 and 8–14).

This warranty documents the responsibility of both the consumer and the machine manufacturer. The warranty explains what the consumer must do to keep the warranty in effect. It also tells consumers exactly what support and backup they can expect from the machine manufacturer. These two points are not the same for every machine. Maintenance and replacement of certified parts may be listed as a responsibility of the consumer.

The part of the warranty that the consumer is usually interested in is the guarantee that the machine will be free from defects in material and workmanship for a certain length of time. The warranty states that any breakdowns or systems problems on the machine will be fixed by the manufacturing company during this specific time period.

A typical warranty will tell the consumer the length of time that the manufacturer will guarantee a warranty. This varies from product to product. For example, a computer manufacturer may extend a warranty to 90 days for parts and labor. A new automobile may have a comprehensive 12,000-mile or 1-year warranty, and an extended limited warranty for six or seven years after the purchase. A certain amount of warranty costs to the machine manufacturer is added to the selling price of the machine. These warranties are in fact paid for by the consumer. Often the warranty is created by the legal department with the cooperation of the service department.

There are situations when a technical operation instruction may add extra costs to the product. By saying too much or by saying the *wrong* thing, the technical writer can place the company in a precarious position. Let's look at three case studies that show how the technical writer can create confusion and add cost to the company warranty program:

Case study 1:

The Universal Lawnmower Company manufactures many types of walk-behind lawnmowers. The company recently released a new model into the marketplace. Sally Brown, Universal's technical writer, wrote the instructions. After she completed the first draft of her instruction, she sent it to the engineering, legal, and safety departments for review and comments.

None of her reviewers found a small mistake that Sally made in her specification chart, namely, that she listed the wrong type of oil to be used. The customer is told to change the break-in oil and add new oil to the engine crankcase after the first two hours of operation. The oil that she tells the customer to use is a type

TERMS AND CONDITIONS OF SALE AND LICENSE OF RADIO SHACK AND TANDY COMPUTER EQUIPMENT AND SOFTWARE PURCHASED FROM A RADIO SHACK COMPANY-OWNED COMPUTER CENTER, RETAIL STORE OR FROM A RADIO SHACK FRANCHISEE OR DEALER AT ITS AUTHORIZED LOCATION

LIMITED WARRANTY

I. CUSTOMER OBLIGATIONS

A. CUSTOMER assumes full responsibility that this computer hardware purchased (the "Equipment"), and any copies of software included with the Equipment or licensed separately (the "Software") meets the specifications, capacity, capabilities, versatility, and other requirements of CUSTOMER.

B. CUSTOMER assumes full responsibility for the condition and effectiveness of the operating environment in which the Equipment and Software are to function, and for its installation.

II. RADIO SHACK LIMITED WARRANTIES AND CONDITIONS OF SALE

A. For a period of ninety (90) calendar days from the date of the Radio Shack sales document received upon purchase of the Equipment, RADIO SHACK warrants to the original CUSTOMER that the Equipment and the medium upon which the Software is stored is free from manufacturing defects. **This warranty is only applicable to purchases of Radio Shack and Tandy Equipment by the original customer from Radio Shack company-owned computer centers, retail stores and from Radio Shack franchisees and dealers at its authorized location.** The warranty is void if the Equipment's case or cabinet has been opened, or if the Equipment or Software has been subjected to improper or abnormal use. If a manufacturing defect is discovered during the stated warranty period, the defective Equipment must be returned to a Radio Shack Computer Center, a Radio Shack retail store, participating Radio Shack franchisee or Radio Shack dealer for repair, along with a copy of the sales document or lease agreement. The original CUSTOMER'S sole and exclusive remedy in the event of a defect is limited to the correction of the defect by repair, replacement, or refund of the purchase price, at RADIO SHACK's election and sole expense. RADIO SHACK has no obligation to replace or repair expendable items.

B. RADIO SHACK makes no warranty as to the design, capability, capacity, or suitability for use of the Software, except as provided in this paragraph. Software is licensed on an "AS IS" basis, without warranty. The original CUSTOMER'S exclusive remedy, in the event of a Software manufacturing defect, is its repair or replacement within thirty (30) calendar days of the date of the Radio Shack sales document received upon license of the Software. The defective Software shall be returned to a Radio Shack Computer Center, a Radio Shack retail store, participating Radio Shack franchisee or Radio Shack dealer along with the sales document.

C. Except as provided herein no employee, agent, franchisee, dealer or other person is authorized to give any warranties of any nature on behalf of RADIO SHACK.

D. **Except as provided herein, Radio Shack makes no express warranties, and any implied warranty of merchantability or fitness for a particular purpose is limited in its duration to the duration of the written limited warranties set forth herein.**

E. Some states do not allow limitations on how long an implied warranty lasts, so the above limitation(s) may not apply to CUSTOMER.

III. LIMITATION OF LIABILITY

A. **Except as provided herein, Radio Shack shall have no liability or responsibility to customer or any other person or entity with respect to any liability, loss or damage caused or alleged to be caused directly or indirectly by "Equipment" or "Software" sold, leased, licensed or furnished by Radio Shack, including, but not limited to, any interruption of service, loss of business or anticipatory profits or consequential damages resulting from the use or operation of the "Equipment" or "Software". In no event shall Radio Shack be liable for loss of profits, or any indirect, special, or consequential damages arising out of any breach of this warranty or in any manner arising out of or connected with the sale, lease, license, use or anticipated use of the "Equipment" or "Software". Notwithstanding the above limitations and warranties, Radio Shack's liability hereunder for damages incurred by customer or others shall not exceed the amount paid by customer for the particular "Equipment" or "Software" involved.**

B. RADIO SHACK shall not be liable for any damages caused by delay in delivering or furnishing Equipment and/or Software.

C. No action arising out of any claimed breach of this Warranty or transactions under this Warranty may be brought more than two (2) years after the cause of action has accrued or more than four (4) years after the date of the Radio Shack sales document for the Equipment or Software, whichever first occurs.

D. Some states do not allow the limitation or exclusion of incidental or consequential damages, so the above limitation(s) or exclusion(s) may not apply to CUSTOMER.

IV. RADIO SHACK SOFTWARE LICENSE

RADIO SHACK grants to CUSTOMER a non-exclusive, paid-up license to use the RADIO SHACK software on **one** computer, subject to the following provisions:

A. Except as otherwise provided in this Software License, applicable copyright laws shall apply to the Software.

B. Title to the medium on which the Software is recorded (cassette and/or diskette) or stored (ROM) is transferred to CUSTOMER, but not title to the Software.

C. CUSTOMER may use Software on one host computer and access that Software through one or more terminals if the Software permits this function.

D. CUSTOMER shall not use, make, manufacture, or reproduce copies of Software except for use on **one** computer and as is specifically provided in this Software license. Customer is expressly prohibited from disassembling the Software.

E. CUSTOMER is permitted to make additional copies of the Software **only** for backup or archival purposes or if additional copies are required in the operation of **one** computer with the Software, but only to the extent the Software allows a backup copy to be made. However, for TRSDOS Software, CUSTOMER is permitted to make a limited number of additional copies for CUSTOMER'S own use.

F. CUSTOMER may resell or distribute unmodified copies of the Software provided CUSTOMER has purchased one copy of the Software for each one sold or distributed. The provisions of this Software License shall also be applicable to third parties receiving copies of the Software from CUSTOMER.

G. All copyright notices shall be retained on all copies of the Software.

V. APPLICABILITY OF WARRANTY

A. The terms and conditions of this Warranty are applicable as between RADIO SHACK and CUSTOMER to either a sale of the Equipment and or Software License to CUSTOMER or to a transaction whereby RADIO SHACK sells or conveys such Equipment to a third party for lease to CUSTOMER.

B. The limitations of liability and Warranty provisions herein shall inure to the benefit of RADIO SHACK, the author, owner and/or licensor of the Software and any manufacturer of the Equipment sold by RADIO SHACK.

VI. STATE LAW RIGHTS

The warranties granted herein give the **original** CUSTOMER specific legal rights, and the **original** CUSTOMER may have other rights which vary from state to state.

8J83

FIGURE 8–13 A warranty statement can be printed on the inside of pages of a manual. (Copyright by the Radio Shack division of Tandy Corporation, *Deluxe Graphics Display Adapter User's Manual,* Cat. No. 25-3047. Used with permission)

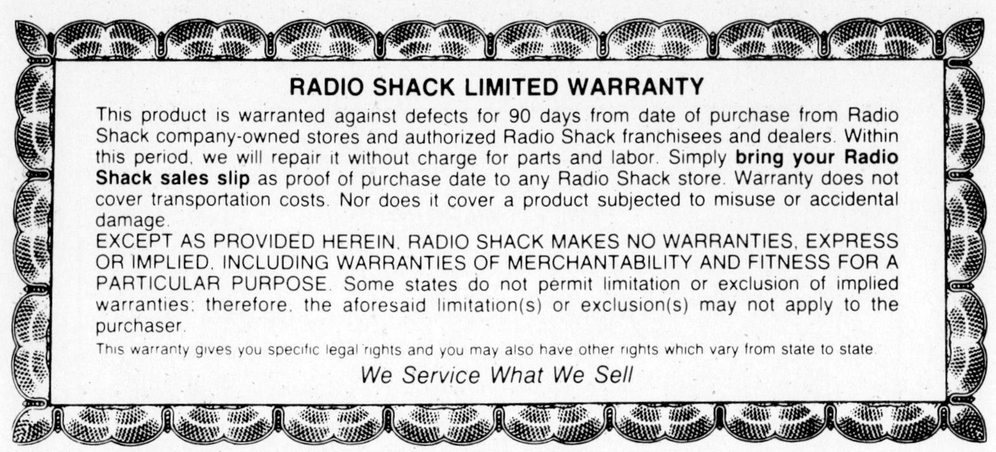

RADIO SHACK LIMITED WARRANTY

This product is warranted against defects for 90 days from date of purchase from Radio Shack company-owned stores and authorized Radio Shack franchisees and dealers. Within this period, we will repair it without charge for parts and labor. Simply **bring your Radio Shack sales slip** as proof of purchase date to any Radio Shack store. Warranty does not cover transportation costs. Nor does it cover a product subjected to misuse or accidental damage.

EXCEPT AS PROVIDED HEREIN, RADIO SHACK MAKES NO WARRANTIES, EXPRESS OR IMPLIED, INCLUDING WARRANTIES OF MERCHANTABILITY AND FITNESS FOR A PARTICULAR PURPOSE. Some states do not permit limitation or exclusion of implied warranties; therefore, the aforesaid limitation(s) or exclusion(s) may not apply to the purchaser.

This warranty gives you specific legal rights and you may also have other rights which vary from state to state.

We Service What We Sell

FIGURE 8–14 A warranty statement can be printed on the back cover of a manual. (Copyright by the Radio Shack division of Tandy Corporation, *Plug'n Talk FM Wireless Intercom Owner's Manual,* Cat. No. 43-207B. Used with permission)

of detergent oil that damages the rings on the pistons of the lawnmower engine. The company's retail outlet stores are being flooded by returned mowers that have piston ring damage. All of the mowers are covered under the company's warranty.

The Universal Lawnmower Company is trying to track down the location of the lawnmowers sold with the incorrect oil specification. Trying to find all of the machines and explain the error to the customers who bought them is an expensive operation. The company, during this search, is responsible for the cost of fixing the damaged lawnmowers at the retail outlets where the mowers are sold.

Many machine manufacturers keep records that tell who buys the machine. A registration card may be put inside the technical operation manual for mass- produced items that are difficult to trace. The manufacturer uses these cards to help locate the machine in the workplace. A manufacturer must be able to "recall" or modify a machine that has been sold when a known defect exists. A registration card as well as a warranty card may be inserted or bound as part of a direction or manual.

Case study 2:

Jim Jacobs is a technical writer for a company that manufactures microwave ovens. His company has developed a new model with a few advanced features on it. Since it is Jim's first attempt at writing technical operation instructions, he is excited. He wants to work alone so that he can take full credit for the results of his work and impress as many people at his company as he can.

In his enthusiastic attempt to give his company a pat on the back for developing such an advanced machine, he makes a costly *implied* warranty. He implies that the microwave oven is maintenance free. He puts this statement at the top of his instructions:

Congratulations for buying the new Model Micr-Z56 microwave oven. It is a remarkable machine. Due to its self-contained design, it *never* needs maintenance. It will last forever! It is the safest machine on the market.

Jim is *trying* to say that the advanced design of the microwave oven is extremely dependable and that all maintenance to the machine *must* be done by a trained service person, and *not* by the consumer. He makes a serious error, however, by trying to resell the microwave oven to the customer. Jim's *implied warranty* costs his company after the original 1-year limited warranty that was intended to cover the machine. He says that the microwave oven will last forever. Now his company is trying to settle legal battles with irate customers who expect their machine to last as long as Jim has implied. Jim's biggest mistake, however, was to tell customers that they have the safest machine that can be purchased. This statement can give consumers a false sense of security and cause them to ignore important safety procedures.

Case study 3:

Warranty problems can also be created by technical operation instructions that are too abstract. Mike Munson writes a technical operation instruction for his company's new garden tiller. A break-in period check and adjustment is extremely important. A necessary item that must be checked is the drive belt that turns the tines of the tiller. In one of the tasks, Mike tells the consumer how to adjust the belt after the first hours of operation. He says:

1. Shortly after beginning to operate the tiller, check the drive belt.
2. Tighten the drive belt as needed.

The problem that some of his consumer operators are faced with is that it is too easy to overtighten the belt. Many of the belts are fraying and snapping. The belts are wrapping so tightly around the drive shaft that they are damaging the tiller gear drive. Mike should have told his consumer audience the exact way to tighten the belt. If the tiller gears aren't put in neutral, then the belt can easily be tightened too much. He should also have told consumers how to measure the belt so that they can get the belt to the correct tightness. Mike might have said:

CHECK AFTER FOUR HOURS OF OPERATION
IMPORTANT: Put the transmission shift in "neutral."
CAUTION: NEVER ADJUST A BELT THAT IS IN OPERATION.

1. Check the drive belt.
2. Belt tension must not be greater than five pounds when it is pulled one inch.
3. To adjust, loosen the drive pulley bolt.
4. Adjust the pulley, and tighten the bolt.
5. Check the tension.

Mike also could increase the usability of his instructions by illustrating the technique (Figure 8–15).

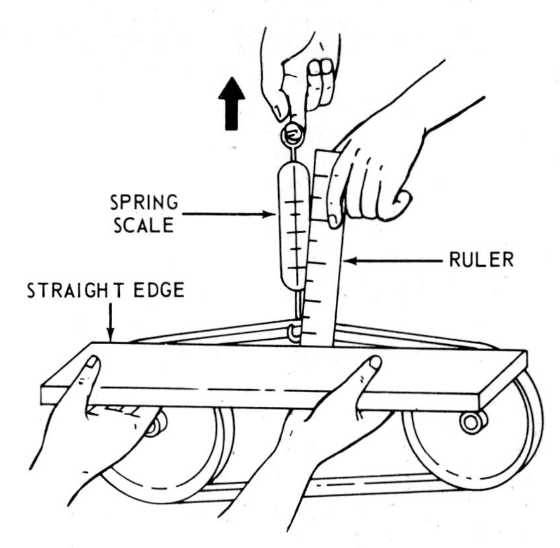

FIGURE 8–15 Illustrations can help reduce ambiguity and create concrete tasks that affect warranty. (Courtesy Deere & Company)

Every word and every piece of artwork in a technical operation instruction is important. Every task and statement of information must be carefully analyzed. Your technical operation instructions must not be the defect that turns your company's well-developed machine design into an inefficient and undependable warranty problem.

Your goal as a technical writer is not to restate the warranty claims, but to tell your consumer/reader how to live up to his or her part of the warranty agreement. Simple maintenance, adjustments, and operation procedures must explain exactly what must be done to derive maximum performance from the machine. The above example is an illustration of an adjustment that is critical. Not all adjustments need this extensive information. If there is a critical adjustment, make sure that you explain the complete process in detail.

Know Warranty Subjects

The warranty document will cover all legal concepts that deal with the consumer and manufacturing company contract. A comprehensive warranty will have the following basic information in it:

The name of the company that provides the warranty.

When the warranty begins and ends.

Limitations. An automobile warranty, for example, may cover a radio for 90 days and a drive train for 5 years. A computer warranty may cover the central processing unit for one year, the keyboard and the CRT (or screen) for 90 days.

Exactly what the warranty covers and exactly what the warranty does not cover. For example, are a computer hard drive and disk drives covered?

Where the machine can be fixed at the manufacturing company's expense.

Geographical limitations.

Any situations that can negate the warranty, such as modifying an engine to perform at a higher peak than it was designed to do.

Who pays for shipping or other transportation costs incurred to return the machine to the manufacturer or an authorized repair center.

Warranties are also often cited in liability cases. If a machine does not live up to the warranty that is included at the time of sale and the consumer is injured, then the manufacturer may be liable. A technical operation instruction must never attempt to restate the basic company warranty. Often the result is an implied or expressed statement that confuses or misleads the consumer and creates problems for the manufacturing company.

UNDERSTAND THE LANGUAGE OF MEASUREMENT AND SPECIFICATIONS

Another language that you will use in your technical operation instruction is the language of *measurement.* Measurement is an important part of the manufacturing process. The people who design and build the machine must know exactly what the tolerances are when they make and fit pieces of the machine together. Interchangeable parts in the assembly line and automated systems must be accurately manufactured to demanding specifications.

Measurement facts in a technical operation instruction must be kept to a minimum. Measurements that may be important to the consumer are the overall dimensions of a large machine. The consumer often must know the physical dimensions and weight of the machine in order to fit it into a work space. Measurements are based on comparisons of a national or international standard. They can be expressed in many different ways, including inches, millimeters, pounds, kilograms, oil viscosity, antifreeze grades, and watts of power output. Important information usually supplied to consumers include the following:

Fuel rating and quality (octane, diesel grade).

Lubrication (weight, quantity, content).

Work limits of the machine (how fast and how slow it can go).

How much work a machine should and should not do, such as how much dirt can an excavator lift, or what load weight can be pulled by a tractor or hitch?

How much fuel and lubrication a machine needs to do the job.

Physical dimensions of certain key parts in case they must be replaced.

Specifications. In a technical operation instruction, measurements are part of the machine specifications. Specifications, however, include the parts ratings that enable consumers to replace parts that need replacing. Specifications are usually listed in a separate section apart from the tasks in a technical operation instruction. Usually they are placed on the last section of a direction or the last page of a manual. For example, a specification in an instruction might begin like the following example (see Figure 8–16):

SPECIFICATIONS FOR THE MODEL
270ZX GRADER LEVELER

LUBRICATION
Oil viscosity.. 15w-30
Oil reservoir size.. 21 oz.
Oil grade ....SE, CC
ELECTRICAL
Front warning signal fuse 6.25z
. . . .

As a technical writer, you must remember one important point when you list specifications. Unless necessary, do not list a specific replacement product such as a manufacturer's name-brand oil. Instead, list the rating or blend of oil. All oil, no matter who refines it, can be used in the machine as long as it meets your company's requirements and is manufactured to the specifications of your machine.

For example, do *not* say:

Crankcase oil.........................International Oil Company, Super Solar Blend.

Rather, list the measurement or specification that can be applied to all oils. As long as the manufacturer makes an oil that meets specifications, customers will have a choice of products. List specifications like this:

Crankcase oil.........................API/SF/or SF/CC or SF/CD

There usually are cross-references for all items

Specifications

450V

Power Head

Engine	Air-cooled, two stroke, single cylinder
Displacement	45 cc (2.75 cu in.)
Bore	43.0 mm (1.693 in.)
Stroke	31.0 mm (1.220 in.)
Carburetor	Diaphragm type
Fuel	Regular (leaded) gasoline and 2-cycle engine oil. See Fuels and Lubricants section for proper mixture
Spark Plug	AM54611
Spark Plug Gap	0.6—0.7 mm (.024—.028 in.)
Spark Plug Torque	14—15 N·m (125—135 lb-in.)
Engine RPM (Under Load)	10500 rpm
Fuel Tank Capacity	0.45 L (15.2 fl oz)
Oil Tank Capacity	0.28 L (9.5 fl oz)

Guide Bar and Chain

Guide Bar Length	406 mm (16 in.) 457 mm (18 in.)
Guide Bar Type	Nonsymmetrical, fabricated, sprocket nose
Chain Pitch	8.26 mm (.325 in.)
Chain Gauge	1.27 mm (0.50 in.)
Chain Type	33 SL
Sprocket	7-Tooth Rim Sprocket
Chain Oiler	Manual and Automatic
Chain Lubricant	Refer to Fuels and Lubricants section

(Specifications and design subject to change without notice.)

Y05;;69SP A 260686

FIGURE 8-16 A specification page is usually in the back of a manual. (Copyright 1986 by Deere & Company, *450V and 800V Chain Saws,* OM-TY20829-F6. Used with permission)

that are not measured against a standard. For example, a bulb that is used in a car turn signal light must be listed in your instruction with a statement similar to the following:

Turn signal bulb AZ1243

The AZ1243 can be cross-referenced with any bulb manufacturer that provides a cross-reference guide. You also can make a statement such as this: "Use the ABC Company bulb number AZ1243 or equivalent." When specifications are used you should give the consumer a choice in determining replacement parts.

Understand Metric Measurements

Machines sold in a worldwide marketplace also must have metric measurement conversions along with the standard inch/pound specifications. When you write technical operation instructions for an English-speaking audience that uses the metric system, such as Great Britain, you must supply the consumer with metric measurements and specifications. You can combine the two measurement systems on the same instruction by putting them side by side. The important point is that the measurements and specifications must be given in the same style throughout the instruction. Do not put the metric equivalent first on one page and then put it second on the next.

A rule of thumb to follow might be to put metric measurements first in an instruction that is written for a machine designed with metric hardware and metric specifications. For standard-designed machines with foot and inch dimensions and hardware such as standard bolts and screws, place the foot and inch measurements first. Examples of this are the following specifications that give the standard ounce (oz.) and the metric liter (L) measurements:

1. Add 21 oz (0.6 L) of 15W-30 oil.

or

1. Add 0.6 L (21 oz.) of 15W-30 oil.

Ounces and liters are dual measurements that are equivalent. Measurements given in a technical operation manual are usually abbreviated to save space. The following dual measurements are common comparisons:

STANDARD	METRIC
miles per hour (mph)	kilometers per hour (kph)
feet (ft)	meters (m)
inch (in)	millimeters (mm)
pounds (lb)	kilograms (kg)
pounds per square inch (psi)	kilo Pascals (kPa)

Abbreviations are usually written in lower case letters such as millimeter (mm) unless part of the measurement is also part of a name such as kiloPascals (kPa). Liter (L) is usually an exception because the lower case "l" can be confused with "1." Kilowatt (kw) is another common exception.

Some things cannot be compared. They are like apples and oranges in a sense that they are completely different. Metric and standard screw and bolt threads are examples of this. Only *like* things can be converted. However, some measurements are universal and do not require a conversion. Examples of these are:

revolutions per minute (rpm)
degrees (amount of rotation)
decibels (db)

Every company should have a style sheet that explains how measurement should be presented in technical literature developed by company personnel. This style sheet should tell you how many decimal points to carry out the conversion from metric to standard or standard to metric. Metrics are extremely difficult to measure more than two places to the right of a decimal point.

Use Common Metric Conversions

1 inch (in) ....25.4 millimeters (mm)
1 cubic inch ....16.387 cubic centimeters
1 foot (ft).............................0.305 meter (m)
1 U.S. gallon (gal) ....3.785 liters (L)
1 U.S. quart (qt) ....0.946 liter (L)
1 U.S. pint (pt) ....0.473 liter (L)
1 U.S. ounce (oz)...................29.573 milliliters (ml)
1 U.S. gallon ....0.8 Imperial gallon
1 pound (lb) ....0.454 kilogram (kg)
1 square inch........................6.452 square centimeters
1 square foot.........................0.093 square meter
1 horsepower (hp).................0.746 kilowatt (k)
1 pound of pressure per square inch (psi) ....6.895 kiloPascals (kPa)
1 degree Fahrenheit (F)..........0.556 degrees Celsius (C)
1 pound of torque force per square inch ....0.113 Newton meters (N.m)
1 mile per hour (mph) ....1.609 kilometers per hour

Understand the Use of Common-Sense Measurement in Instructions

The heterogeneous mass-consumer audience is made up of people from all walks of life. They cannot be expected to have sophisticated calibration or measurement instruments to keep their machines in specification. For example, it would be useless to tell the consumer to gap a spark plug 0.1265 inches. First of all, there are no instruments outside of a metrology laboratory that could read this measurement on a spark plug. Your consumer/operator normally has access to gauges that measure only two places to the right of the decimal, such as a spark plug feel gauge.

The basic measurement tools of the consumer might be limited to the following:

A 12-inch ruler or tape measure that is accurate up to 1/16 inch. The human eye without the aid of magnification can only read a measuring device up to 1/32 inch accurately.

A measuring cup accurate to 1 ounce.

Feeler gauges accurate to hundredths of an inch.

A protractor accurate to whole degrees.

A weight scale accurate to ounces.

These measuring devices are good for quick and approximate measurements. Any adjustment or service task that requires a measurement more accurate that the consumer can make with available instruments must be done by a service technician equipped with the proper measuring devices.

A deviation from the above guideline can be made if the technical operation instruction is written to a homogeneous audience. In this situation, the consumer may have special knowledge that allows him or her to use more accurate measurement tools such as optical flats, calipers, micrometers, and other higher amplification measurement instruments. These more sophisticated instruments are not found in the "average" consumer's garage or tool box. Hence, you must be aware of the exact measurement devices that your audience can use.

The final concept of measurement language that you must become familiar with is "tolerance." You can define tolerance as a variation. Tolerance is the difference between the minimum and maximum size that a part must be manufactured in order to be used as a replacement on the original machine. Usually tolerances are found in instructions written for a homogeneous audience. Tolerance is measured as a plus or minus. If the part fits between the allowable maximum and allowable minimum measurement, then it can be used.

CHAPTER SUMMARY

- Safety is the most important word in the vocabulary of the technical writer.
- The technical writer has a dual responsibility: to instruct the consumer in safety, and to help reduce liability claims against the company.
- The technical writer must try to force the consumer to be conscious of safety messages at all times during machine operation.
- There are many machine experts who must be collaborators with the technical writer on safety messages.
- The technical writer is in position to continuously review the entire company safety program for inconsistency and incomplete safety information.
- Product liability means that the consumer has a legal right to attempt to recover damages from a manufacturer if a consumer/operator believes that the manufacturer is at fault.
- Liability can result from a failure to warn or instruct.
- Liability laws have developed individually in each state of the Union.
- Liability can be expensive to everyone. Liability costs must be included in the selling price of the machine.
- There is a liability crisis for the machine manufacturer.
- There are two media for safety messages: the decals and signs on the machine, and the technical operation instruction.
- Safety messages in an instruction can be grouped together in an independent safety section, or they can be placed at the point of use.
- Safety messages can be created at different levels for maximum impact upon the consumer. Danger to the consumer is the most important level.
- Safety messages must be used consistently throughout the technical operation instruction.
- Every machine-powered system has some inherent dangers.
- Warranty is a contract between the machine manufacturer/seller and the consumer.
- Warranty tells the consumer what responsibility the manufacturer has to ensure the machine works properly. It also tells consumers what responsibility they must assume in order to keep the warranty in effect.
- Warranty can become the cause of a liability case.
- Measurements are necessary in an instruction. They become specifications.
- Standard measurements in a technical operation instruction used in a worldwide market must have metric-dimension equivalents.
- The technical writer must not include sophisticated measurements in a technical operation instruction for a heterogeneous audience.

EXERCISES FOR CHAPTER 8

The Languages of Technical Writing

1. Select five machines.
   a. List the safety messages on each machine in order of importance to you. These are the decals and signs attached to the machine.
   b. Are these safety messages useful?

c. Do you think that the technical writer overreacted when placing these safety messages in the instruction? Can you see a legal necessity for all of the decals and signs?

2. How many measuring tools can you use?
   a. Compare this list with lists from other members of your class.
   b. What are the most common measuring devices that your classmates can use?
3. Compare a warranty for an inexpensive machine and one for an expensive machine. An example is a hand-held hair dryer and a car.
   a. What is the length of time that the warranty covers?
   b. What is the responsibility of the consumer in the care of the machine?
   c. What are the major topics covered by each warranty?
4. Collect as many examples of safety messages as possible from technical operation instructions.
   a. Arrange them in order of degree. Which ones are more important? Why?
   b. Divide the messages into general ones and ones that are point-of-use warnings.
5. Find at least two examples of specifications in technical operation instructions.
   a. Are these specifications complete?
   b. Are they useful? Can you interpret all of the entries?

9

Write the Instruction

The information in this chapter will enable you to:

- Put the different stages of technical writing together into a technical writing process.
- Decide what the content of a direction or a bound manual should be, based on the needs of the consumer.
- Budget time based on priorities and available resources.
- List the steps that can be used for creating an instruction and explain them in detail.
- Determine how to test the readability and judge the usability of an instruction.
- Set up a review process of your instruction by the manufacturing company experts.
- Write a technical operation instruction.

INTRODUCTION

The technical operation instruction is a set of logical tasks that explain how to operate a machine. It also is a medium that warns the consumer about possible hazards associated with misuse and misapplication of the machine. Simple machines may require instructions that contain only a few steps and tasks. These one- or two-page instructions are called *directions*. Conversely, some complex machines may require that the instruction be broken down into many separate sections. When many pages of an instruction are *bound* together under the same cover they are called a *manual*.

No matter how many pages are required to support the machine properly, they must tell consumers everything they need to know in order to control and operate the machine safely and efficiently. In order to write complete instructions, machine experts from all areas must be brought into the technical communication effort between the consumer and the manufacturing company. Instructions are the single voice of the manufacturing company both in the marketplace where the machine is sold and the workplace where it is used.

The complete process of writing technical operation instructions brings together data from many departments of the manufacturing company that makes the machine. Instructions must reflect the collective effort of both the technical writer and the people responsible for the final product. Poorly written instructions that do not include input from all areas of manufacturing may undermine the safety of the machine. Incorrect and incomplete instructions can turn an efficient machine into a headache for the company's service and warranty personnel. Machine warranty and service costs can add a substantial cost to the final price of the machine. The ability of a manufacturing company to produce competitive machines depends upon keeping the cost of the machines as low as possible and the quality of the finished product as high as possible.

The job of the technical writer is to translate and arrange data into an instruction that is easy to read and understand by the consumer. As a technical writer, you must develop a writing style and a writing strategy that allow you to complete instructions on time in order to meet production deadlines. The technical operation instruction must be available when the first machine is assembled and packaged for shipment into the marketplace. The technical operation instruction has become as important as any metal or plastic part on the machine. It is as important as the engine of an automobile or the hydraulic pump of an industrial robot.

The main goal of this chapter is to help you bring together all of the separate but major elements from the preceding eight chapters that can be used to write "visual and active" instructions for industry into one technical writing *process*. The chapter that follows shows *one way* that can be used to put a technical operation instruction together.

Just as it is impossible to give a formula for writing a novel, poetry, or advertising copy, it is impossible to reduce technical writing to a few unchangeable rules. Like all types of writing, technical writing must be approached as an attempt to "creatively" complete a writing assignment that can be used by as many people in the consumer audience as possible.

The following information provides a hands-on method that you can use to develop and write visually active instructions. Like the preceding chapters, it is intended to be a guideline of basic writing techniques and elements that can be applied to the many technical writing situations you may encounter either on the job or in the classroom. The fascination of technical writing may well lie in the fact that every machine has a different mechanical "personality." Each one is interesting not only in the aspect of how it is designed but also how it has been created in stages throughout the manufacturing process.

The concept of this book is predicated on the idea that the simple instruction is the basic building block of all industrial technical literature. The professional technical writer on the job in a small to medium-sized company usually begins by writing simple one- or two-page directions. The second step usually is the construction of larger and more detailed manuals. Because writing operator's manuals is usually the "natural" second step or advancement in a technical writer's career, this chapter will also provide information that can be used to develop content in a bound set of instructions. While a one-page technical operation direction may contain complete "how to operate" information under one section, the manual must be skillfully divided into many chapters or separate sections.

Above all, the information contained not only in this chapter but also in the entire book does not pretend to answer every technical writing question that has been, or will be, asked about technical writing subjects. Rather, it is an attempt to approach the business of technical writing from a unique perspective. It attempts to prepare the individual to meet the challenges of technical communication in our high-tech and visually oriented world with a "problem-solving ability" that can only be developed through an interdisciplinary approach to the study of technical writing fundamentals. A technical writer who concentrates solely on technology, or solely on communication, is only studying half of the discipline. Modern technical writers must have the ability to determine how a machine works and translate that information into simple how-to-operate instructions.

PREPARE TO WRITE

There is a practice that can help you prepare yourself for the job at hand. Before you begin collecting and filing data, and before you interview the machine experts, you should "take stock of yourself." If your strength and training lie in the area of communication, then you must concentrate on preparing yourself to understand machine systems concepts and basic technology. If technology is your main area of study, then you must begin to prepare yourself for the art of communication. Without the two disciplines, you will not be able to write the best possible instruction. You must honestly accept and try to compensate for your weaknesses. You must also understand and build upon your strengths. Students can make use of their training and build a strong technical writing no matter whether their study has been in science, applied science, liberal arts, or any of the wide variety of curriculums offered in a college or university.

Above all, when you are given your writing assignment you must write objectively. You must not consider yourself as anything but the voice of the manufacturing company. The technical instruction is the best example of the need to express the facts and not to impress the audience. You must see yourself as a member of this audience and strive to keep this in mind throughout the technical writing assignment.

Budget Your Time for Maximum Results

From the first eight chapters of this text you can see that the technical writer spends more time in preparing to write than the actual act of putting words and pictures together on paper or in a computer file. Each assignment may be different, depending upon the complexity of the machine and your knowledge about how the machine works. An estimate of the division of preparation and writing time for the technical writing process breaks down into these categories:

20 percent researching the documents and other information generated during the design and manufacturing stages.

15 percent firsthand inspection of the machine and asking questions on a random basis.

10 percent interviewing the machine experts in formal interviews.

10 percent organizing the information.

15 percent collecting or creating artwork.

30 percent audience research, outlining, writing, reviewing, revising, and preparing the copy for print.

These divisions of labor can be concretely drawn with actual writing experience. An exact division, however, may not be possible. Your writing assignment may require you to overlap all of these separate jobs in one day. Usually the task of technical writing is not a strict division of the categories of labor that are listed above. An "average" workday may find you spending the morning walking along the assembly line, watching the workers assemble the prototype machines, selectively asking questions of the line managers, and drawing rough sketches of the machine systems. You may find that a machine is available for a short photo session in mid-morning. As the technical photographer, you must take as many pictures as possible before you lose the machine to the test and quality-assurance technicians.

You may spend a half-hour before lunch sorting the material that you collect into your filing system. After lunch you may find yourself in a formal interview with the design engineer asking questions about engineering changes that you "heard" were going to occur.

Part of your afternoon may be spent in the test area where the technicians are cycling a prototype machine and writing down their quantitative results. You may find it necessary to check an engineering drawing to make sure that you know the name and function of an unfamiliar machine part. After all of your research, you end your workday by updating old information and adding new information to your files and revising the rough draft of your instruction outline. As a technical writer, you may find that your workday constantly changes and that you must be prepared to take advantage of the opportunities presented.

Double-check all information. The technical operation process and the success of the technical writer in preparing a highly readable and usable instruction depends to a great extent on the accuracy of your files. Like a mirror, your files must reflect the machine that is being fabricated and assembled on the shop floor. If you use inaccurate data, then your technical operation instruction will be outdated before it is printed and attached to or packaged with the machine. Moreover, you must budget your time so that you can double-check every source no matter what or who it is. A failure to take the time to make sure information is as accurate as possible can cost you valuable time in the writing and printing process. You will find that it is expensive and time-consuming to make changes to your technical operation instruction manuscript after you develop the final manuscript that you send to the printer.

The deadline. As a technical writer who has been given a writing assignment, no matter whether it is in industry or the classroom, you must always be aware of your deadline. Like the writers for magazines and newspapers, you must be able to schedule your time so that the job is complete on schedule. As a technical writer, however, failing to meet a deadline in industry can be much more serious than missing a newspaper deadline. Because the technical operation instruction or operation manual is considered an integral part of the machine, a manufacturing company often will not assemble the machine unless the documents are available.

Failure to supply technical operation instructions on time will prevent the manufacturing company from closing the machine shipping carton. In a mass-production process, thousands of open boxes may clog the shipping area of the factory because they cannot be sealed without the instruction. If the machine is made and packaged without the instruction, it probably will not be shipped out the back door of the manufacturing company. Instructions are a legal necessity. Thus, the consequences for not meeting a deadline for an instruction or manual can be costly to the entire manufacturing company and to the professional survival of the technical writer.

Plan Your Strategy

Because time is your greatest enemy, you must formulate a militaristic type of strategy against it. After you are given a writing assignment, minimize wasted time by developing a writing schedule. Write down all important dates on a desk calendar or on a computer time-organizer program. Do not forget items that may cause you a delay in the technical writing process. The only way you can defeat time and produce the best possible technical operation instruction is to organize and set your priorities. Some important entries in your calendar or organizer program might include:

The day the prototype is assembled on the shop floor. A prototype is a machine that is built to test the efficiency and quality of the manufacturing process before mass production of a machine begins. You may be able to learn much about the internal workings of the machine if you watch it being assembled.

The days that machine experts who are instrumental in designing and manufacturing the machine will be absent. Like everyone, these key people may go on vacation or become involved in projects outside of the company during your writing assignment. You can easily and indirectly identify these people and find out their work schedule through contacts with the various department secretaries and other key support people.

The prototype machine test dates. By making videotapes or photographs of the machine while it is being tested, you may be able to learn much about how to operate the machine from firsthand experience. When you observe a machine, divide its work output into cycles. A work cycle for an automatic clothes washer, for example, is the operation of the machine from the time the water enters for the wash sequence until the final rinse cycle. Concentrate on the control variations that can be made during a cycle. Each variation may require a separate task in the technical operation instruction that explains how to make the change.

The date the first draft of the manuscript is needed. This is the first time that your instruction artwork and words will be completely put together.

The date the final review draft, including all up-to-date information, is to be completed.

The review time that the key machine experts, including the design engineers, attorneys, safety personnel, and manufacturing people, need to review your instruction. They will check for accuracy of information and the completeness of the tasks. Small instructions may take them a few days, and larger ones may take many weeks. You must be sure that the machine experts know when you must have their review. You must keep a close observation on the instruction as it travels from expert to expert.

The date you must have the final review copy revised. Revision time in relation to the size of the instruction must be taken into account.

The assembly time needed to put the instruction in final manuscript form. This will be sent to the printing company, which will mass-produce the instruction.

The time that you need for company travel, vacation, and other time away from the writing assignment. You may find that writing assignments overlap and that one instruction will interfere with a second instruction that you are simultaneously writing. Be sure that you take this into account.

The day and hour that the first machine is scheduled to come off the assembly line, the instruction must be printed. When you plan a strategy, start with this date and work backwards.

Manage time wisely. Practice time management both on the job and in the classroom. Your success in technical writing, whether it is to get a superior grade in classroom competition or earn your salary, depends heavily upon how well you can plan your writing strategy and your schedule. Without a detailed outline of what you must do and a record of what has been done, some of the small pieces of the jigsaw puzzle or writing assignment may be lost or forgotten. A technical operation instruction must include every piece of useful information no matter how insignificant it may seem.

Sharpen Your Language Tools

Technical operation instructions must meet the needs of consumers who use them. Instructions must be written to the level of the audience. They must reflect the age, intelligence, reading ability, comprehension, education, special training, and, if possible, disabilities of the consumer audience. The technical writer must try to *target* the consumer group as accurately as do the marketing experts who attempt to predict population shifts and product demand.

Failure to warn consumers of potential dangers created during the operation of the machine, and failure to instruct consumers in language and words geared to their intelligence and ability level, can result in product liability cases. Product liability is costly both to the manufacturing company involved and to consumers. Thus, you must write your technical operation instructions carefully and cautiously, examining each word and picture to make sure that both word and pictures symbolize exactly what you are trying to say to your audience.

The types of audiences. There are two basic types of consumer audiences that you must attempt to identify. The first one is the homogeneous audience that you are able to measure. You know each consumer's ability to understand tasks and information for a specific machine. Thus, you can determine beforehand the education and training of these people because they have met a "standard" of competency prior to assuming their job title or operation role with the machine. A homogeneous, or stratified audience, may also be small enough to allow you to test and evaluate the members before you write the instruction. In any event, the homogeneous audience is a group of consumers or machine operators that you are able to identify and make assumptions about based on their job level or other measures of reading and understanding ability, training, and overall intelligence.

The heterogeneous audience. The second, and usually the largest, group is the heterogeneous audience. Often this audience is so large and dispersed that any significant measurement of the members is impossible. A diverse and heterogeneous audience is one that can never be concretely evaluated. Members of this group may include dentists, lawyers, domestic engineers, college students, machine assemblers, and the hundreds of other people divided into age, sex, and occupation categories who may be experts in their particular profession. These people may be highly skilled in some area; however, they may not have the ability to control a particular machine. It is impossible to put labels on and assume anything about this unknown audience.

Safety I.Q. Although it may be possible to determine the ability of a homogeneous audience to use an instruction, it may be impossible to evaluate completely the safety knowledge of that audience. Technical knowledge and knowledge of certain procedures necessary to operate a machine must not be confused with the ability of the audience to understand safety hazards associated with the machine. Safety information must always be complete and well written. You must never assume that your audience will know the safest procedure simply because members of the audience have machine knowledge. The results of misuse and misapplication must always be explained in detail to the audience before members find themselves in a dangerous situation.

Write to the lowest common denominator. You are at a disadvantage when you attempt to customize a technical operation instruction for an audience that cannot be determined by normal measurement means. Another problem is that a homogeneous audience may have individuals who are better trained and have more ability to operate the machine than other individuals. For example, graduates from the same college or university program may have completely different competency levels in the same field.

One answer to this predicament is to write to the lowest common denominator, or the character sketch of the person who you think has the most difficulty reading and understanding the written instruction that supports your machine. If you do this then you will reach a larger percentage of the audience compared to the technique of writing to the "average" or middle person in the audience. By writing to this mentally pictured individual, who is at the bottom of the audience in terms of machine aptitude, you will be completing the task that is before you. You will be writing a technical operation instruction geared to the largest percentage of people possible.

Words can be rated. The key to writing instructions for either of these audiences is to select the best words possible. Word choice is extremely important if you are going to match your technical operation instruction to the audience or consumer who may buy your company's machine. This is possible because a person's vocabulary at times can be predicted. Formal education relies to some extent on teaching a selected vocabulary to the individual. Lawyers, health care workers, engineers, service technicians, or other "trained" people are given certain terms and definitions as part of their formal study.

Education, age, social background, hobbies, and other cultural influences can shape the words that a person uses. For example, a chemist may use different words from that of an engineer. Lawyers and doctors spend large sums of money to attend schools to learn the language of their professions. A shortwave ham radio operator will develop a specialized vocabulary that may sound foreign to the person who knows nothing about this particular hobby. Each of these people will look at your machine and your technical operation instructions through a different set of communication symbols or words.

Vocabularies are layered. We speak and communicate with different words because they are learned and developed in layers. Children begin communicating with a survival vocabulary that enables them to express their immediate needs and wants. Their basic language, no matter what culture they are born into, would be the same. This shared basic language stays with the individual throughout his or her life. This vocabulary is the foundation upon which the child begins to interact with the outside world with verbal recognition.

As children develop and interact with family and society, they add specialized words to their basic vocabulary. Their language becomes increasingly complex as they progress through the educational system and become involved in their work and activities. Words are added as people develop their values and needs in a competitive worldwide society.

Just as words are added to a person's vocabulary, so can they be stripped away. This is the process that you, the technical writer, must also go through before you can attempt to reach your audience. You must strip away the complex vocabulary that you have developed and use as much as possible the basic words that are familiar to everyone. As you need them, specialized words can be added to your vocabulary depending upon the composition of your audience. This is an especially important self-evaluation if your audience is composed of people who have been educated, trained, and who live in a completely different society from the one you have experienced.

Usually the word with the fewest syllables is your best choice. Foreign derivatives other than German root words are usually the most complex and hardest words to understand. A good rule to follow is to use the right word for the situation as long as it is the shortest, most descriptive word that you can choose.

Instead of saying "acquire," say "get."

Instead of saying "discontinue," say "stop."

Instead of saying "utilize," say "use."

Instead of saying "in conjunction with," say "and."

Make Sentences Short, Active, and Visual

Along with choosing the shortest words, you can make your technical operation instruction more useful if you use, active-voice verbs such as *push*, *pull*, *put*, *press*, *turn*, *open*, and *start*. These, active-voice verbs create action in the mind of the reader of your instruction. Active first-person verbs are also easier to read and translate into motor actions by the consumer of machine operator.

Another technique that you must use to make your technical operation instruction active and visual is to choose *concrete* nouns. Concrete nouns help cast concrete images in the mind of the reader. It is better to choose words such as *hammer*, *open-end wrench*, and *nail* than it is to use words such as *tool*, *wrench*, and *fastener*. Nouns, like *nail* and *hammer*, create a more concrete image in the mind and leave the audience with little doubt as to what you are trying to say. Concrete nouns do not allow the consumer or machine operator to interpret your meaning. They keep the imagination of the consumer in check and eliminate the necessity for the consumer to interpret the meaning of your word symbols.

Active verbs and concrete nouns are the main ingredients that you must use to build sentences for the instruction. In order to develop a workable visual technical style, you must perfect and use uncomplicated sentence structures. The three basic sentences you can use to write any instruction are:

The command sentence.

The simple sentence.

The compound sentence.

The command sentence. The *command sentence* is the one most used in the instruction. This structure is also called the *imperative sentence.* You will write all of your tasks in this sentence format. This sentence demands with authority that your reader completes the task before going to the next one. The structure of the command sentence is:

(You) Turn the knob.
("You" is understood. "Turn" is the verb. "Knob" is the direct object.)
Push the button.
Open the door.
Change the filter.
Drain the crankcase.

The simple sentence. The *simple sentence* is used to write safety-warning information and other important facts that cannot be put into the form of the command sentence. A simple sentence has the subject-verb-object format. For example:

Battery acid is a hazard.
("Battery acid" is the subject. "Is" is the verb. "Hazard" is the descriptive object.)
Air causes cavitation.
Incorrect oil overheats.
Gasoline can explode.
Gears form pinchpoints.

The compound sentence. The *compound sentence* is a variation on the command sentence. It is two command sentences that are connected together with a conjunction. Use this type of sentence to tell the consumer to do two tasks simultaneously. For example:

Hold the flip cap up and add fuel.
("You" is the understood subject. "Flip cap" is the object. "And" is the connector. "Add" is the second verb and second action that simultaneously must take place. "Fuel" is the second object of the compound action. It is the thing that must be accomplished by the simultaneous action of the consumer.)
Press the button and lower the blade.
Hold the start button in and set the timer.
Remove the plug and drain the crankcase.
Start the engine and release the choke.
Cycle the system and fill the reservoir.

WRITE WITH PICTURES

The second part of a visual technical writing style is the use of pictures or artwork. By skillfully using artwork, you can eliminate unnecessary words from the technical operation instruction. Photographs, line drawings, and other illustrations make the instruction easier to "read" and understand. Without artwork, tasks must be written with many descriptive words that tell you where the action takes place and how you must get there. Seldom will a technical operation instruction be written without an illustration.

There are a few techniques that can be used to choose active pictures, just as there are techniques for choosing visually active words. First of all, you must make sure that the quality of the artwork is as good as possible. This means that halftones or photographs must have sharp contrast. Also, make sure that cluttered backgrounds are eliminated from the composition of the picture.

Show action in pictures. To help make a line drawing or halftone active and alive, show people doing the task. Do not simply show an adjustment screw and say "Turn the screw." Instead, show an operator with a screwdriver making the adjustment. Show people controlling the machine; this will provide the consumer and machine operator with a better concept and mental image of what must take place. Use heavy arrows to show motion if a part of the machine moves or should be moved. Any movement in the artwork will give it life and show action (Figure 9-1).

Call outs. Use call outs to help connect together the artwork and the words in the task. Number and letter call outs can be purchased in a variety of formats including "rub-off" and "pressdown" sheets. These call outs also include leader lines, arrows, and other signs that point out a specific place on the artwork where consumers must focus their attention. Without these visual symbols, artwork would not be an effective communicator of ideas. Words and leader lines are also used to call out items on artwork. It is easier to translate nomenclature on art if numbers are used as pressdown call outs and the legend is printed under the art or referred to in the text (Figure 9-2).

In order for the consumer to better understand your instructions, begin your artwork sequence for a task with a general or overall view of the machine. This overall view orients the consumer and shows what you are going to describe or instruct. This

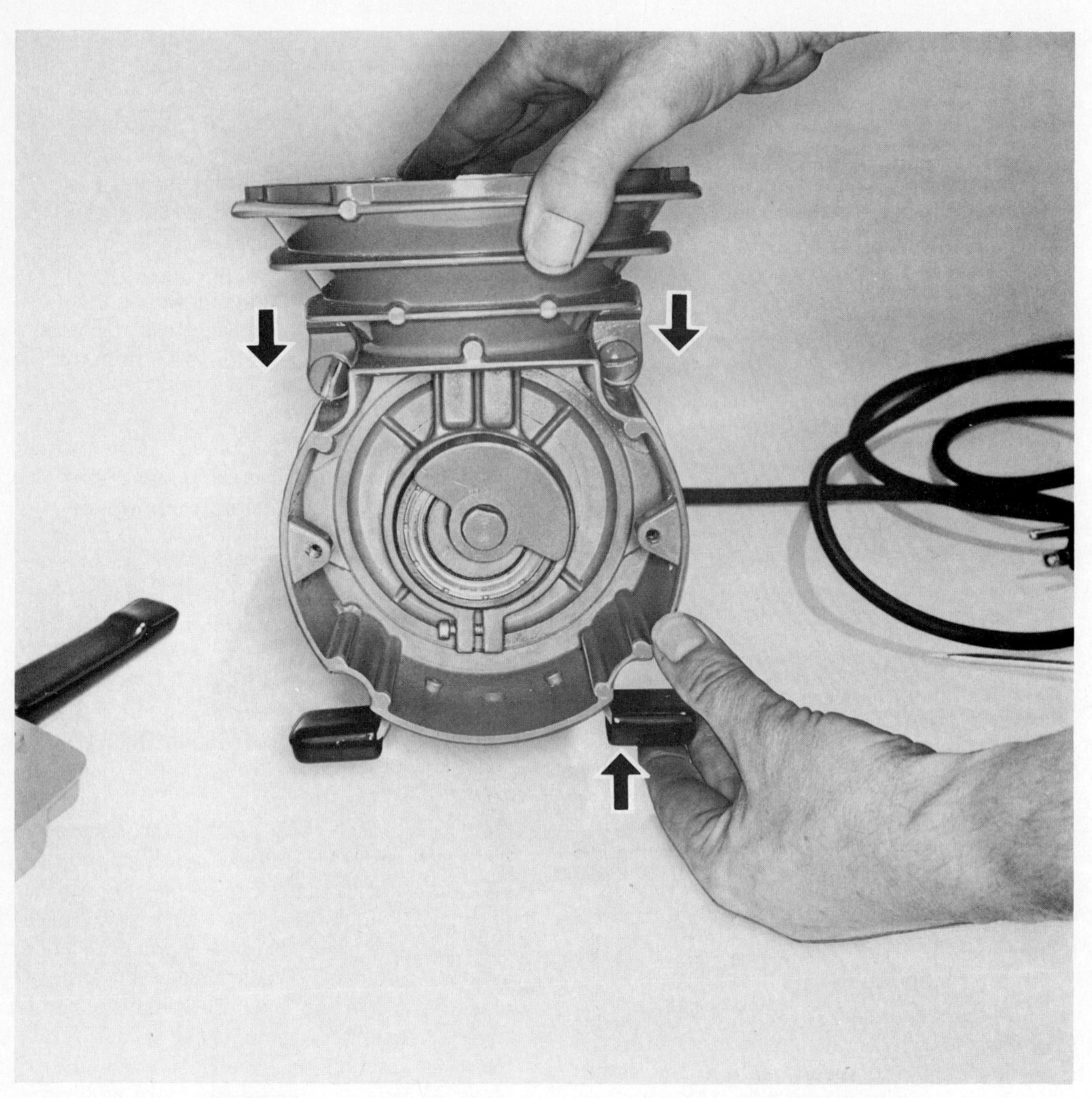

FIGURE 9–1 Action is created by "arrows" and people doing the task.

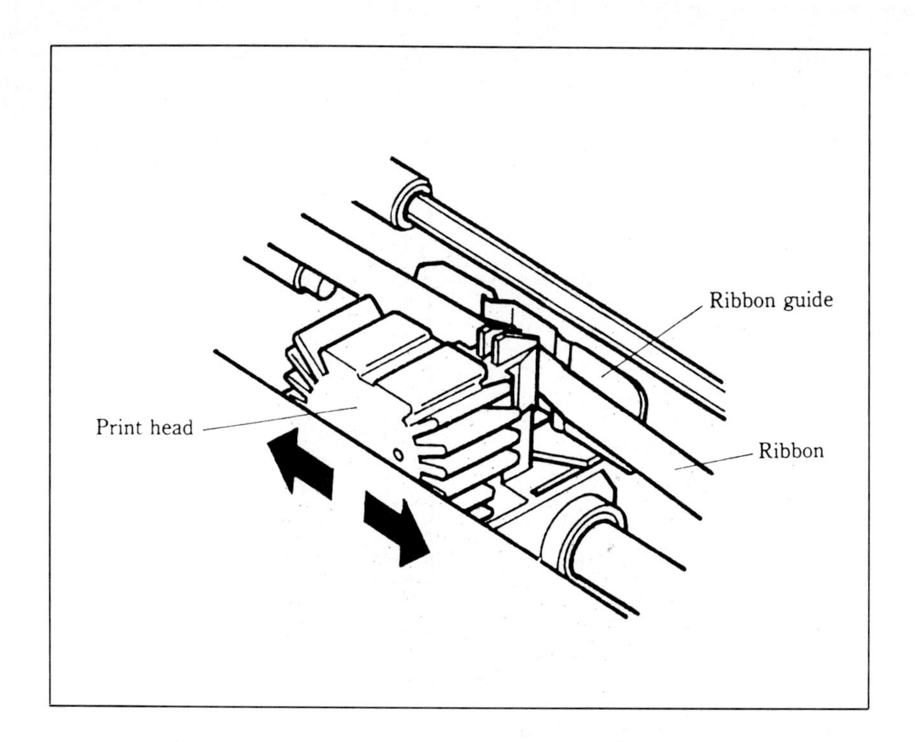

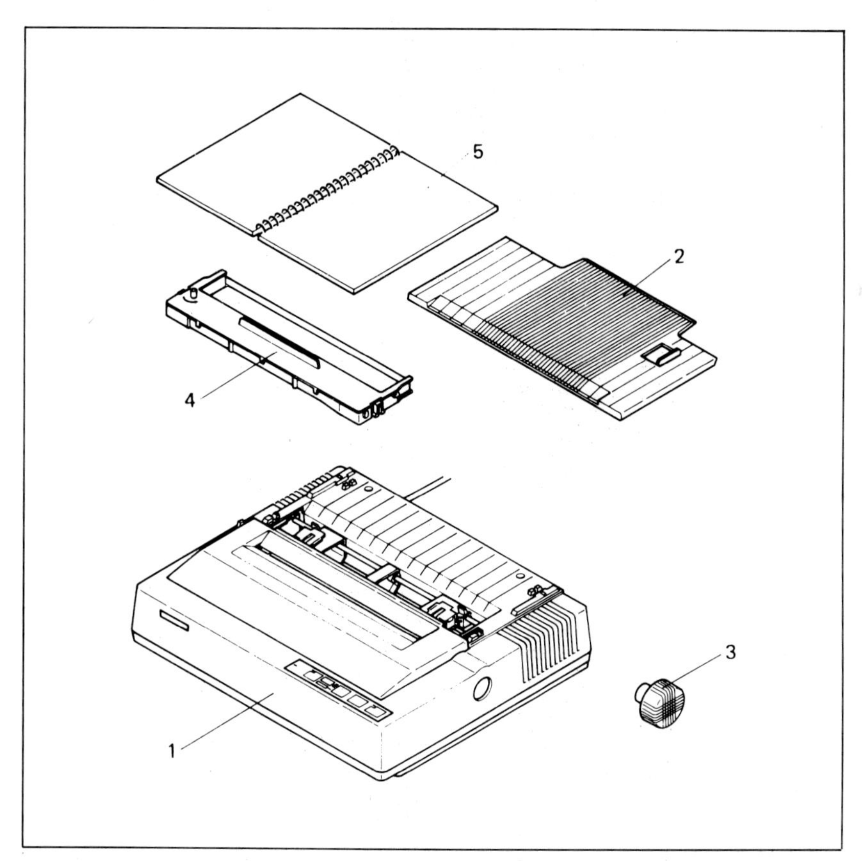

FIGURE 9–2 Both words and numbers can be used to call out items on artwork. (Copyright 1986 by Star Micronics Co., Ltd., *NX-10 Users Manual,* PN 80820122. Used with permission)

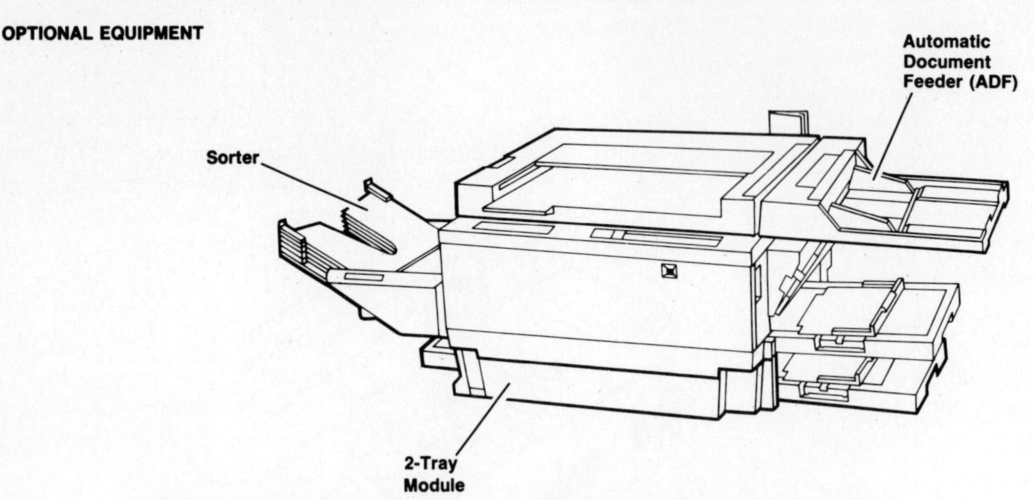

FIGURE 9–3 An overall view of a machine. (Copyright 1985 by the Xerox Corporation, *The Xerox 1038 Reduction/Enlargement Copier User's Guide,* 600P88316. Used with permission)

piece of artwork illustrates the main features of the entire machine and the major points that will be the control points or location of action in the instruction. By moving from the general to the specific, you will be taking the first step to making your technical operation instruction logical. The overall view is extremely important in manuals that require extensive operator movement around the machine in order to complete the tasks that are presented (see Figures 9-3 and 9-4).

Spatial logic is an important part of the instruction. Like verbal logic, where tasks and information are given in numbered steps, artwork must be organized to help orient and move the operator to the points of control on the machine. Use artwork to move from the general overall view to specific points of action where adjustment or control is required. Spatial logic must be used to keep the operator in the safest and most convenient place on or at the machine. This type of logic is especially important for larger machines that require extensive technical operation instructions.

Show safe procedures. Another important concept is that you must depict people doing the task as being as safe as possible. Artwork such as photographs are a powerful communicative tool that suggests many things besides what is being covered in the task. Always use the photograph or drawing to depict the operator of the machine in proper protective clothing and in the correct position. Artwork must be considered a major part of any overall safety message in the technical operation instruction.

DEVELOP AN OUTLINE

Before you begin to write the instruction or manual, you should develop a *contents outline*. This is a list of major items that you must include in the finished document. Without this general organization of ideas, you may find yourself unnecessarily repeating words and artwork. The contents outline includes all major, secondary, and third-level headings that signal logical breaks of information.

The development of an outline should not be done by trial and error. The outline is a mental sketch of all the machine tasks and safety messages listed in headings. An outline gives your instruction the basic framework that you can use to build your tasks upon.

The only time that you may not need to develop your contents page is when an industry, government, national, or international standard has been created that demands certain information be placed and arranged in certain instructions or manuals. If standards are available, they may cover only the basics of content, and you may be required to apply these basics to your situation and develop them further.

Because you are the communication expert for your company or for the particular machine that you are writing about, it is more efficient if you initiate the contents outline if one does not exist. Your first step may be to operate the machine firsthand. Run the prototype machine through the standard operation cycle if possible. Put yourself in the place of your audience. You must constantly ask yourself what information should be included so that you and your audience can control the machine properly, maintain it on a daily basis, or adjust it so that it runs efficiently. Study the test results for out-of-the-ordinary operation "quirks" or characteristics of the machine that might not show up in the machine parameters or standard operating procedures.

Your second step in developing a contents outline is to evaluate your safety messages carefully. You must include all safety warnings and notes that are essential to help keep the operator and the machine safe during operation. Visually picture these safety messages in place under the various headings. Part of the safety message program in your technical writing procedure must be to project yourself into the workplace where the machine will be used. What must your company tell the operator about misuse and misapplication of the machine? These questions can only be answered by the safety and legal experts within your company. Note, however, that you must only *assume* the role of the operator and must place these questions before the company experts.

Create your contents list with the level of headings that you plan to use in your final instruction. A rule of thumb is that an instruction or direction that is not bound usually does not need extensive breaks

PREPARING TO TYPE

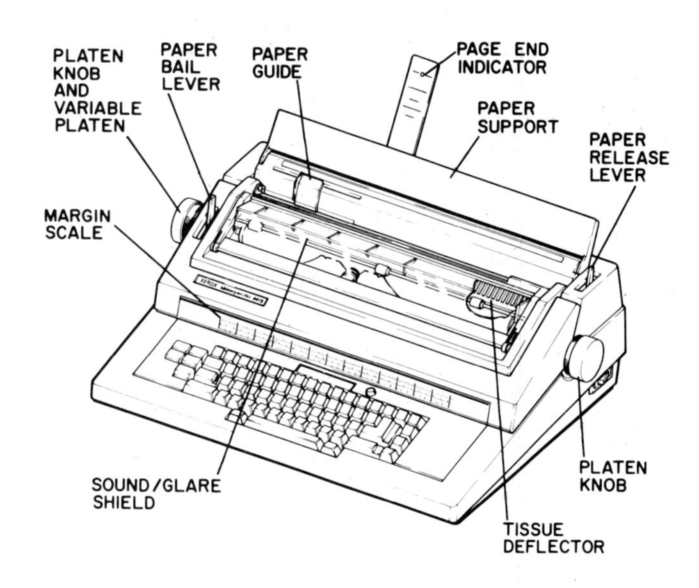

6015 MEMORYWRITER 2-3

FIGURE 9-4 An overall view at the beginning of a major section. (Copyright 1985 by the Xerox Corporation, *Xerox Memorywriter 6015 Training Manual,* 610P72804. Used with permission)

in the text. The reason for this is that consumers should not encounter difficulty finding what they want to know about in only a few pages.

A manual, on the other hand, may be used as a reference in order to retrieve information throughout the operation of the machine. An extensive manual may require many breaks in the text. This can be compared to a floppy disc that stores information on a computer. In order for the computer to retrieve information efficiently, the disc must be "formatted." Formatting divides the disc into sections in order to help facilitate the storage and retrieval of information.

A "rough" working contents page for an extensive instruction must include the safety information and a simple outline of the contents. Changes can be made to this instruction as you develop your information. Here is an example:

THE MODEL 6YN SNOW BLOWER

I ASSEMBLY

II STARTING THE BLOWER
- Control points
- Adding fuel and lubrication
  - *Safety points*: Gasoline handling information.
- Greasing wear parts
- Startup
- *Safety points*: Rotating auger. Stay clear of during operation. Do not attempt to unclog with spark plug connected. Clear area of trash. Make sure people are clear of area.

III MAINTENANCE
- Daily
  - *Important points*: Use correct lube and gas.
- Weekly
- Monthly

IV STORAGE

V SPECIFICATIONS

VI OWNER'S REGISTRATION CARD

VII WARRANTY

Not only does the basic outline give you a framework you can use to add information to, but it can also serve as a table of contents for your operator's manual. Larger manuals need a contents page added at the front of the publication. The contents page is usually added to these bound instructions so that readers of the technical information can go to the section that they need without having to read the entire manual everytime they operate the machine and have questions. A contents page in a technical operation instruction is used like a textbook table of contents.

Both the size and the information contained in a technical instruction contents page is determined by the *complexity* of the product. For example, a hand-held calculator with few functions may require the support of a few operation tasks. A hand-held calculator with many functions may require a manual with many pages and many divisions of the operation instruction. Some machines may require more in-depth contents than do others. The physical size of the machine is not necessarily reflected by the contents page.

For a better understanding of what contents can encompass, look at the following "tables of contents" from documents that are used to support specific machines. These contents include a wide variety of machines of all shapes, sizes, and functions. Each table of contents is designed for a specific application in order to meet the reference needs of the consumer and operator for that machine. They do not include the sketch of safety information that is given in the outline above. You can use contents pages from existing machines *as a guide when you create your working rough outline.*

When you write an instruction for similar machines, always remember that these sample outlines may not be adequate for your particular needs. Always develop specific contents pages and manual content for each type of machine design.

The first example is used with permission from the Ford New-Holland Company. It is taken from the copyrighted *Operator's Manual, Loader Backhoe General LB-620, Issue 4-86, Form No. 0 (LB-620)-2-2M-486P.* The numbers on the right are page numbers in the manual where the information is located (see Figure 9–5).

The second example below (Figure 9–6) is used with permission of Speed Queen, a Raytheon Company. It is taken from the *Operating Instructions for Home Laundry Automatic Washer.* In this example, the contents page is divided into three levels of headings with an emphasis on the safety section at the front of the manual.

Another type of machine that requires an operator's manual is the automobile. The next example (Figure 9–7) is used with permission of the Ford Motor Company, Ford Parts and Service Division, Training and Publications Department. It is taken from the *Mercury Sable 1987 Owner's Guide.*

FIGURE 9–5 A loader backhoe. (Courtesy Ford New-Holland)

CONTENTS

Table of Contents

FIGURE 9–6 A contents page for a home laundry washer. (Copyright 1987, Speed Queen, a Raytheon Company, *Operating Instructions for Home Laundry Automatic Washer Model NA 3311-3059,* Part No. 31575 4/87. Used with permission)

4 CONTENTS

FIGURE 9–7 A contents page for an automobile operation instruction. (Copyright 1986 by the Ford Motor Company, *Mercury Sable 1987 Owner's Guide,* FPS-12059-87A. Used with permission)

Products that used to be installed by professionally trained people often become so refined that they often can be installed by the amateur home do-it-yourselfer. Some instructions must be written for both the home carpenter and the professional. An example of this is the manual for a garage door opener. The following is used with permission of The Chamberlain Group, Inc., A Unit of Duchossois Industries, Inc. The contents page is from the *Chamberlain Garage Door Opener System 550-6 & 350-6 Owner's Manual, Models 550—1/2 HP, 350-1/3 HP, 114A1003, Copyright 1987.*

CONTENTS PAGE

Some new machines require extensive safety and introductory information because the technology is new and not as well understood as with other machines. An example of this is the microwave oven. In order to cook food, the consumer must adapt an entirely new approach to food preparation. As our technology advances, it sometimes is necessary for the technical writer to acquaint the consumer with the basics. The following example is used with permission of Amana Refrigeration, Inc. The material is from the company's *Use & Care Manual, Amana Commercial Radarange Microwave Oven, Models RVS7 and RVS10, Part No. A1056307*, copyright 1987.

CONTENTS

The final example is used with the permission of the Xerox Corporation, Multinational Service and Technical Support/Multinational Service Education. It is from the company's copyrighted manual entitled *Job Preparation and User Guide for the Xerox 8200, 600P83889* (see Figure 9-8).

The contents of the manual depends mainly upon the complexity of the machine, the consumer's familarity of the machine technology, the audience, and the number of controls that are on the machine. Once you have decided what you basically must put into the direction or operator's manual, you can begin to solidify the structure of your document. A tentative contents list can be routed to the various machine experts for their opinion. The legal, safety, engineering, and manufacturing experts can save you time by giving you an evaluation of what you are going to include before you write the instruction.

A direction may contain all or some of the above information under one or two main headings. No matter how large the instruction is, it must be *complete* and contain all major topics necessary to the operator for a safe and efficient operation of the machine.

Contents

FIGURE 9-8 A contents page for an office copying machine. (Copyright 1981 by the Xerox Corporation, *Job Preparation and User Guide for the Xerox 8200, 600P83889. Used with permission.)*

Develop the Rough Outline Into Storyboards

Once you have determined the basic contents page or rough outline of the instruction, you can begin to arrange your information with spatial and verbal logic. A working visual outline gives your technical operation instruction a strong visual quality when you complete your writing assignment. A visual outline saves you time by letting you group, arrange like information, and rearrange your ideas. It is one of the simplest tools that you can create, and it is the most time-saving tool that you can use when you "write" a visual instruction.

Good visual working outlines are a direct result of a good filing system and a professional approach to the collection of information. If you can see each topic in an isolated file, you can better grasp exactly what you must tell the consumer. If all safety information is together, then you can get a clear mental image of the unsafe situations that can develop during the operation of the machine. A good filing system is a source of building blocks that you need when you begin to assemble the working outline in pictures.

Your simple written working outline can be developed with simple drawings or sketches. This outline is your first attempt to put your information together, and it must be created so that it can be easily revised. As you analyze the steps of each instruction, you may find that you have confused the task sequence, or have placed the operator out of position through a piece of artwork that does not correctly move the operator to the task at hand.

The complete outline of the manual can be created on storyboards. Storyboards begin with simple sketches of the part or item that you are explaining. Remember that when you put together the instruction, you must think in mental pictures if you are going to think clearly and completely. The storyboard is your mental picture of the machine that you create as you research your files and follow your rough outline that gives you your division of ideas.

After the storyboard is constructed with simple drawings, you can add the text to the art. By thinking first in pictures, you can eliminate many descriptive words that you otherwise might be forced to use without the right picture. On the other hand, your picture, or drawing, helps you explain your words. After the basic tasks are added to the storyboard, you can add the safety information before the tasks that they refer to at the point of use. With pictures, words, and safety and information messages on a storyboard, you have all of the information in front of you in the form of a "rough visual outline."

Adapt These Basic Steps to Your Visual Writing Process

The following information is a listing of the steps that you can follow in order to create visual technical operation instructions. Use this outline to construct your personal approach.

1. *Create an outline.* Develop a simple outline from sketches of the control points of the machine. Begin with the heading or title of the task that you are going to explain. Keep headings contained in the direction or manual active and parallel in construction. For example, headings that are *not* parallel and active might look like this:

   I Take out the battery.
   II Battery needs to be cleaned.
   Safety hazards.
   III Connect battery correctly.

   An active set of headings would look like this:

   I Remove the battery.
   II Clean battery.
   Wear safety clothing.
   III Connect the battery.

2. *Develop a storyboard.* Develop a storyboard with simple sketches that show the points that you want the consumer to focus upon.
3. *Develop a rough word sketch.* Add the words near the picture that the words help clarify. These are simple ideas in words that can be revised at a later time when you are concentrating upon the accuracy of your instruction.
4. *Add safety messages.* Add the necessary safety messages to the storyboard. Look at safety from two angles. First, where do point-of-use messages go? Do you need a separate safety page because of industry standards or your company's legal requirements? Second, review all safety information to determine if your instruction is complete. You may need more tasks in order to warn the customer about safety problems that you may forget on your preliminary outline. Remember, your instruction must not be guilty of the *failure to warn* the operator of potential dangers.
5. *Create a checklist for art.* Complete your artwork requirements. Copy the storyboards and use your rough sketches either to buy art from a graphic artist, supervise a photo session, or take photos for your project. Make a master checklist of all art so that you do not overlook a shot that you need. It may be extremely expensive and even impossible to set up the machine and retake photographs that you forgot at the original photo session. Remember, arrows and illustrations of people doing the task can add to the active tone of your instruction.
6. *Select words carefully.* Reread all the words that will be on the draft of your instruction. Analyze each word. Determine how well each word communicates your ideas when put into a sentence. Analyze each word to find the shortest, most active, or concrete word possible.
7. *Finish artwork.* Add the call outs, arrows, and nomenclature to the artwork to close the gap between the visual picture and the words. Words and artwork must work together.
8. *Lay out a rough draft.* Create a rough draft. Type or print out from a computer program the words in columns small enough to fit side by side on a standard 8½- × -11-inch piece of paper. Reproduce your original art and paste the reproduction on the paper. *Do not use original artwork on your rough draft.* Artwork can be reduced on many office copy machines to the approximate or exact size that you will use on the finished layout.
9. *Desk-check With Direct Feedback.* Route the rough draft to the machine experts who assist you as you gather information. Be selective. Do not waste time by routing the rough draft instruction to people who are not the machine experts. Attach a routing slip to the rough draft to keep a record on file of the people who review your information. This routing slip will also tell the machine expert who is next in line to review your material. Have each expert use a different colored pencil or pen when writing comments. You will know exactly who made what comment or correction and you will be able to follow up on the correction more efficiently. Remember, keep the reviewer list as small as possible, and only allow the key people in your company to make comments on your copy. An example of a routing slip is given below. You may not need as many experts as in this example.

REVIEW FOR THE OPERATOR'S MANUAL FOR THE 174 LAWN TRIMMER

SIZE OF THE MANUAL—8 1/2″ WIDE × 11″ HIGH.

PAGE COUNT IS APPROXIMATELY 20 PAGES.

PLEASE ROUTE TO:

DATE OF REVIEW

JIM ARNOLD, ENGINEERING DEPARTMENT ______

BILL STEWARD, PRODUCTION ______

NANCY LEONARD, LEGAL ______

ROBIN MAST, SAFETY ______

JACK HIGGINS, TESTING ______

SALLY JOHNSON, WARRANTY ______

MICK ROUSSEAU, ADVERTISING ______

NOTE: PLEASE SIGN THE DATE IN THE SAME COLOR THAT YOU USED TO REVIEW THE INSTRUCTION. PLEASE USE SEPARATE COLOR FOR EACH REVIEW. PLEASE ROUTE QUICKLY. DUE DATE FOR THE COMPLETION OF THIS REVIEW IS 2 AUGUST.

10. *Collect direct feedback.* Desk-check your information by setting up role-playing situations. Allow as many people as possible a chance to read your instruction and operate the machine.
11. *Corrections.* Make corrections to both the artwork and the text according to the results of your desk-check.

Desk-Check Your Draft Copy

You must check the accuracy and content of your technical operation instruction when it is in the rough-draft stage. One way is to let the machine experts review the contents of your material at the rough-draft stage. Another way to make sure that your technical operation instruction is accurate and complete is to desk-check the contents by allowing individuals to role-play or field test the tasks with the machine in front of them. As a technical writer, you should be role-playing your instructions continuously as you write your instructions.

Your role-playing sample group should be as diverse as possible if you are writing to a heterogeneous audience whose capabilities are not measurable. If you are writing to a homogeneous group of people whose abilities can be measured, be sure that the desk-check is made with people who have the same skill and background as your potential audience.

A video recording of these people as they operate the machine with nothing but your instruction is a good way to record reactions, hesitation points where information may be unclear, and unsafe situations that consumers may put themselves in. Always try to record as much body language as possible that the role-player displays. Analyze reactions at each task.

By using the direct feedback of the machine experts and the sample consumer reaction to your instruction, you can adjust your material before it reaches the final print stage. Charles E. Witherell sums up the importance of a desk-check exercise:

> A product user can be an expert in his field and still be totally unfamiliar with terminology of the industry that manufactured the product and those operating manuals he must be able to comprehend to safely use the product.
>
> For these reasons it is highly desirable, and practically essential for some products, for a manufacturer to "test" its operating and service manuals, in addition to its products. This is done through selecting individuals representative of the typical customer or product user, but having no expertise or knowledge within the field of technology of the product, and observing their ability to comprehend key points in the manual; particularly an understanding of details critical to safe operation of the product, system or equipment. Such tests often reveal unanticipated difficulties and frustration of users failing to grasp information that the manual's writers thought would be readily understood.
>
> Information, descriptions, warnings and written procedures so obscure that the average user can understand them only with difficulty or not at all cannot be regarded as effective and , therefore, do not really satisfy communication standards prescribed by regulations and other requirements.[1]

Readability formulas. There are a few readability formulas that have been used to determine who will be able to read the instruction. Readability formulas are usually based on the premise that a person with a certain grade level is able to read longer sentences with more syllables than people at a lower grade level. Use of the formula usually measures the length of words, number of syllables in the words that are chosen, and the length of sentences.

[1]Charles E. Witherell, *How to Avoid Products Liability Lawsuits and Damages: Practical Guidelines for Engineers and Manufacturers.* (Park Ridge, N.J.: Noyes Publications, 1985), p. 184.

While these formulas might be useful for the mass-media writer such as the journalist, they are not particularly useful to the technical writer.

If the guidelines for developing short, active sentences with active present-tense verbs and concrete nouns are followed, then these formulas will not be necessary. Readability formulas do not tell you how well your written words and *artwork* communicate. How well a technical operation instruction informs and warns the consumer is the major concern of the technical writer. For example, the following two sentences will have the same readability level in every system that measures compound words and syllables:

Oil the gear, and rotate it one turn.
Rotate one turn it and oil the gear.

A Case Study

The following case study is an *example* of how a technical operation manual can be created by using the 11 basic steps presented earlier. It is a step-by-step look at how an instruction can be created block-by-block with the visual writing techniques covered in this text. It is presented only to show the method and approach that can be used both on the job and in the classroom to create a clear and concise set of tasks for a heterogeneous audience. Although the actual writing sample given is not one that shows how a complete manual can be created, it does show the development of one main heading of the instruction. By writing each heading in the same manner, the entire manual can be created.

Jill Davenport is a technical writer for a company that makes consumer products used in domestic lawn care and commercial landscaping. Most of her customers are landscape hobbyists, homeowners, and other people who do not operate her company's machines on a daily basis. She knows that it is impossible to attempt to measure this mass audience. Her readers may be professional people who sit behind a desk during the week. They may be young adults working during the summer. They may be of any age and background.

Mike Packard is in charge of developing the overall technical writing group strategy. He gives Jill an assignment that includes the writing of complete technical operation instructions for a new *chain saw* that their company is about to market. Jill works on the assignment for a month. She has collected information and has developed a rough outline that has been reviewed and modified by the company experts.

Jill knows that she must write a manual with many sections in order to break up the information for easy reference by the consumer. Chain saws are not a simple machine. She divides the major topics and major parts of her operator's manual into the following:

FRONT COVER
INSIDE FRONT COVER
LEARN THE CONTROLS
OPERATE THE SAW
- Complete prestart procedures
- Mix oil and gas
- Start the engine
- Stop the engine
- Store the saw properly

USE SAFETY PROCEDURES
- Handle gas properly
- Wear proper clothing
- Handle fuel properly
- Properly start the saw
- Protect against chain kickback
- Keep the chain tip guard in place
- Handle the saw properly

TROUBLESHOOT PROBLEMS
SERVICE DAILY, WEEKLY
- Keep the chain lubricated
- Clean the saw

CORRECTLY STORE THE SAW
SPECIFICATIONS
- Note size or physical dimension
- Choose the proper fuel
- Know the weight of the saw

COPY YOUR SERIAL NUMBER
INSIDE BACK COVER
BACKCOVER

Jill knows that she can change the order of these sections and add and subtract minor sections for each of them. In order to keep her job on deadline, Jill begins with the section that she has thoroughly researched and one that she is confident that she can finish. She spends much of her time at a test plot in the back of the manufacturing company where trees are being trimmed, felled, and debranched with a prototype of the chain saw. She decides to begin with the section OPERATE THE SAW and subsection "Start the engine." As long as she has a working outline, *she can write any section and subsection that she is prepared to complete.*

Her first step is to review completely her notes on safety and machine operation. Because she is

FIGURE 9–9 Jill uses an overall view to begin her main OPERATE THE SAW section. (Courtesy Deere & Company)

writing about a saw that is powered by a gasoline engine she knows that there will be standard safety warnings concerning fuel handling. She also has collected industry safety standards that she must put into the instruction at the point of use. Jill is aware that the test lab experts who have been cycling the machine have determined that there are two definite start procedures. One is for a cold engine and one is for a warm engine.

Storyboard. After a complete check of her notes and a final double-check on a few items that she thinks might be redesigned in the near future, Jill prepares to write the "Start the engine" section of the instruction. Jill's writing begins with an extremely rough outline of sketches in a storyboard. Jill has already obtained the overall picture of the saw that she will use in the front of the main section, entitled OPERATE THE SAW. Jill will add call outs to this piece of artwork after she has completed the writing of the OPERATE THE SAW section. She does not want to miss any important parts on this overall view that she covers in the section (Figure 9–9).

Jill creates the first part of her rough storyboard, which she will use as a foundation for her visual instruction, with a series of roughly drawn pictures (Figure 9–10). From this rough picture outline, she can determine exactly what pictures she will need to develop a complete set of tasks. After the pictures have been created, she adds a rough written outline. From this, she can see where her safety messages will fit into the total picture of the task. With the storyboard complete with pictures, words, and safety message location, Jill can begin to fine-tune her information and develop a complete and accurate write-up of how to start the saw (Figure 9–11).

Jill then reviews her outline with the data in her files. After arranging all of her information in the best place possible and making sure that her instruction is verbally and visually logical, she begins to fill in the visual detail in her instruction. She types a master art list that she can give to the company photographer. She is confident that, with her sketches and her supervision, the photographer will take the sharpest, clearest photographs possible. She has discovered that she will need more pictures than she had originally used in her rough outline. She also will wait until she works with the photographer to determine the best possible *angle* that the photographs will be taken.

The rough layout. While the photographs are being taken and processed, Jill writes the complete tasks and safety information. She has a good mental picture of each photograph that she has requested, and she begins the task of making the pictures and words work together to project the technical information that she must give to the consumer. As she finishes the photographs with call outs and arrows, she begins the procedure of creating a rough layout. She reduces her art on the company copy machine to the approximate size that it will be on the finished instruction (Figures 9-12 and 9-13).

This rough layout does not necessarily show how the final instruction will be printed. The rough layout only shows the tasks, safety messages, and artwork *together* as they will be in the finished instruction. Jill uses her computer to generate printed copy for the text. By doing this, she can electronically transmit her words to the next step in the printing process.

She uses tape to fix her artwork and text that she has generated from her computer printer to the page. Jill knows that there are many machine experts who will review her work before it is printed. She holds this subsection of the instruction until she completes the entire OPERATE THE SAW section. When she is ready to send this material to the reviewers, she will make two copies of it in order to divide her review list in half so as to reduce some time from the process.

Jill will make use of feedback from some of the company workers who are unfamiliar with chain saws. She will ask them to operate the saw by using and following her instructions just as they are written. Jill will also operate and role-play the operator's part as she follows her own instructions word for word. According to Jill, it is better to be honest with herself while she still has the opportunity to revise the text and pictures. She does not want to wait for the customer who purchases the saw to use an instruction that may be unclear and incomplete.

After Jill collects and reviews all of the feedback, she meets with a few of the machine experts in order to clarify a few points that they have commented upon. She makes her corrections to *everyone's* satisfaction. Her next step is to turn her rough draft into a manuscript, which will then be sent to a printing company where it will be printed. Chapter 10 covers the process of preparing a rough draft for the printing process.

FIGURE 9-10 Jill creates her rough outline with pictures on a storyboard.

FIGURE 9-11 Jill adds tasks and safety messages.

Ⓐ

START THE ENGINE

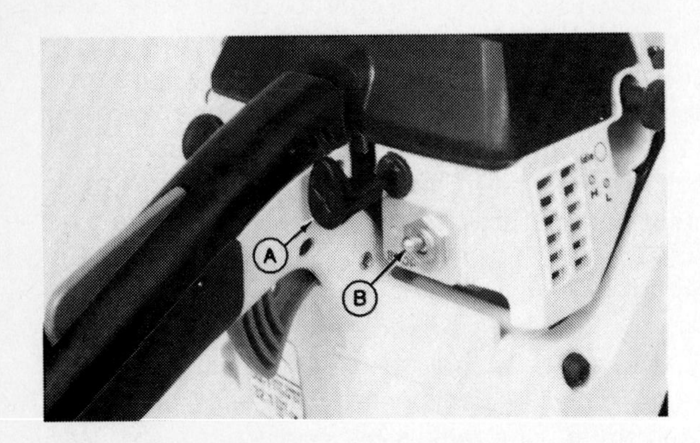

WARNING: Run saw only outdoors and not in an enclosed area. Exhaust gas from the exhaust is a deadly poison.

1. Cold engine: Pull knob (A) out.
   Warm engine: Do not choke.
2. Push switch (B) up.

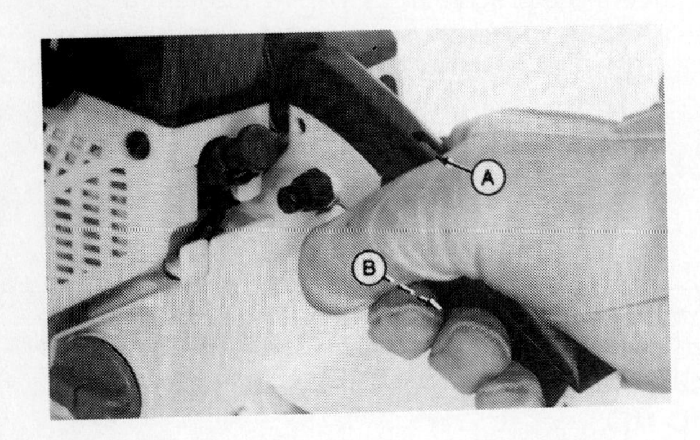

3. Press in throttle interlock (A), and depress trigger (B).

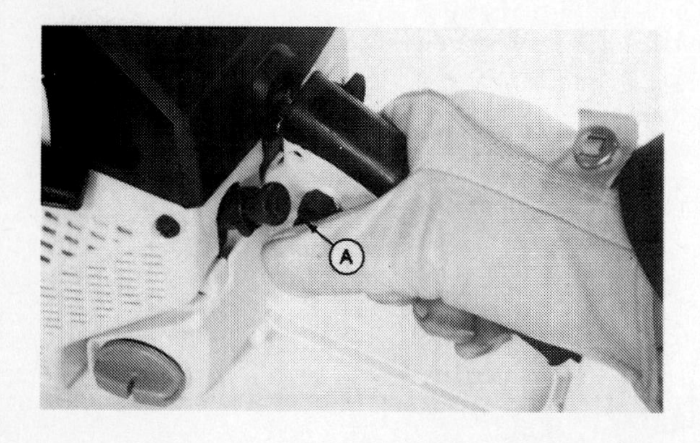

4. Push in throttle lock (A), and release trigger.

FIGURE 9–12 Jill creates a rough draft from the storyboard. (Artwork courtesy Deere & Company)

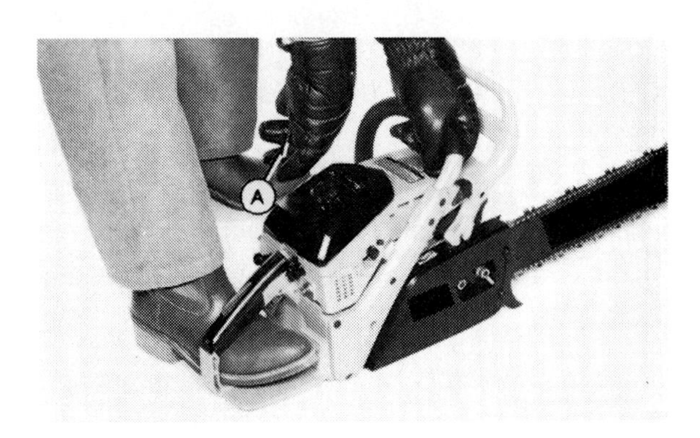

Important: Pull only 2/3 of the rope out at each pull.

5. Hold front handle with left hand.
6. Put right foot on rear handle.
7. Pull recoil starter (A).
8. Cold engine: Push in choke knob.

Note: Allow engine to idle until it runs smoothly.

9. Squeeze trigger (A) to engage chain. The chain will move.

FIGURE 9–13 The rough draft—page 2. (Artwork courtesy Deere & Company)

CHAPTER SUMMARY

- Technical operation instructions can be one or two pages in length. These are called *directions*. Manuals are many pages of instructions that are bound together.
- The technical operation instruction can only be created with the collaboration of many machine experts.
- Many sources must be researched before a technical operation instruction can be written.
- The technical operation instruction is the basic foundation of professional technical writing.
- Technical writers will be more successful writing instructions if they begin their assignment by objectively identifying their strengths and weaknesses.
- Time must be budgeted. The technical writer must plan a strategy to defeat time. The job of technical writing includes more skills and time-consuming work than simply writing the technical operation instruction.
- The deadline, or due date that the instruction must be on the shop floor for the production of the machine, must always be kept in mind.
- Use a desk calendar or a computer program calendar to keep track of important and conflicting dates associated with the project.
- A usable technical operation instruction must be written to the audience. Sex, age, education, training, and other social aspects

must be evaluated when the audience is determined.

- Words can be added to the individual vocabulary in layers.
- Write technical operation instructions in layers according to the audience's vocabulary.
- In order to write visually, make sentences short, accurate, and active. Use active verbs and concrete nouns.
- The *imperative* is the most common sentence structure found in the technical operation manual.
- Active pictures or artwork can give action and life to the instruction.
- The first step in the writing process is to decide upon the content of the instruction. Each machine may require a different content. Safety messages and warnings about possible unsafe situations must be a major part of each instruction.
- Break larger instructions down into sections or chapters, and then subsections or subchapters. Information will be easier to find if headings break up the information and tasks.
- The size of the direction or manual is not determined by the size or weight of the machine.
- Technical operation instruction size is determined by the complexity and newness of the machine technology, legal requirements, industry standards, and the professional judgment of the technical writer.
- A table of contents at the beginning of the instruction lists the divisions of information and tasks.
- The steps involved in creating an instruction include the following: The making of a rough outline; the making of a storyboard outline; the addition of simple headings and sentences to the storyboard; the addition of safety messages and tasks related to safety; the completing of artwork and text; the production of finished art; the layout of the rough draft; a desk-check of the material; and the correction of the rough draft according to the feedback that is received.
- The technical writing process must include a review of a rough draft by the machine experts, who must check for accuracy and completeness.
- The instruction can be desk-checked by people unfamiliar with the technology in the machine.
- The technical operation instruction manuscript must now be prepared for the printing process.

EXERCISES FOR CHAPTER 9

Write the Instruction

1. Mentally *create* a make-believe machine that is based on existing or futuristic machine power systems. This make-believe machine can be a small toy or a large piece of equipment. An example would be a robot that can clean windows.
   a. Create a series of sketches that outline everything that you need to say about your machine.
   b. Put call outs and arrows on your pictures and add the necessary words.
   c. Add text and safety messages. Explain how to use this imagined machine safely and efficiently. Make up operation procedures as you write.
   d. Create a rough draft of your final ideas.
2. Your fellow students will be your audience of experts for this rough draft. Make enough copies of the instruction that you wrote in question number 1 to give to every class member.
   a. Criticize each instruction on the use of artwork, word choice, and overall neatness. Remember, this is a rough draft.
   b. Correct your instruction to reflect your feedback from the experts.
3. Either select a poorly written technical operation instruction and rewrite it in your own visual style, or create an instruction complete with artwork for a machine that you are familiar with.
   a. Put your material in draft form.
   b. Desk-check the instruction and evaluate the usability of the instruction.
   c. Does desk-checking help you evaluate accuracy, readability, or usability?

10

Prepare the Instruction for Print

The information given in this chapter will enable you to:

- Better appreciate the complexity of the printing process.
- Understand why it is expensive and time-consuming to make corrections to an instruction during the printing process.
- Choose a layout for the pages of an instruction.
- Understand the terminology of selecting type specifications.
- Select the proper paper and binding for your bound instruction.
- Prepare a price quote that will help you select a company to print your instruction.
- Complete the manuscript for printing.
- Transmit files of text from the personal computer to the typesetting equipment.
- Create a control sheet to keep track of the instruction during the printing process.
- Use proofreader's marks.
- Obtain copyright forms.

INTRODUCTION

The technical writer's writing assignment may not end when the rough draft of the technical operation instruction is revised. Many times, the technical writer is both the editor and the publisher of the technical document. This situation has gradually developed as manufacturing companies have felt the increasing cost of wages and salaries, the expenditure of capital for state-of-the-art automated high-technology manufacturing equipment, and cost of raw materials. Worldwide markets and worldwide competition have challenged manufacturing companies further to reduce "overhead" costs while increasing product quality.

Many work forces have been trimmed. Many jobs that once were done by two or three people are now done by one person with many skills. For this reason, the technical writing position in many manufacturing companies often has been combined with the editing and publishing job descriptions. The technical writer, especially in a small or mid-sized company, may be given the responsibility for total production of the instruction. Thus, not only must the technical writer-editor-publisher know how to write the instruction but he or she must know how to prepare it for print.

In order to be successful as a technical writer in industry, the technical writer must know the basics of the prepress or preprint steps of manuscript preparation. Any mistakes in the preparation of an instruction can be extremely costly. A machine that is mass-produced by a modern manufacturing company may require tens of thousands of instructions during its life on the shop floor. One small mistake can be reproduced many times over.

The technical writing process is as much of a collaboration between the technical writer and the company machine experts as it is a cooperation between the technical writer and the printing company that mass-produces the instruction. The technical writer who is in charge of the editing and publishing of the instruction must be able to work closely with the people who keyboard, typeset, and produce the finished technical operation instruction. Without a good working relationship with the printing company personnel, the technical writer will find it very difficult to complete the printing of an instruction quickly and mistake free.

As with each machine that requires a unique set of instructions and warnings, each printing company may require special manuscript handling by the technical writer. Printing companies usually vary in the quality of the finished product, printing methods, and the cost of labor and materials. Some printing companies can do better jobs on certain "types" of instructions than others. One printing company may be more dependable and meet deadline dates more readily than other printing companies in the area.

Every department in the manufacturing company that produces a product will be limited by a budget. The technical writing department or group is not an exception. Budgets for particular instructions usually include the entire writing and printing process. For this reason, printing costs must be kept to a minimum in order to meet budget guidelines. Small details in the printing process that are overlooked often add up to large overrun costs.

To stay within budget restraints, it may be necessary and acceptable to sacrifice *some* quality in the final product. It is *unacceptable,* however, to sacrifice the readability and usability of the instruction because of cost limitations. If the instruction is not usable, then it does not fulfill its purpose. If because of poor print quality or instruction layout the instruction fails to tell the consumer in clear and accurate terms how to operate the machine safely and efficiently then the money spent to develop and produce the instruction is wasted. Low-quality technical operation instructions that are sent with the machine as a support in the marketplace will become more of a liability than a benefit to the company.

This chapter will examine the *basic* areas of printing that the technical writer must know in order to get the best quality of instruction at the lowest cost. *Cost* is a key word to the technical writer. A mistake in manuscript preparation or unexpected results from a printing company can add large sums of money to a project that otherwise would have been under budget. By understanding the basics of the printing process, technical writers can communicate exactly what they expect from the people who will publish the instruction for the mass-consumer audience. Above all, this chapter conveys the message that the technical writer must be aware of how the instruction is going to be printed *before* the final manuscript is developed.

UNDERSTAND THE TERMS OF THE PRINTING PROCESS

A Case Study:

Jill Davenport, a technical writer at a consumer products company, is given the assignment to publish her instruction on the chain saw that she completed. Until today, she has only been responsible for writing the technical material and not preparing it for the printing process. Publishing is a new and challenging area of technical communication that she approaches with enthusiasm. She knows that the machines and techniques used in modern printing companies are developing at a fast rate. She believes that this part of the job will be exciting and that she will have the opportunity to learn much about the printing industry.

She contacts one of the larger printing companies where her department buys printing. She does this because she has been told that the printing company has a wide variety of hardware that she should become familiar with. Mike Block, the printer "rep," sets up an appointment for her to tour the plant facilities. Because Mike's company has the capabilities to do all of the "prepress" work on her instruction, Jill will be able to see the entire prepress and printing process in one visit.

Mike meets her at the receptionist desk and explains that his company is divided into three main sections. They are:

The sales, scheduling, and other front office departments.

The "makeready" area.

The pressroom.

Mike begins his tour in the sales and scheduling area. Jill is introduced to the people that she will be talking to at the company. Personnel in this area are in charge of selling, planning, and customer relations. They are the people who will furnish her with a schedule of any future job that she has in work. A print schedule is extremely important to Jill. It tells her when to expect the printer's proof that she receives, when it must be returned to the printer, and when the printing company can furnish the final printed instruction in quantity. The people in the front office are her contacts if she has a scheduling or time problem.

An important group of people for Jill are the experts who will use her "quotation" or explanation of what she wants to print in order to give her a price bid on the job. Jill will supply the same quotation to each of the major printers that her company has selected. Each printing company will bid or offer her a price for completing the printing job. Jill will use the bid to select which printer will print her instruction. Price is a major consideration when Jill chooses a printing company (see Figure 10–1).

Makeready. Jill and Mike's next stop is the printing company's makeready room. Jill knows that the jobs that are done here also can be purchased from graphic arts companies that do nothing but prepare "mechanicals" of instructions and other manuscripts for printing. However, she doesn't know exactly what takes place in this area. Mike tells her what she will see on this part of the printing company tour. The makeready jobs include:

Input of the text or words into a format that the electronic typesetting machine can use to set the type and create a negative.

Photography of the artwork and the creation of a negative.

Stripping of the artwork and text negatives into a "flat."

Platemaking for the plates required for the printing presses.

The words of an instruction will begin the step-by-step procedure through the makeready area in two places. If the text is input on a personal computer and the information is properly stored on a floppy disc, then it can be taken directly to the typesetter. If the text is typed from a mechanical type-

FIGURE 10–1 Each instruction that Jill publishes will be bid upon by many printers. (Courtesy Bawden Printing, Incorporated)

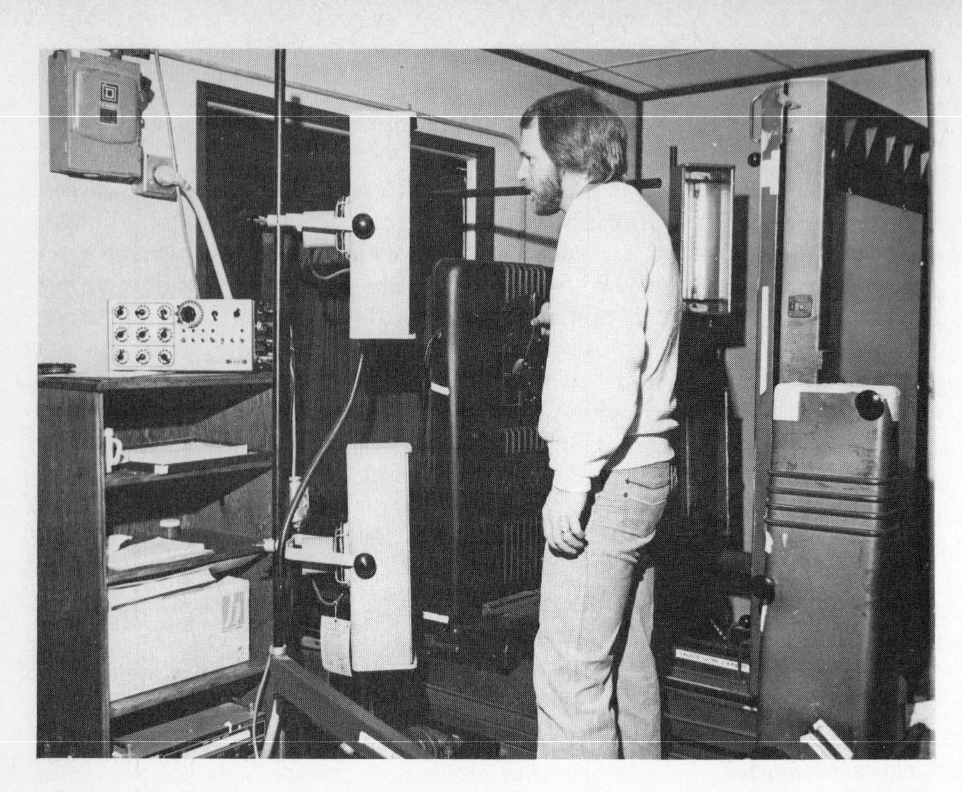

FIGURE 10–2 The composition camera creates negatives of art and text. (Courtesy Bawden Printing, Incorporated)

writer and exists only on paper, then it must be keyboarded into program commands that the typesetter can use. Mike shows Jill how the program is transmitted to the area of the electronic typesetter where the text is "set."

The typesetting machine at Mike's company translates the program commands into usable flashes of light. As the light reflects through a small wheel that has the letters of the alphabet and special symbols cut out of it, light-sensitive film picks up the exposures at an extremely rapid rate. In this particular typesetting machine, this "negative" is produced in a "galley" or a continuous sheet of negative paper that contains line after line of the instruction without a break.

Mike further explains this step of the process by saying that the galley that is being made is a *line negative.* That is, what will be the black ink of the words is clear on the negative paper, and the white background is dark. On the other hand, the black and white photographs, or artwork, provided with the instruction copy are being shot in a "process camera" in the next room (Figure 10–2).

Halftones. The composition camera is specially designed for graphics art reproductions. The black-and-white photographs are called "continuous tone" according to Mike. When they are mounted temporarily on a board and photographed again by the process camera, they become "halftones." Halftones are created when a "screen" of tiny dots is placed between the photograph and the film. This is done, explains Mike, because the photographs will not reproduce during the printing of the instruction if the small dots are not on the negative. After they are screened, the photographs are called halftones because part of the continuous tone is removed by the screen.

Mike gives Jill an example. Newspaper comic strips are printed with a coarse 85-line screen because of the pulp texture of the newsprint paper and the printing process. This means that there are 85 lines of dots per inch on the screen that is used. You can see the dots if you look closely at the cartoon. A newspaper is printed on rough paper stock and usually requires such a coarse screen to be legible. Paper quality, ink, printing methods, and the expected quality of artwork in Jill's instruction determine what screen to use. Mike explains that most instructions are printed with photographs that have been screened at 85, 100, 120, and 133 lines per inch.

The printing method will have some impact on which screen Jill wants for the photographs. For example, she might choose a 133-line screen for an "offset" printing process. It will cost Jill more money to print an instruction as she increases the screen; however, the print quality should also increase. An increase in screen density will require more ink as the artwork is reproduced.

Flats. The next stop for Mike and Jill is the tables where the artwork negatives and line negatives of text are "stripped" together into complete pages of the instruction. Each page is then put into windows and taped to a stencil. This step of the process creates "flats." Flats are used to make plates. More than one page is put onto a flat because it is less costly to print more than one page at a time.

Flats are created with 4, 8, 16, and 32 full-sized 8½- × -11½-inch pages. Mike points out that, for this reason, it is often less expensive to print a 16-page than a 12-page instruction. A 12-page instruction takes special handling, but a 12-page instruction can be reduced in cost by adding four blank pages at the end of the tasks and specifications in order to make a 16-page book (see Figure 10–3).

The proof. An important step in the printing process for Jill happens after the flats are created. They are used to create a "proof" that Mike sends to Jill. Proofs are often called *blue lines, brown lines, silver prints, blues,* or *Vandykes.* However they are made or whatever they are called, proofs are a valuable double-check of the instruction to make sure the instruction is correctly laid out before it is printed.

Jill must review and mark up this proof with the necessary corrections using proofreader's symbols. Mike explains that she should also mark corrections with symbols that show the cause of the correction. For example, if the printing company made the error, then Jill should mark the correction with a "PE" (printer's error) or similar mark to show that the mistake was made by the printer. If Jill adds new material because the data for the instruction changes, or because she made the mistake, then she should mark the corrections with "WE" (writer's error) or similar mark that the printer understands.

This is important to Jill because her company must pay for all revisions that she makes to the proof. If the printer makes the error, then the cost should not be charged to Jill's company. Changes, especially additions, can be extremely expensive at this point. Not only must the corrections be input, set in new negatives, and stripped on the page, but the instruction may need to be put into new flats. After the proof is returned to Mike, he reschedules the complete revision process. All of the steps for creating flats must be redone.

The flats are then used to create "plates." Metal plates are treated chemically and put inside a platemaking machine. As light passes through the negatives in the flats, a chemical reaction occurs. The plates are created so that the light and dark areas stand out in contrast. The chemical-coated light-sensitive plate develops after it is exposed to light much like a photograph. Mike tells Jill that there are two basic plates that his printing company makes for its two types of printing processes. They are:

FIGURE 10–3 The negatives of the text and artwork are brought together in "makeready." (Courtesy Bawden Printing, Incorporated)

FIGURE 10–4 The Letterpress method prints pictures and letters directly on paper.

Plates with a raised print area for the "letterpress" method.

Chemically treated plates that are flat for the "offset lithography" method.

Mike takes Jill to the pressroom to see the two types of printing presses used to print technical operation instructions for his clients. One is a *letterpress*. For this type of press, a metal plate with letters, dots, and symbols that stand up from the plate is used. As the light source behind the flats hits the chemically treated plate, the background begins to be eaten away with acids. A fixative is applied to the plate to stop the process. For this reason, the letterpress type of printing is often called *relief printing*. A basic example of this method is the rubber stamp that Jill can purchase to stamp her address on an envelope. Ink is applied directly to the plate and the plate is pressed against the paper (Figure 10–4).

Offset lithography. The second press that Mike shows Jill is the *offset*. The offset press plate is exposed to light and chemically changed on the surface. The image or words and artwork areas accept ink and reject water. The white areas accept water and reject ink. Mike explains that his company uses the offset method to do the bulk of their direction and manual printing (see Figure 10–5).

FIGURE 10–5 The Offset method prints from a flat plate that is treated to accept or repel ink.

The term "offset" is derived from the way the ink and the plate image are transferred to the paper. There are three cylinders in offset printing. One cylinder is a mount for the printing plate. Another cylinder surface accepts the inked image. The third is the impression cylinder that presses the paper against the inked-image cylinder. Mike explains that this method is usually less costly than letterpress for the body of the instruction.

Two types of presses are used to print instructions. One is the *sheet-fed* press that prints one sheet of paper at a time. The second type is the *role-fed* press that is used for the larger manuals. From practical experience, Mike tells Jill that the role-fed press is the less costly press to use; however, he feels that the sheet-fed press can produce a better quality job (see Figures 10–6 and 10–7).

The instruction is printed on one of these presses at Mike's company. If the instruction is an operation manual, the body pages are printed at the same time in flats of 4, 6, 8, 16, or 32 pages. A cover that is a *heavier* stock is printed separately from these flats. The heavier cover is printed in a flat of four pages, which includes the following:

Front cover, which usually shows the company name, logo or symbol, manual number, and title of the manual. The copyright symbol often is displayed on the front cover if the manual is copyrighted.

Inside front cover, which usually gives introductory information about the product.

Inside back cover, which sometimes contains the warranty, machine serial number registry, or other important information.

Back cover, which is used with the front cover to protect the inside information.

After the printing plate is mounted to the printing press and the paper is printed front and back, the paper is folded and cut. Two flats or two plates are used to print the pages front and back. These flats are set aside until the remaining flats are printed. After all of the impressions are made of the flats, they are collected and fed in sequence through a binding machine if the page count is below 92 pages. Here they are bound together with the cover. The cover and inside pages are printed on a slightly larger page than is necessary. The entire manual is then "trimmed" to separate the pages of the manual and to cut them down to a uniform size (Figure 10–8).

Mike further explains that the printing process, like technical writing, varies between compan-

FIGURE 10–6 A role-fed press is fast at printing and folding signatures. (Courtesy Bawden Printing, Incorporated)

FIGURE 10–7 A sheet-fed press prints one signature or plate of flats at a time.

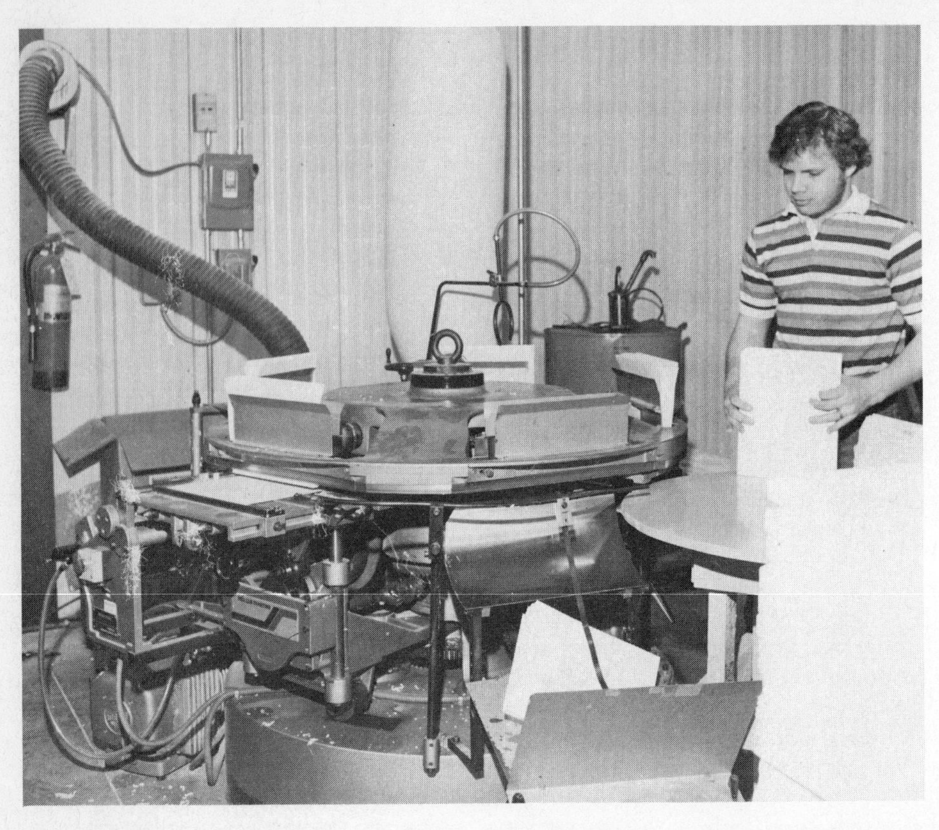

FIGURE 10-8 A binding machine glues the backbone of the book and attaches the cover. (Courtesy Bawden Printing, Incorporated)

ies. Not all presses print the same. Not all inks look the same. Paper that is used to accept the image is not all made by the same manufacturer. There are four things that Mike wants Jill to keep in mind at all times when she deals with a printer. They are:

> Always ask for print samples when talking to a printing company about a particular job or "run."
>
> The higher the quality of reproduction, the more the work will cost.
>
> At every printing company, corrections to the instruction caused by the client add to the total estimated or "bid" print cost.
>
> Make sure that the printer knows exactly what is required of the finished product before the instruction is released for printing.

Mike gives Jill a checklist that she can use to compare cost versus quality between his company and the competing printers who may try to obtain Jill's printing contract. He also gives her print samples with paper, ink, and press specifications to compare with other printers' samples. Because Jill's company has done business with other printing companies, she can obtain printed copies from past print jobs that have been done by each printer. Mike's quality checklist includes these items:

Printer Quality Checklist

Check for these basic items when comparing samples of printing:

1. Ink rub off. Manuals must be printed with enough quality to last the life of the machine. They will be constantly referred to during the operation of the machine. Ink must be permanent.
2. The quality of halftone reproduction. Not every printing company will obtain the same results from a 100-line screen. Quality may vary considerably.
3. Poor ink drying that causes ink to transfer to the page next to it.
4. *Ghosting,* or images that tend to get lighter for no reason.
5. *Show-through* to the next page. If you hold a page up and can read the text on the backing page, then you are looking at show-through.
6. Foreign particles in the ink or on the paper.
7. Paper curling due to improper drying, poor quality paper, or too much pressure on the press.

Jill begins her research into the business of preparing her instruction for the printing process. Mike tells her that her success as a publisher depends upon what she can learn about the inside workings of a printing company. Mike gives her a list of important items that she should research before she begins her search for a printer. She must know the basics of:

Page layout specifications.

Type specifications, and how to use type to emphasize important information.

How to select paper and binding for each manual.

How to prepare a quote to allow the various printers that Jill's company deals with to compete for the printing of her manual.

How to lay out a manuscript properly.

How to transmit files electronically to the typesetter.

How to create a control sheet to keep track of her manual at all stages of production.

How to mark the printer's proof with proofreader's marks.

How to copyright the instruction to protect art and text from being plagiarized.

Jill begins her research by reading a few graphic arts books that explain how to prepare copy for printing. Mike explains to Jill that she must know how to complete the above list no matter what printing company or composition method is used. Mike explains that these skills will be essential to have when Jill's company decides to develop a computerized desk-top publishing system to create mechanicals or layouts of instructions "in house" or within the company.

PRINT THE INSTRUCTION

Develop a Consistent Page Layout

For you, the technical writer, the physical arrangement of the technical information on the page is important for two reasons. First, the layout will determine how easily the consumer can perceive the material. Second, the physical layout affects the size of the instruction. Often, some machine manufacturers issue one- or two-page unbound directions along with operation instructions that are squeezed into the two pages without regard to how well the consumer can read and use what is printed. Manuals that are bound sometimes have so many different types of layouts in them that they become confusing to the consumer who is referencing the information.

The key to designing instruction pages for the needs of the consumer is to keep the layout *consistent.* If information is always presented in the same way, then consumers will not be confused as they try to follow your tasks. The readability and usability of your instruction will be influenced as much by the overall design as by the words and pictures that you choose. There are a few common-sense guidelines that can help you develop a usable page layout of your instruction.

How you lay out your instruction will be determined by how many pieces of art you have, what emphasis you want to put on your words or artwork, and where the instruction will be attached or put inside the machine. As you determine what layout you will use, you must keep in mind that the layout must be kept "open" so that your instruction can be easily revised. Every machine that is manufactured must be constantly reengineered in order to stay ahead of the competition. Reengineering means that you must revise your instruction by adding information, deleting out, or replacing it.

If you have an open layout design with white space, then you will have blank space enough to "add" information without having to completely restrip and remake printing plates every time a correction to the instruction is made. An open design also enables you to delete information and leave open space without making consumers feel that some of their instruction has been accidentally omitted. A manual or direction with white space, or unused space, if it is skillfully used, is easier for consumers to read.

The three basic ways that information can be consistently presented are:

The vertical style.

The horizontal style.

The cartoon or "chessboard" style.

The vertical style. This is a popular layout when the tasks are relatively simple and easily explained. Artwork is usually kept to a minimum, and words tend to take over. However, it may be difficult for readers to concentrate on the pictures in this layout while following the text. The vertical style layout is a good layout to use if descriptive or other informative information must be presented. It is not necessarily the best way to present tasks to the consumer, however.

The horizontal style. Text can be isolated to the left or right of the artwork in the horizontal-style layout. This type of layout may take more space than the tighter layout of the cartoon series. Artwork, however, can be used to support multiple tasks. A problem may arise if too many call outs are added to the artwork. Usually, company style sheets limit the number of call outs to four or five symbols unless the illustration is a major overall view of the product that shows the location of all major items.

An emphasis upon artwork is created if the artwork is on the left, and the text is on the right. The text is primary if it is on the left, and the artwork is on the right.

Since we read from left to right in our culture, this may be the easiest layout for the consumer to follow. However, it seems to be a more difficult layout to use if the two arrangements of art and text are mixed. Position of art and text in translated manuals can be varied depending upon the direction in which the consumer reads.

The cartoon style. This by no means is a derogatory connotation of an often used layout. In the cartoon style, pictures are arranged like a comic book or cartoon. This method can save space in lengthy manuals; however, care must be taken to ensure that it is open enough for revisions and changes to the text and artwork. This layout may be difficult to follow in some situations if the instruction is not numbered sequentially. Cartoons can read from left to right, or from top to bottom. The problem with cartoon layout is that one piece of artwork may not be able to support more than one or two tasks. Many illustrations often must be repeated. The artwork tends to demand the most attention in this layout. Care must be taken not to overcrowd the page (see Figure 10–9).

There is no best way to lay out every type of technical operation instruction. A layout depends upon what you want to emphasize, how much space you have to work with, how readable and usable your instruction must be, and how much money you can spend on paper and printing.

Layout also may depend upon the size of your instruction. The standard page size is 8½ inches wide by 11 inches high. If you look at instructions from many products, you will find that their size differs as much as do the machines that are being explained. An 8½- × -13-inch two-page (one sheet) direction can be packed inside a wristwatch case if it is folded five times. An operation manual for a car may be 7½- × -4-inches in order to fit into a glove compartment. Each physical dimension of the instruction may require a different layout specification.

It is important to note that most designers of technical operation instruction layouts allow the art and the text to determine the length of the manual. For a contrast, look at the rough of an advertising page. Each page is a vignette. Each page must be designed to be appealing to the eye of the reader. In this case it is important to "fit" type and artwork into blocks. Type is "sized" to fit. Artwork may be of different sizes depending upon the effect that the layout artist wants to create (Figure 10–10).

The technical writer must look at each task as an important bit of information. Technical information in the instruction must not be "fitted" for a particular space at the expense of readability and usability. The instruction must be laid out simply and consistently so that each task is a vignette. Information must be easy to read, easy to find, and the safety messages must stand out from the regular text.

Important page dimensions. The final manuscript that is sent to a printer must be dimensioned. Measurements required by the printer include the *overall* size of the instruction. You also must include the following dimensions illustrated in Figure 10–11:

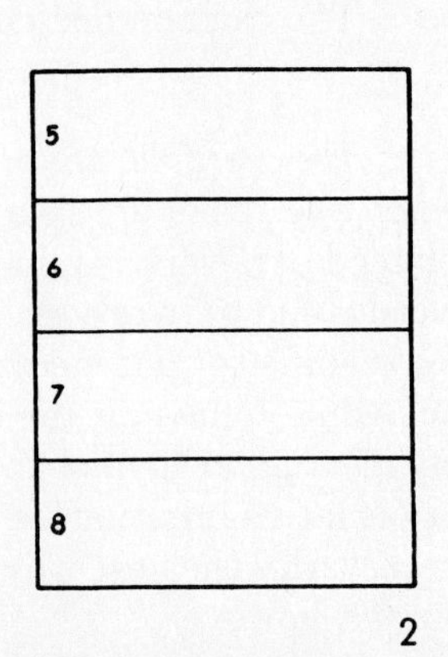

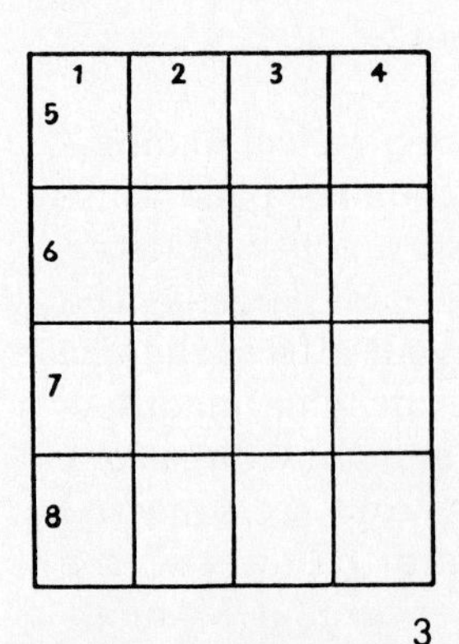

FIGURE 10–9 (A) Vertical format; (B) Horizontal format; (C) Cartoon format.

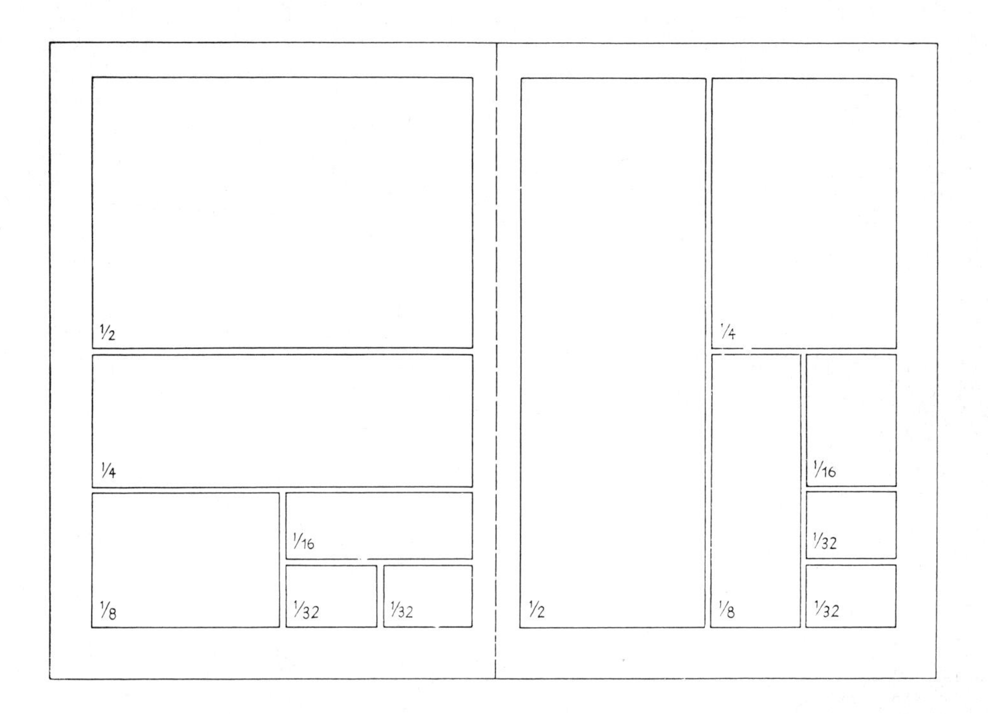

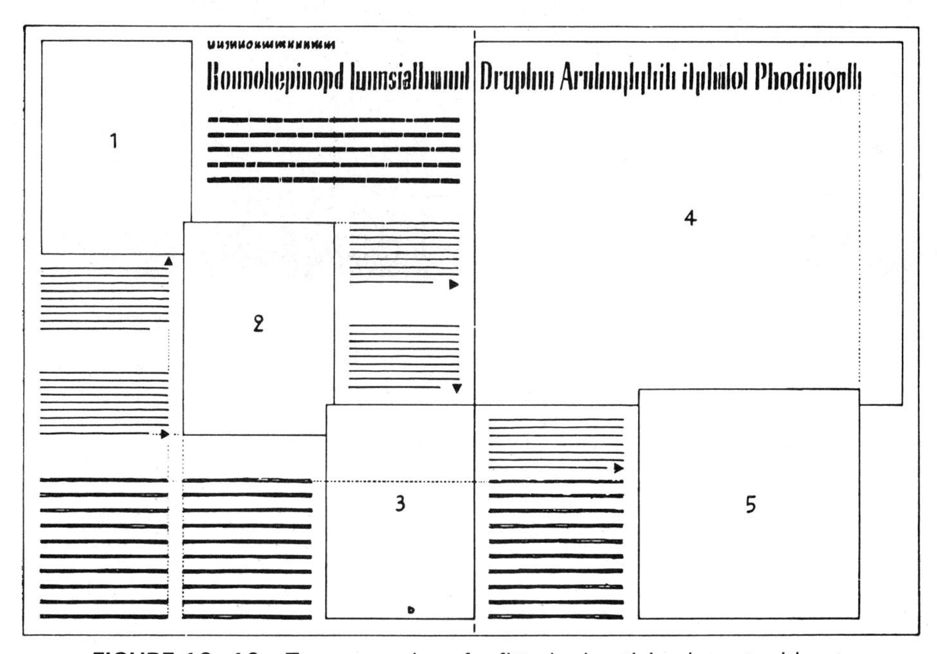

FIGURE 10–10 Two examples of a fitted advertising layout with art and text sized for overall page appeal.

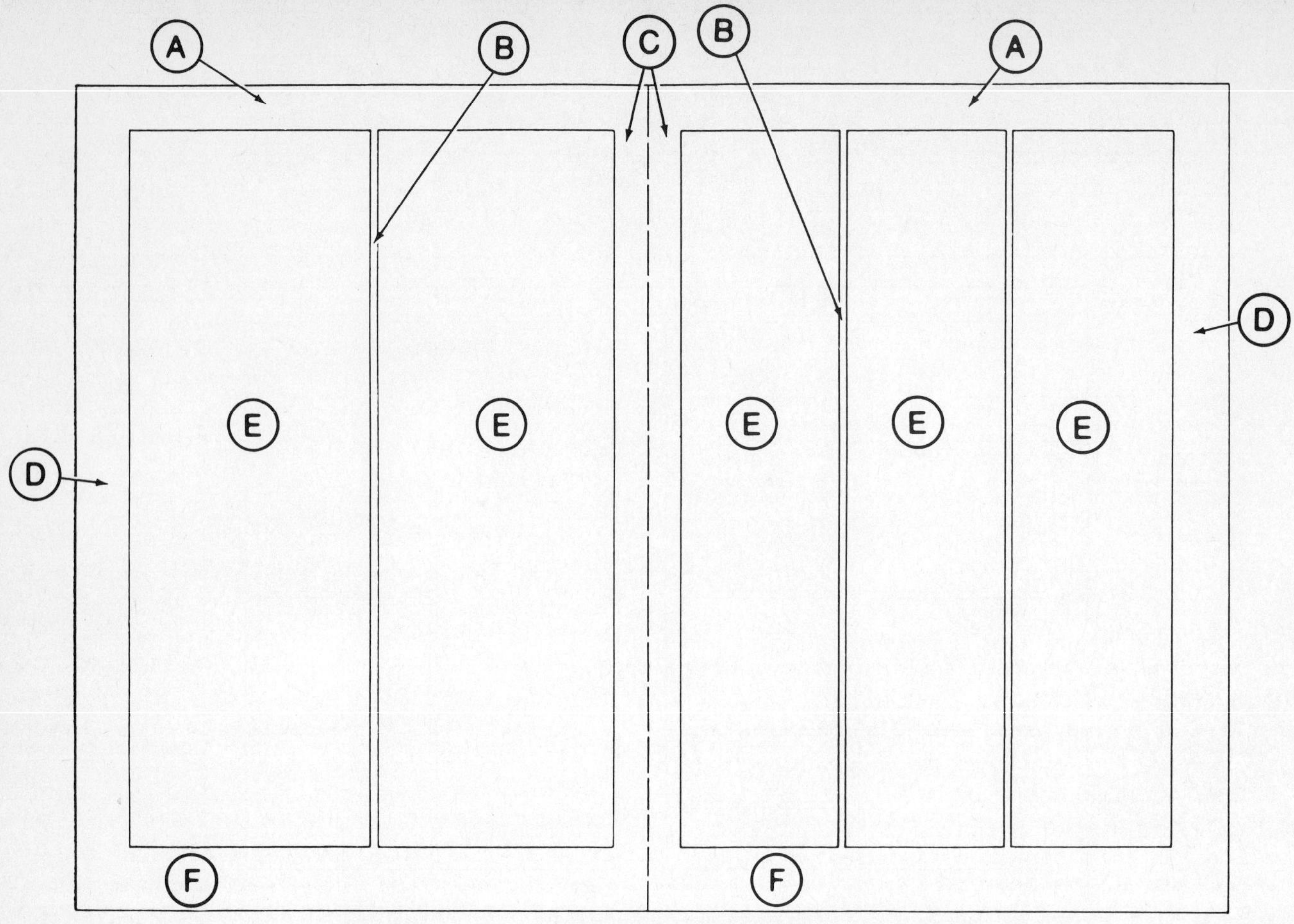

FIGURE 10–11 Two facing pages with varying formats.

A. *The top margin.* How far must the printer go down from the top of the page to begin the text and art?

B. *The alley.* If there is to be a space between the columns of type and text, how wide will it be?

C. *The gutter.* How far will the columns be placed from the binding? The dotted line in Figure 10–11 is the binding that separates two facing pages.

D. *The outside margin.* How much space is there between the column of text and art and the edge of the page?

E. *The column.* How wide is the text and art on the page?

F. *The bottom margin.* How far must the text and art be from the bottom of the page?

The above example shows two pages with two separate layouts. Usually, one layout will be used throughout a manual or direction by the technical publisher.

Type-block arrangements. Another aspect of page layout is the choice of line arrangements. There are three basic types of line arrangements used in the printing of technical operation instructions. These are:

Lines blocked with flush-paragraph formats. These lines are arranged like this:

Lines that are blocked with indented paragraphs:

The most used arrangement of text lines is the flush left and "ragged," or "rough right," format. Many studies of readability cite the alignment of the left column and the ragged variety of the right column as the easiest line arrangement for the eye to read:

The length of a printed line can affect the eye movement of the reader. A widely accepted line length of 3-¼ inches to 3½ inches, or 40 to 50 characters, may be the easiest to read. Column size depends on the layout you choose. Lines are usually flush left, or aligned with the left margin of the page. The right side of the line is sometimes left ragged or not set flush. The ragged 3-5/32-inch column may be a good place to start when you lay out the rough draft of your instruction.

Choose and Mark the Typeface and Character Size

Under most circumstances, the type style used in an instruction remains the same throughout. There are many type styles including some of the most popular ones used in instruction printing: (Megaron) or Helvetica, Bookman Light, Century Regular, Schoolbook Regular, and Times Regular. The choice of which type to use is the responsibility of the publisher of the instruction. Each printing company may have slightly different typefaces for the same type style. Ask the printers to see a "font" or a collection of upper- and lowercase letters of each type style before you choose the type from a particular printing company (see Figures 10–12 and 10–13).

You must keep in mind that the typeface must be simple and clear and must not include decorative lines. Readers can be as confused by elaborate or out-of-the-ordinary typeface styles as they can be with inaccurate or incomplete information. Type must be chosen for usability and not decoration.

Each typeface can be varied as to size. Size is measured in *points.* For example, the overwhelming size of type used in technical operation instructions is 10 points. This measurement is made from the bottom, or "descender," of the characters such as "p" "y" and "g" to the top, or "ascender," of characters such as "h" "T" and "L." There are 72 points in one inch. Point rulers are available at graphic arts supply stores to help you measure type size (Figure 10–14).

Another measurement that is made in points on copy is the "leading," or space between the lowercase letters such as "y" and "g" in one line and the top of the characters such as "T" and "L" in the line directly below. This space must be wide enough to allow the reader-consumer to separate lines easily and yet not lose his or her place as the eye scans from one line or task to another. A common leading is 2 points. If a 10-point size type is used with a leading of 2 points between the lines of the text, then you must mark your copy with a **10/12 leading sign.** A 10-point type with a 3-point leading is **10/13.** Readability depends to a great extent upon the arrangement and type size. Too much leading is as detrimental to readability as not enough.

The space between each character is also measured in points. It is important to ensure that this space is wide enough to permit the consumer to see and read each character of type. It is important for you to see printer samples of character spacing, leading, and character size before you specify exactly what you want for your type size. You must make certain that you know what is being talked about when type samples are submitted to you. A regular or 1-point spacing is a common measurement for this part of the type format.

As you can see, the space on a page is as important to the overall usability and readability of the technical operation instruction as are the typeface and the images. Another major measurement of space in typesetting is the "quad" or gap between words. To the average reader, this is not an important measurement unless the space is so close that words are run together. A quad is the square of the point size of each character. For example, a quad for 10-point type is 100 square points, or 10 × 10 points. Quads can be measured in:

Em quads.
En quads.
3-to-the-em space.
4-to-the-em space.
5-to-the-em space.

An average word spacing in a technical operation instruction is a 3-to-the-em quad.

Megaron (Helvetica) Light 10 pt/12 pt.
ABCDEFGHIJKLMNOPQRSTUVWXYZ
abcdefghijklmnopqrstuvwxyz

Megaron (Helvetica) Light 12 pt/14 pt.
ABCDEFGHIJKLMNOPQRSTUVWXYZ
abcdefghijklmnopqrstuvwxyz

Megaron (Helvetica) Light 14 pt/18 pt.
ABCDEFGHIJKLMNOPQRSTUVWXYZ
abcdefghijklmnopqrstuvwxyz

Megaron (Helvetica) Light 18 pt/20 pt.
ABCDEFGHIJKLMNOPQRSTUVWXYZ
abcdefghijklmnopqrstuvwxyz

Bookman Light 10 pt/12 pt.
ABCDEFGHIJKLMNOPQRSTUVWXYZ
abcdefghijklmnopqrstuvwxyz

Bookman Light 12 pt/14 pt.
ABCDEFGHIJKLMNOPQRSTUVWXYZ
abcdefghijklmnopqrstuvwxyz

Bookman Light 14 pt/18 pt.
ABCDEFGHIJKLMNOPQRSTUVWXYZ
abcdefghijklmnopqrstuvwxyz

Bookman Light 18 pt/20 pt.
ABCDEFGHIJKLMNOPQRSTUVWXYZ
abcdefghijklmnopqrstuvwxyz

Century Regular 10 pt/12 pt.
ABCDEFGHIJKLMNOPQRSTUVWXYZ
abcdefghijklmnopqrstuvwxyz

Century Regular 12 pt/14 pt.
ABCDEFGHIJKLMNOPQRSTUVWXYZ
abcdefghijklmnopqrstuvwxyz

Century Regular 14 pt/18 pt.
ABCDEFGHIJKLMNOPQRSTUVWXYZ
abcdefghijklmnopqrstuvwxyz

FIGURE 10-12 Type samples.

Century Regular 18 pt/20 pt.
ABCDEFGHIJKLMNOPQRSTUVWXYZ
abcdefghijklmnopqrstuvwxyz

Schoolbook Regular 10 pt/12 pt.
ABCDEFGHIJKLMNOPQRSTUVWXYZ
abcdefghijklmnopqrstuvwxyz

Schoolbook Regular 12 pt/14 pt.
ABCDEFGHIJKLMNOPQRSTUVWXYZ
abcdefghijklmnopqrstuvwxyz

Schoolbook Regular 14 pt/18 pt.
ABCDEFGHIJKLMNOPQRSTUVWXYZ
abcdefghijklmnopqrstuvwxyz

Schoolbook Regular 18 pt/20 pt.
ABCDEFGHIJKLMNOPQRSTUVWXYZ
abcdefghijklmnopqrstuvwxyz

Times Regular 10 pt/12 pt.
ABCDEFGHIJKLMNOPQRSTUVWXYZ
abcdefghijklmnopqrstuvwxyz

Times Regular 12 pt/14 pt.
ABCDEFGHIJKLMNOPQRSTUVWXYZ
abcdefghijklmnopqrstuvwxyz

Times Regular 14 pt/18 pt.
ABCDEFGHIJKLMNOPQRSTUVWXYZ
abcdefghijklmnopqrstuvwxyz

Times Regular 18 pt/20 pt.
ABCDEFGHIJKLMNOPQRSTUVWXYZ
abcdefghijklmnopqrstuvwxyz

FIGURE 10–13 Type samples.

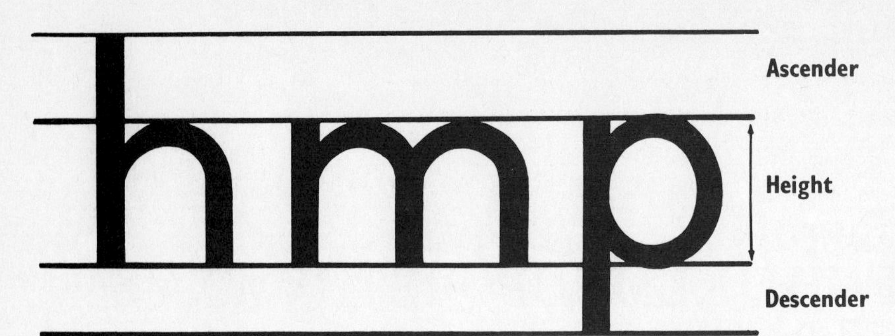

FIGURE 10–14 Type is measured from the ascender to the descender.

Use Type to Emphasize Information

There are many reasons why the layout is important in an instruction. The main one includes the setting off of warning and safety information. Usually safety warnings and related information are set off from the text of the technical operation instruction by using a **boldface type** (BF) or an *italic typeface* (IF), and by varying the justification of the line. Each font, or type style, can be printed in a variety of thicknesses and specific configurations.

Another way to emphasize warnings and other important information is to indent the line that the information is printed on. In typographic terms, the text of the instruction can be justified left and the warnings indented 10 ems. In other words, all tasks are aligned left except the warning and important information begun a few spaces later. This arrangement makes the consumer stop and contemplate the information because it is *emphasized.*

Hierarchy headings also can help ease the task of reading an instruction. A primary head that begins a section of an instruction may be set in 18-point type with all caps, or capital letters. A secondary head may be set in 12-point type just slightly larger than a 10-point body type in caps and lowercase letters. A subhead may be printed in 10-point boldface with caps and lowercase letters. Each heading is a key to where the consumer is in the instruction. At every major head, for example, the consumer knows that he or she is beginning another major part of the operation of the machine.

Select Paper and Binding

Every item or specification on the printed page affects the usability of the instruction. This includes the ink, the paper, the white space, and the arrangement of information. You saw the major types of arrangements of information earlier in this chapter, but you must also be aware of other details of art and text that can make your technical operation instruction easier to read.

The color of the paper and the color of the ink must be chosen to match the application. White paper with black ink is best for an instruction. There are, however, at least eight or nine shades of white paper. Yellow, off-white, gray, or tan paper, for example, will not give you as good a reproduction of halftones as will a good white uncoated book paper. Paper must match the printing process. Paper is graded by weight and finish.

The weight of paper is measured in *pounds.* It is the weight of 500 sheets of paper. Paper size varies depending on the use. A sheet of standard paper for the inside pages of a manual measure 25 inches × 38 inches. Hardcover stock for manuals measures 20 inches × 26 inches. A 50-lb coated cover stock paper is not the same as a 50-lb standard bond stock. Paper also varies in quality to some degree from manufacturer to manufacturer for the same grades. Always obtain a sample of the paper stock that your instruction is going to be printed on before you accept a bid from a printer. Paper may be 40 percent to 50 percent of your total print cost—choose paper carefully.

Paper quality. The most important qualities that must be considered when specifying paper is opacity, smoothness, and ability to absorb light. *Opacity* determines how much "show-through" a paper has. If you look at a piece of paper that has been printed on both sides, you will see an example of opacity. If the images of the artwork and the text on the other side of the paper can be seen, then the paper has low opacity. Show-through is distracting and makes the text and artwork harder to read. Rough paper, depending on the printing process, does not give as sharp an image as smoother paper. Paper that absorbs light creates darker halftone artwork. High-opacity paper also adds to the cost of the finished instruction.

Along with paper specifications, you must

choose the least expensive "binding" that can hold the pages of a bound manual together if you have more than two pages of copy. The four major types of binding include:

Wire stitch.
Mechanical wrapped.
Perfect.
Edition, or hardcover.

Wire stitch. Hardcovers are usually eliminated because of cost, which can run from five to ten dollars per manual for a hardcover. Saddle-wire stitch is the most popular for medium-sized manuals. This type of binding is usually found on manuals up to 92 pages. The wire staples are placed on the backbone of the manual. A side-wire stitch also can be used; however, it is harder for the consumer to keep the wire-side-stitch book flat when it is open. Side-wire stitch is most economical and usable for smaller manuals of approximately 24 pages or less (Figure 10–15).

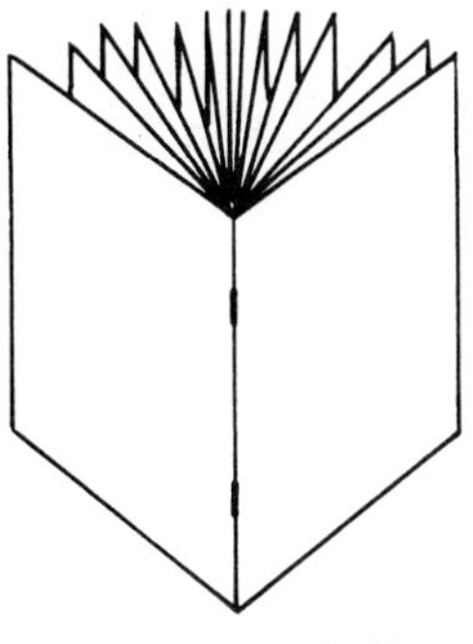

Saddle-wire stitching.

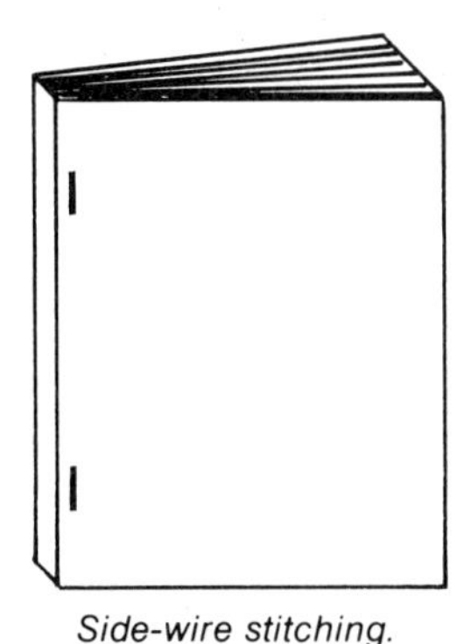

Side-wire stitching.

FIGURE 10–15 Top: Saddle-wire stitch binding. Bottom: Side-wire stitch binding.

Mechanical wrap. A mechanical-wrapped backbone is used on manuals that must be easily referenced and also lie flat when they are used. A common example of this type of binding is the spiral notebook used in the classroom. The metal or plastic clips or spirals add cost to the publication. Most of the work required to do this type of binding is often done by hand, which can add significantly to the cost of the final product (Figure 10–16). Most computer manuals have a mechanical wrap binding.

Perfect bound. Perfect binding is often used for manuals of 96 pages or more. This depends, however, upon the binding equipment that is in the print shop. Perfect-bound manuals are trimmed on the backbone, and abrasive material is used to scuff up the inside pages of the manual. The backbone is then glued and the cover is wrapped around the inside pages (Figure 10–17).

The most important point that can be made is that you must get the most quality possible for the money that you have to spend on your instruction. There is a limit as to how much quality is needed. You must look at how the instruction is going to be used, and then determine the paper, type size, line screen for the artwork, layout of the page, binding, and so forth. Every decision you make affects the final cost and quality of the instruction. *You must balance the quality of your materials with the use of your manual. Technical operation instruction must be printed to be used. Often the highest-grade paper and the most expensive type of binding are not necessary for the application.*

Prepare a Quote

A Case Study:

Jill Davenport is ready to complete the final manuscript that she will send to the printing company. Before she begins to put her layout and text and art on paper, she develops a *quote.* The quote allows her to send the vital specifications of her technical operation instruction to the various printing companies that she has selected. From this information, the printing companies will make a "bid" for the printing job. They will tell Jill exactly how much they will charge her for completing the printing of the chain saw manual. Jill must realize that this must be a fair bid for all of the printing companies that are trying to get the job. They must be given the same quote sheet and time limit for responding. Jill develops a specific quote for the chain saw manual.

Helvetica is a contemporary typeface of Swiss origin. Although typefaces without serifs were used in the nineteenth century, it was not until the twentieth century that they became widely used. Helvetica was introduced in 1957 by the Haas typefoundry and was first presented in the United States in the early 1960's. Although Helvetica has a large x-height and narrow letters, its clean design makes it very readable. Sans serif types in general have relatively little stress and the strokes are optically equal. Because there is no serif to aid the horizontal flow that we have seen is so necessary to comfortable reading, sans serif type should always be leaded.

18 POINT HELVETICA 8 POINT LEADED

ABCDEFGHIJKL
MNOPQRSTUV
WXYZ&abcdefg
hijklmnopqrstuvw
xyz1234567890
$.,"-:;!?

72 POINT HELVETICA

74 DESIGNING WITH TYPE

ABCDEFGHIJKL
MNOPQRSTUV
WXYZ&abcdefg
hijklmnopqrstuv
wxyz123456789
0$.,"-:;!?

72 POINT HELVETICA ITALIC

RTWhaego

Characteristics

Thicks and Thins. Evenness in the strokes
Serifs. No serifs
Stress. No noticeable stress
Other Sans Serif Faces. Futura, News Gothic, Venus, Univers, Folio

FIVE FAMILIES OF TYPE

Mechanical-bound book.

FIGURE 10–16 Mechanical binding.

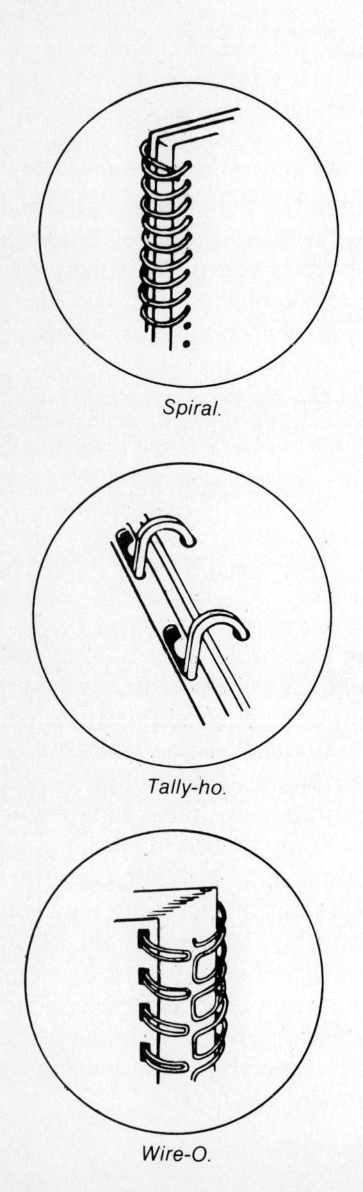
Spiral.

Tally-ho.

Wire-O.

FIGURE 10–17 A perfect-bound, or glued, manual backbone.

The letter that she will send to all of the companies is put together on her word processor so that she only changes the address for each printer:

Jill Davenport
(Company title)

Printing company address

SUBJECT: QUOTATION FOR PRINTING THE 15B CHAIN SAW MANUAL

Dear ________:

I would like to have the following bid by December 15 for the printing of the chain saw manual.

SPECIFICATIONS FOR THE JOB:

Size—8½″ × 11″ trimmed.

Number of pages—48 including covers and two blank pages before the inside back cover.

Type—Helvetica 10-point body; safety and important messages 10-point Helvetica bold; headings 12-point and 24-point. 10/12 line spacing. 1-point character spacing. Word gap 3-to-the-em. Indent safety and important information to 24 points from left margin. Line format is ragged right.

Manuscript—Supplied on floppy disc for type setting.

Ink—Black, per the sample included with this quote.

Layout—To be furnished with the manuscript.

Artwork—60 continuous tone, and 35 line drawings.

Paper stock—50# *white* per the sample included with this quote.

Binding—Saddle-wire stitch, 2 wires.

Proof—Furnish 1 proof 15 January.

Quantity—10,000 copies.

Due at the factory—February 1.

Manuscript transmitted—December 20.

Cost per each additional 1,000 copies—

____

Prepare bid with separate costs per each item of: 1. Cost of alteration charges; 2. Paper; 3. Shooting art and text; 4. Stripping art and text; 5. Printing; 6. Packaging; 7. Shipping.

Net 30 days billing on all work.

For every two (2) working days that this job is late, 10 percent of the total bill will be subtracted if the delay is caused by the printer.

Sincerely yours,
Jill Davenport

Enclosure: Ink on paper sample.

An analysis of this quote shows that Jill knows the pages that her printed instruction will have. She has used a manual similar to her quoted manual specifications to estimate the size. She has set her computer program for the size of type, line spacing, and line dimensions to match the final printed copy. By laying out her type and dimensioned artwork on a rough draft, she can estimate total page count.

Because she does not have an industry standard to guide her, she creates her own manual design. She decides that the cover will be the same stock of paper that she will use for the body of the instruction. This means that the front cover, inside front, inside back, and back cover can be counted as part of her 48 pages. Jill knows that she must keep her page counts in increments of 4, 8, or 16. She has added two blank pages at the end of the book before the back cover to make 48 pages overall. She will send a copy of her company's standard cover to the printer with the name of the machine when she submits the manuscript. The printer will use this standard design to typeset all cover material.

After a week passes, Jill calls the representatives of each printing company to make certain they have all of the information that they need in order to give her a fair bid. She makes sure, however, that she does not give one printing company an advantage by telling the representative something that she does not tell all of the other printing company reps.

Prepare the Manuscript

On December 15, Jill opens all of the bids that she has received for the chain saw printing job. Four of the printing companies have responded, and two have not. Jill selects the company that can deliver the printed book on the due date at the lowest price *with the best quality for the money.* She contacts those companies that sent a bid and tells them whether or not they have won the job. She is ready to complete all that remains to be done to the manuscript before sending it to the printing company.

To help her develop each page layout, she has created a master page dimension sheet. She can copy this layout and use it for every page of the manual. She has chosen a horizontal format *for this particular machine.* With this design, she can put artwork on the left or right of the text. Her layout dimensions are:

> *The top margin* is ½ inch or 36 points to the heading (A). The start of every major section has a 24-point heading, and every page in that section thereafter has a 12-point boldface "running heading."
>
> *The heading block* where the headings are located is ¾ inches or 54 points deep.
>
> *The gutter, alley, and outside margins* are ½ inch or 36 points wide.
>
> *The art and text columns* are 3-¼ inches or 234 points wide.
>
> *The bottom margin* is ¾ inch or 54 points deep. The page number is centered ⅛ inch or 9 points from the bottom of the art and text columns. Pages are numbered consecutively. Page one always starts on the right-hand side of the manual backbone. All even numbered pages are on the left side of the backbone when the manual is open (see Figure 10–18).

Jill can fit four pieces of art on the page if she limits the size to 2 inches or 144 points deep. She can, however, still fit larger overall views into the page without redesigning her basic format. She uses her art proportion scale to fit into position the artwork that she has taken and ordered from graphic art companies (Figure 10–19).

She creates some "white space" to give her instruction better readability and to leave her room for future revisions by placing every main heading at the top of a page. She knows that every manufactured machine must change in order to keep up with the competition in the marketplace. Her manual must keep up with these changes as rapidly as they are made. Jill could have chosen other layouts that can work as effectively as the horizontal layout. She has desk-checked this particular layout by allowing people to use the manuscript as a guide while they operate the machine. The layout is easy for them to read and use. Jill decides to stay with this arrangement.

In order to give those engaged in the printing process a blueprint of exactly what she wants, Jill prepares her manuscript by laying out the art and text on her page dimension sheets. This manuscript shows where the art and text must be placed, where to put the headings, and the page numbers. She must also indicate each typeface change with a mark and instructions. She uses a sideways caret to denote a typeface change, and she puts the type specification to the left of the line. This helps those who are either inputting her text or using the inputted text from her computer to double-check her commands. There is no need to mark type after the first line of type change. For example, she does not need to mark every line of 10-point regular type after she has signified the type change. She marks her type specs in *blue pencil. Red pencil* is reserved for corrections and additions (see Figures 10–20 and 10–21).

She also tells the printer how to handle her artwork. In Figures 10–20 and 10–21, which shows two selected pages from her manual, Jill has told the printer to "shoot" or reproduce her artwork 100 percent. It must not be blown up or reduced. She also has put the *specific art number on each window.* The numbers such as CS125 correspond to the "cut" numbers that she has issued for the artwork. With these numbers she not only can locate the artwork to the instruction window but she can also store and recall it whenever she needs to use it. Jill has cropped her artwork correctly, and she will send it to the printer with the manuscript.

Jill will see her manual in a "proof" before it is printed. During the proof of the section on chain saw operation, she notices a mistake in the first task on page 20. She closes the space between the words "completely" and "out." She must be able to mark correctly any revisions to the proof. After she sees the proof and returns it to the printer, the corrections are made and the manual is printed (see Figures 10–22).

Note: The sample chain saw manual is presented to show the process that can be used to create an active visual instruction. It is not intended to be used in the operation of a chain saw. For chain saw operation procedures,

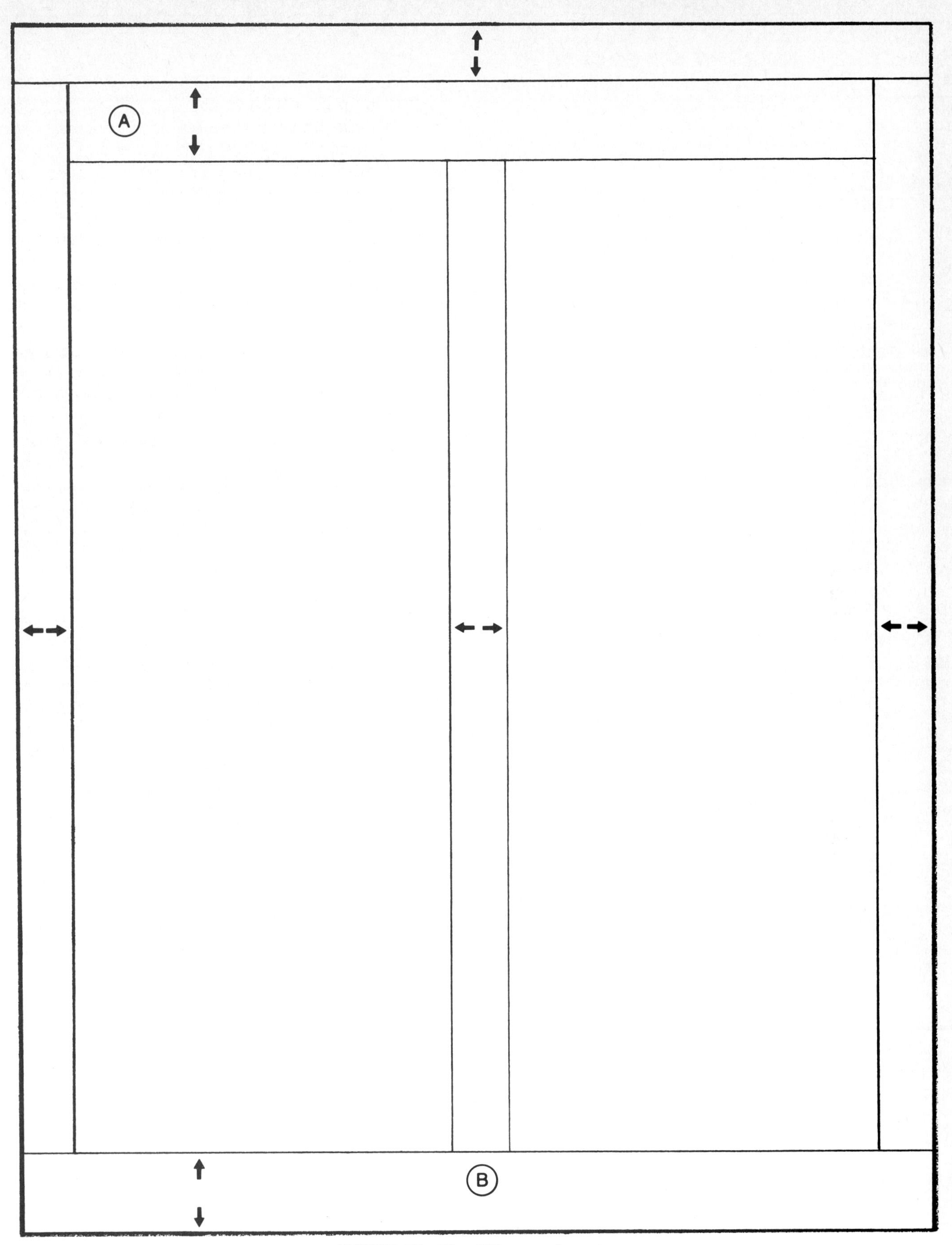

FIGURE 10–18 Jill's page layout.

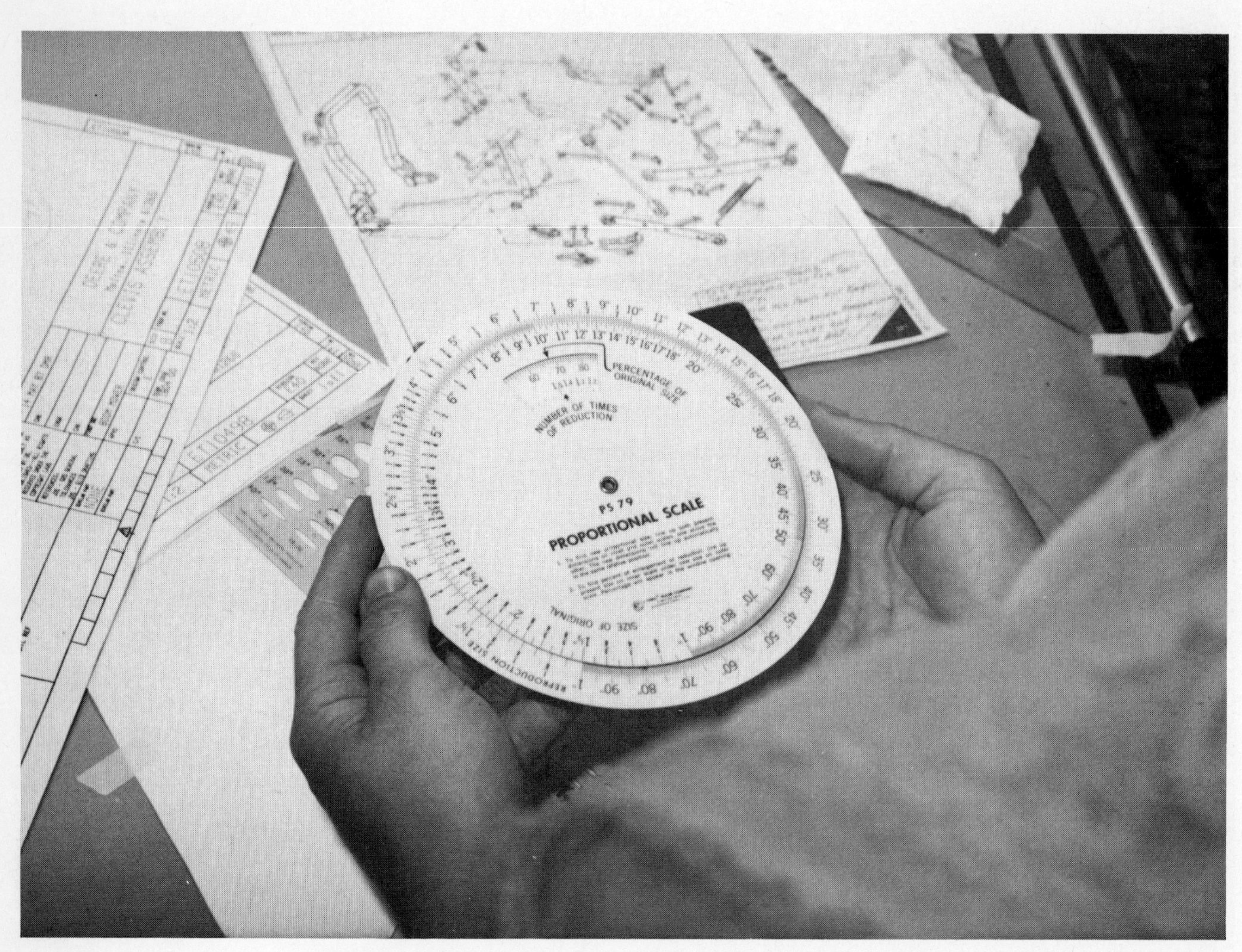

FIGURE 10–19 Jill proportions her art to fit the given space.

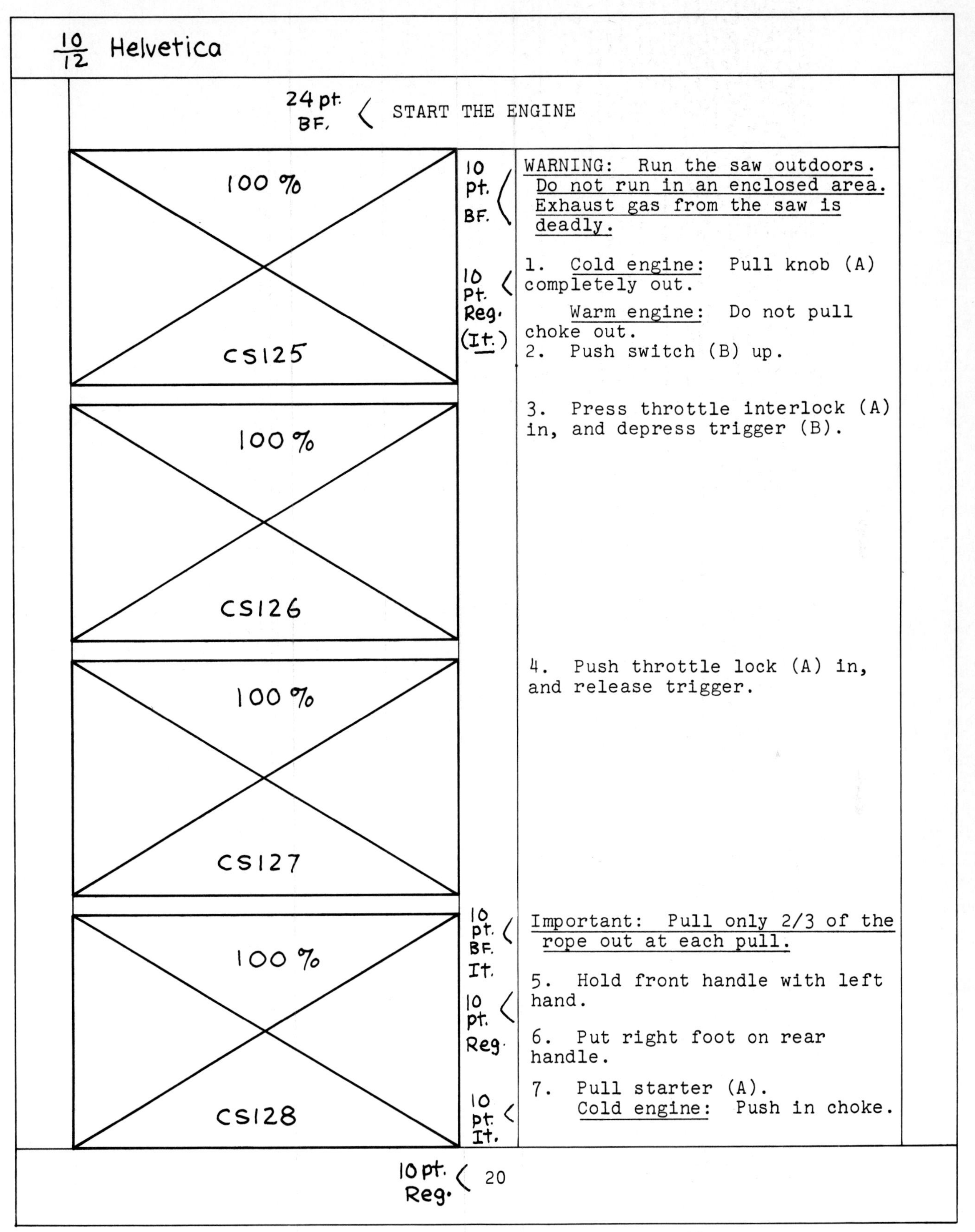

FIGURE 10–20 The final manuscript prepared by Jill.

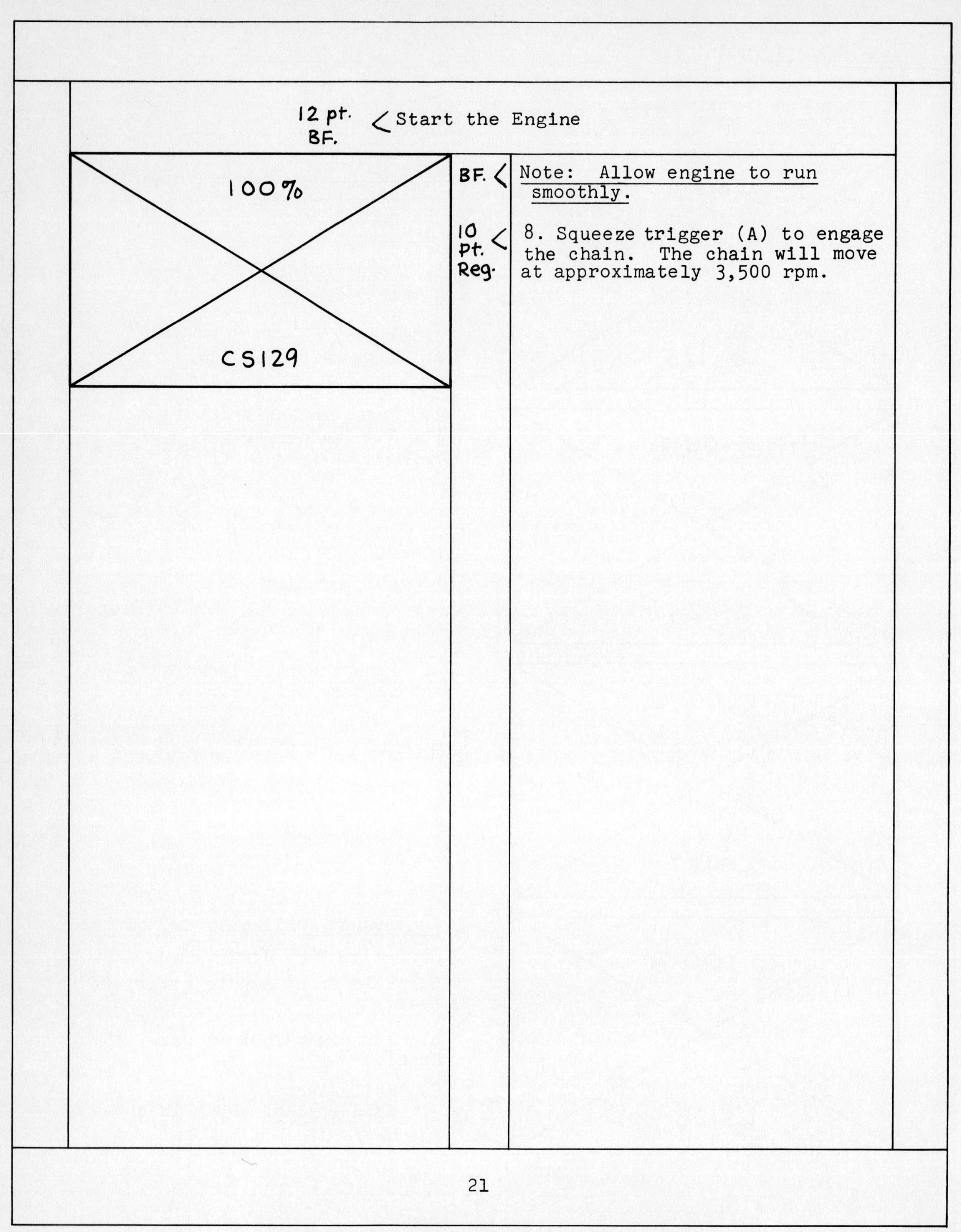

FIGURE 10–21 The final manuscript prepared by Jill.

START THE ENGINE

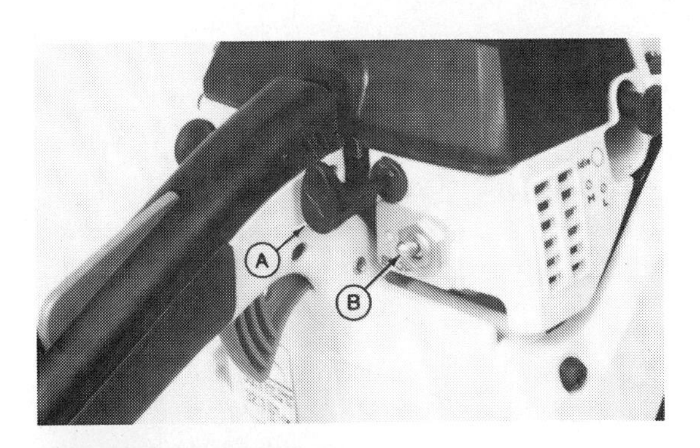

WARNING: Run the saw outdoors. Do not run in an enclosed area. Exhaust gas from the saw is deadly.

1. *Cold engine:* Pull knob (A) completely out.
   Warm engine: Do not pull choke out.
2. Push switch (B) up.

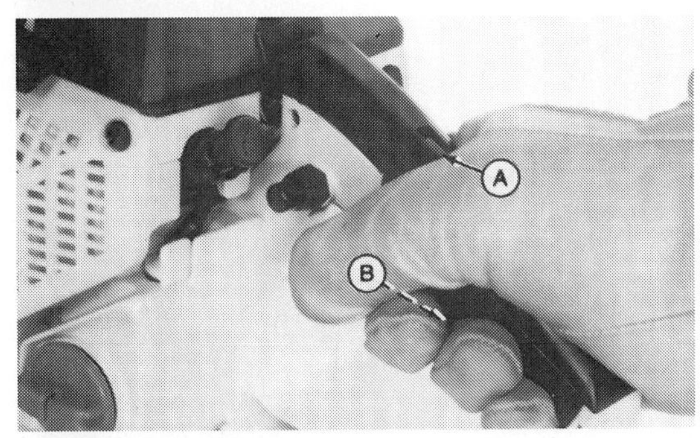

3. Press throttle interlock (A) in, and depress trigger (B).

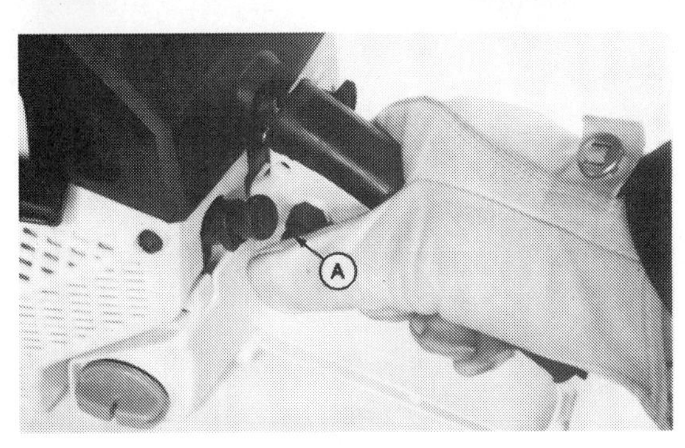

4. Push throttle lock (A) in, and release trigger.

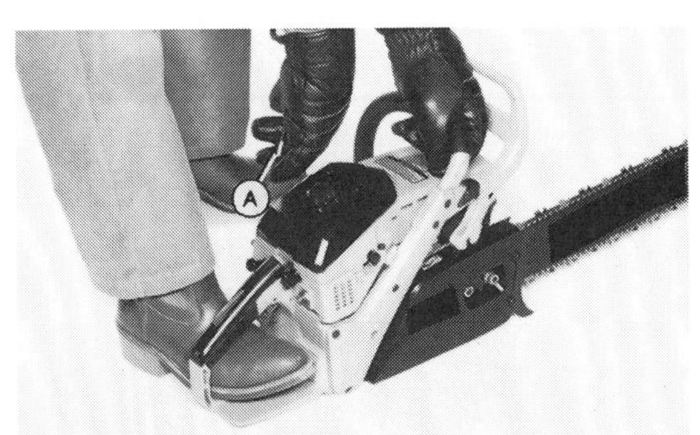

Important: Pull only 2/3 of the rope out at each pull.

5. Hold front handle with left hand.
6. Put right foot on rear handle.
7. Pull starter (A).
   Cold engine: Push in choke.

FIGURE 10–22 The final printed copy of Jill's manuscript.

(continued)

FIGURE 10-22 (cont.)

START THE ENGINE

Note: Allow engine to run smoothly.

8. Squeeze trigger (A) to engage the chain. The chain will move at approximately 3,500 rpm.

refer to the specific chain saw manufacturer technical operation instruction.

The artwork for these chain saw page examples is furnished by **Deere & Company,** Moline, Illinois. They are used with permission.

Transmit Files Properly

As a technical writer, you may find yourself in a situation where you have input your text on your computer program. This inputted information often can be fed into the electronic typesetting machine. This saves time and money that must be spent to pay someone to input your material into electronic codes. If you have input your information on a program, there are a few points that you must be aware of before you attempt to have your electronic program match the input hardware.

Be sure all corrections to the text of your instruction are made before you send it to the typesetter. Corrections to your material after it is "set" is a needless expense if you can make the changes while you are preparing your manuscript. Careful proofreading on the computer screen also saves valuable time at a point when time is an extremely valuable commodity.

There are two ways that your text can be sent to a typesetter. One way is the floppy disc. Most word-processing programs allow you to save your input in more than one program. Usually the files of your text must be "saved" in ASCII assembly language. Check with your typesetter before you set a transmission time. The electronic hardware that turns these programs into printed words must be instructed with a compatible language that it can understand and use.

Another way that can be used to transmit files to the typesetter is with a telephone "modem" hookup. By using telephone lines instead of the mail, you may save time in the transmission of your files. One day, or even a half day, can make the difference between getting the technical operation instruction to the assembly line on time and missing an important deadline.

Our modern electronic equipment is extremely fast and accurate. However, it is not very forgiving. One small mistake in a file can cause a transmission failure over a modem and a failure in transmitting floppy disc materials to the typesetter. Other important points that you must clarify with your typesetter in order to get a clean transmission include the following checklist:

How many "return" keys must you hit between each task? Usually there are two returns between them.

How many "space bars" should you use between sentences? Usually one space is required.

How must "indents" such as the bodies of warnings and other important information be set off from the normal text?

How must you align copy?

Must your word-processing program have a "word wrap" capability? Usually a sophisticated typesetter must have an automatic wrap ability that begins a line of type automatically after the preceding line.

How many characters must be in each file of your text? You must establish a minimum and a maximum.

Must you prepare copy in columns? This is called "galleys." Multicolumnar material will be set side by side during the composition process.

How do you close a file?

Do you need a marked-up hard copy with the disc? Some attributes are not transmitted by certain programs. Many special characters must be put into the program *after* it is transmitted.

For all transmissions, you must be consistent. Use the figure "1" when you write "one." Do not use "1." Do not use an "O" for a zero (0). Type the main heads, secondary heads, and subheads on your program the way you want them to appear in your final typeset material. Be sure to set a pattern for heads and other special formats and stay with that pattern. Be sure to check these items as well as the technical specifications of your modem and the typesetter's modem before you attempt to transmit.

Like Jill Davenport, you must furnish a "hard" or marked-up paper copy of the material that you electronically send to the typesetter. This marked-up copy has all of your type, page layout, and other specifications that are important for the people who will create the final page makeup.

Create a Control Sheet

If you are the publisher of a manual or direction, it is extremely important for you to know where the job is at all times. The best way to do this is to create a *control sheet* that allows you to check off each stage of the printing process. An example of a control sheet is:

CONTROL SHEET

FOR THE 15B CHAIN SAW MANUAL

DUE	DUE TO PRINTER
ROUGH DRAFT	XXXXXXXXXXXXXX
ARTWORK	XXXXXXXXXXXXXX
FIRST DRAFT	XXXXXXXXXXXXXX
REVIEW BY MACHINE EXPERTS	XXXXXXXXXXXXXX
QUOTES FOR PRINTING	
MANUSCRIPT	
PROOF	
FINAL PRINTING	
AT THE ASSEMBLY LINE	XXXXXXXXXXXXXX

Use Proofreader's Marks Correctly

When the printing company gives you a proof to review, you must mark all corrections in red. Designate which corrections are the printer's fault and which corrections have been made by you. A printing company has the right to charge you for any changes that you make to the proof. Abbreviation marks must be made in the left column as well as a corresponding mark at the point of correction (Figure 10–23).

Copyright the Instruction

Like any written document, the instruction can and should be copyrighted. This copyright protects your company from unauthorized use of your artwork and text. You cannot copyright facts or ideas; however, you can copyright your information to ensure that the hard work you put into it is not usurped by someone who will reprint and sell your document for a profit.

You must use Form TX for a nondramatic literary work. A nonrefundable filing fee of $10.00 must accompany this application as well as two complete copies of the best edition. This will ensure that your instruction will be copyrighted for 75 years from the publication date. Few machines will last as long as the time period of a copyright. It is your responsibility to use a copyright notice in your publication to inform others that your company's work is protected under the copyright law. For a "copyright information kit" write to:

The Copyright Office
Library of Congress
Washington, D.C. 20559

You will receive Form TX, a booklet entitled *Copyright Basics,* and a short summation of the 1976 copyright law. Although the Copyright Office receives more than 500,000 applications annually, the copyright process is usually a fast one. If problems arise, the staff will help you clear up your difficulty.

Unpublished instructions that are written for a limited audience or that are never going to be published can also be copyrighted.

POSTSCRIPT

Technical writing is one of the most interesting and rewarding jobs in our modern manufacturing plants. As a technical writer, you can move into many related areas because of the fact that the technical writer must develop an extensive amount of product knowledge. You can also find a rewarding career as a writer or publisher as you write and develop your company's technical operation instructions. You may experience a feeling of accomplishment by making the difficult simple and by explaining to others how to operate your company's technology as efficiently and safely as possible.

In your development as a technical writer in the classroom or on the job, you may be exposed to various types of "technical" documents that are issued by manufacturing companies. This book concentrates on the writing of technical operation instructions because these are usually the first documents that the technical writer creates in an industrial setting. However, you also may be asked to write and produce the following:

Assembly manuals. The assembly manual explains how to put a machine together. Often this information is included in the operation instructions that accompany the machine. An example of this is the sophisticated toy that you may be required to put together before operating. Manuals that deal completely with setup are often used on machines that are more complex. Often larger machines are shipped in pieces to reduce shipping costs and to ensure the machine arrives as damage-free as possible at its point of destination.

Technician's manuals. Manuals that explain

EXPLANATION	MARGINAL MARK	ERRORS MARKED
Take out letter, letters, or words indicated.	⸿	He opened the windoow.
Insert space.	#	He opened thewindow.
Turn inverted line.	@	He opened the window.
Insert letter.	e	He opned the window.
Set in lowercase.	lc	He Opened the window.
Wrong font.	wf	He opened the window.
Broken letter. Must replace.	x	He opened the window.
Reset in italic.	ital	He opened the window.
Reset in roman.	rom	He opened *the* window.
Reset in bold face.	bf	He opened the window.
Insert period.	⊙	He opened the window
Transpose letters or words as indicated.	tr	He the window opened.
Let it stand as is. Disregard all marks	stet	He ~~opened~~ the window.
Insert hyphen.	=/	He made the proofmark.
Equalize spacing.	eq #	He opened the window.
Move over to point indicated. [if to the left; if to the right]		[He opened the window.
Lower to point indicated.	⊔	He opened the window.
Raise to point indicated.	⊓	He opened the window.
Insert comma.	^,^	Yes he opened the window.
Insert apostrophe.	v'v	He opened the boys window.
Enclose in quotation marks.	" "	He opened the window.
Enclose in parenthesis.	( )	He John opened it.
Enclose in brackets.	[ ]	He John opened it.
Replace with capital letter.	cap	he opened the window.
Use small capitals instead of type now used	sc	He opened the window.
Push down space.	⊥	He opened the window.
Draw the word together.	⁐	He op ened the window.
Insert inferior figure.	^2^	Sulphuric Acid is HSO$_4$.
Insert superior figure.	v2v	2a + b^2 = c
Used when words left out are to be set	out, see copy	He window.
The diphthong is to be used.	æ	Caesar opened the window.
The ligature of these two letters is to be used.	fi	He filed the proof.
Spell out words marked with circle.	spell out	He opened the 2d window.
Start a new paragraph.	¶	door. He opened the
Should not be a paragraph. Run in.	no ¶	door. He opened the window.
Query to author. Encircled.	(was?)	The proof read by.
This is the symbol used when a question mark is to be set. NOTE: *A query is encircled.*	?	Who opened the window
Out of alignment. Straighten.	=	He opened the window.
1-em dash.	1/M	He opened the window
2-em dash.	2/M	He opened the window
En dash.	1/N	He opened the window
Indent 1 em.	□	] He opened the window.
Indent 2 ems.	□□	] He opened the window.
Indent 3 ems.	□□□	] He opened the window.

Proofreader's marks.

FIGURE 10–23 Most common proofreader marks.

how to diagnose, tear down, and fix machines are extremely important. Usually they are written to the level of the trained service technician. An example of this is the automobile technician's manual. Even though these manuals may be available to the general public, they explain many procedures that require the training and equipment of a trained service professional. These service manuals are also called "shop manuals."

Shop-time manuals. Shop-time manuals are written to give the service technician an estimated time that a job will take. This is important for billing purposes and for estimating the money that it will cost to do a particular job. Without these manuals there would not be a consistent charge made to each customer for the same work. The consumer can receive an estimate in writing for a particular type of service.

Passports. These manuals are similar to an individual passport that contains a physical description of the traveler. Some importing countries require passports that not only include operation instructions but also certificates of safety checks and work. Examples of certification are pressure-vessel certificates, which are issued by the pressure-vessel manufacturer and which explain how the part was made and tested. Passports are not a common publication; however, they may increase in importance as the worldwide market continues to develop.

Parts catalogues. Parts catalogues are manuals or direction-size lists and pictures of parts that are used in the machine. The consumer and service technician can use parts catalogues to order replacement parts for the machine.

The goal of this text is to give you, the student of technical writing, the tools that you can use to write the technical operation instruction. By using this skill and know-how, you may be able to open doors of opportunity in the fast-paced and challenging area of industrial technical writing and communication.

CHAPTER SUMMARY

- The technical writer in some smaller or mid-sized companies may also be the technical editor and publisher.
- The publisher of technical instructions must be familiar with the printing process in order to get the best printing job for the money.
- The printing process is complex and labor-intensive. It includes the combined efforts of the front office, makeready, and pressroom experts.
- The first step in the printing process is the inputting of text in order to interface the words with the typesetting machines.
- Continuous-tone photographs must be shot with screens to help in the transfer of ink to paper.
- Flats that are used to make printing plates are stripped up in increments of 4, 8, or 16 sheets. Plate type may vary as to the printing method used.
- Offset lithography is an important method of printing that is often used to produce copies of manuals and directions.
- Print quality and cost vary from printer to printer, and often from print job to print job.
- To cut down on revision costs, develop a consistent page format that allows changes to be made without affecting every proceeding page in the manual.
- The three main styles of layout for page formats are the *vertical, horizontal,* and *cartoon* design.
- Ragged-right lines may be the easiest lines to read under most circumstances.
- Type on a manuscript must be marked as to style or font, and size or points.
- By varying type size and thickness, you can make information stand out of the body of the instruction.
- Paper and ink quality must be chosen to balance the printing budget limitations and also the application of the manual.
- The most popular and inexpensive binding is usually the wire-saddle stitch. Computer user's manuals are often bound with mechanical wire spirals that allow the manuals to lie flat when opened.
- A quote is a means by which the publisher of an instruction can allow many printers to bid fairly and accurately on a print job.
- A manuscript includes all important type specifications, word spacing, and arrangement of information not covered in the quote.
- Before you input your text into a computer

program, be sure the program and your input methods interface with the typesetting equipment.

- Use a control sheet to keep track of every step of the writing and production of the instruction. Make sure that each step is on time. One late due date can affect the final print date.
- Copyright both published and unpublished instructions to protect from plagiarism and misuse of words and pictures.

EXERCISES FOR CHAPTER 10

Prepare to Print

1. Write and layout an instruction for a common household appliance. Create a storyboard, a rough draft that your instruction checks, and a final manuscript. Mark all specifications on the manuscript that a printer must have in order to print your instruction. Take halftone pictures and/or create sketches of line drawings and crop and size them so that they work with your text.
2. Write and prepare for publishing a manual on a make-believe machine. The manual should be one that you have designed from the basic power systems used to design modern technology. Be creative: Remember, technical writing is a creative problem-solving process that often demands resourcefulness and imagination in order to create a technical operation instruction.
3. Contact a local computer and software dealer. Find out what new advancements are available in:
   a. Desk-top layout of instructions.
   b. Artwork scanning machines.
   c. Laser printer output.

Glossary

Artificial language: Languages that have been created for a particular application such as jargon and legalese. These languages are understandable by a limited audience. Words are "coined" to fit a situation.

Artwork: This term refers to the many types of illustrations used in instructions. Artwork includes photographs, line drawings, and charts that visually support a main idea or task in an instruction.

Audience: The audience includes everyone who *potentially* may use the instruction to operate the machine.

Bid: A bid is a listing of all costs that a printing company will charge for a print job.

Binding: Binding is the method used to hold the sheets of an operation manual together. Binding can be done with glue (perfect bound) or wire (side- or saddle-wire stitched), or be bound mechanically, such as with spiral wire.

Boldface: Type that is thicker and darker than the regular type of an instruction.

Call out: A call out points to and identifies a particular item on a piece of artwork. Call outs can be letters or numbers. Lines or arrows connect these call outs to the item.

Caveat emptor: "Let the buyer beware." This statement sums up a marketplace that does not put the responsibility of machine efficiency on the manufacturer. Once the machine is purchased, the buyer must assume complete responsibility for safety and operation.

Caveat venditor: "Let the seller beware." This statement sums up a marketplace where the manufacturer is totally responsible for certain machine efficiency and safety points. *Caveat venditor* is the opposite of *caveat emptor.*

Chart: A chart is a grouping of information that is visually and easily used.

Communication model: A communication model explains how communication takes place. The main parts of the model include the sender, the medium, and the receiver.

Concrete noun: A concrete noun creates an exact visual image in the mind of the receiver. Concrete nouns symbolize exact things.

Consumer: The consumer buys the machine in the marketplace. The consumer in this book also is synonymous with the operator of the machine.

Continuous tones: A photograph that is composed of chemically created shades of black-and-white tones.

Control point: The place in a system that the operator can use to adjust the output or amount and type of work done by a system. Each system must have at least one control point.

Control sheet: A control sheet lists all of the stages of instruction production and publishing. Control sheets tell the technical writer where the instruction is at all times. It will show if the instruction is early, late, or on time for a particular stage of the process.

Corpus callosum: That part of the brain that connects the left and right hemispheres.

Departments: Every manufacturing company is divided into specialized sections, or departments, of the manufacturing process to help create efficiency.

Desk-check: The check of an instruction before it is put into a manuscript form or before it is sent to a printing company to be reproduced.

Diagram: A diagram is a visual connection of all parts of a system to show how they go together. Diagrams can be abstract to show only the idea of the part, or they can be pictorial and illustrate the parts of the system in detail.

Direction: A short, unbound instruction that tells how to operate the machine safely and efficiently.

Edge up: The photograph that lacks contrast can be edged up to bring out details.

Electrical system: An unbroken connection of similarly rated parts that carry electrical power from the power source or prime mover to the point of work. The electrical system is the most popular system that can be found in a machine.

Engineering diagrams: Drawings made by the engineering department to show how a system is connected.

Engineering drawings: Commonly misnamed blueprints. Drawings are illustrations of machines or machine parts that are "dimensioned" to show exactly how they should be made or assembled.

Fabrication: The machining of machine parts and the making of machine components before the assembly of the machine begins. Planers, drills, grinders, presses, and lathes are a few of the machines used in the fabrication stage of manufacturing.

Feedback: Feedback is a measurable result of a system output. Feedback allows the system operator to evaluate how the system should be adjusted to attain maximum results. Direct feedback in a technical communication model results from a role-playing session or a review by the machine experts. Indirect feedback can be obtained by reviewing warranty and liability claims and suits against a particular machine.

Flats: Before a plate can be created for the printing press to print an instruction, a flat must be made of the negatives of the pages. Flats are created in sequences of 4, 8, 12, 16, and 32 pages.

Fog index: An artificial method used to measure the readability of the instruction. Fog indexes usually measure the number of syllables in the words of an instruction and compare them in a ratio to an established reading level.

Format: The dimensions of a page. Format is a consistent pattern that allows for the layout and arrangement of information.

Front office: The division of a manufacturing company that creates the concept of the machine, the directions on how it must be made, and how it must be put together.

General assembly: An overall view of the machine that shows the arrangement of machine parts such as control points and other major points that are explained in the instruction.

Halftone: During the process of making negatives for the printing process, a screen is placed over a continuous-tone photograph. The screen, or halftone, is made up of tiny dots or squares that break up the "tone" of the photograph so that it can be printed with greater contrast.

Heterogeneous audience: An audience that cannot be completely analyzed as to each member's training, technical knowledge, familiarity with the machine, or reading and understanding ability. A heterogeneous audience is usually large and undefined.

Homogeneous audience: A stratified group of people who share the same education, training, and background in a certain area. This group of people can be analyzed for their machine knowledge.

Hydraulic cylinder: This is a common point of work for a liquid-under-pressure system. Power is converted to back-and-forth movement.

Hydraulic motor: The motor is a common point of work in a liquid-under-pressure system. Motors convert power into torque, or circular motion.

Hydraulic pump: A pump creates hydraulic oil flow.

Hydraulic system: A closed connection of similarly rated parts that transfers power into fluid flow.

Imperative sentence: A command sentence that is used to give tasks to the reader of an instruction. The subject (you) is understood in this type of sentence, thereby allowing the verb to be the first symbol that is interpreted by the consumer.

Input: Text or words of an instruction must be keyboarded into the printing system before the words can be "typeset" into letters that will make up the finished instruction.

Instruction: An instruction contains everything that the consumer must know in order to operate a machine safely and efficiently. Instructions tell "how to operate, control and maintain" a machine on daily basis.

Interchangeability: The basis for the mass-production manufacturing process is the ability to make machine parts in close tolerance so that they can be interchanged on any similar machine. This concept has widened markets and allowed machines to be produced at lower cost so that they are affordable by more people.

Liability: The responsibility of a machine manufacturer to ensure that the safest machine possible is manufactured and sold.

Limited vocabulary: A conscious effort to limit the number of words used in an instruction to only the most basic descriptive terms possible. Limited vocabularies are also created by using the same terminology throughout an instruction.

Line art: A silhouette of a machine or part drawn to show only the outline of the subject.

Lowest common denominator: The writer's mental perception of the person who is least familiar and least able to operate a machine. This person represents the lowest level of the audience as far as machine ability and understanding of the technology embodied in the machine. If an instruction is written to this person, then the instruction will be usable by the highest percentage of the audience as possible.

Machine: A machine is designed and manufactured with at least one powered system. Machines convert power into useful work.

Machine expert: A person who is highly skilled in a certain area of the manufacturing process. Responsible for a particular stage of machine design or production.

Machine system: A system is composed of connected parts that carry and transform power into work. A useful system will have at least one control point, internal protection, and a point of work.

Manual: A manual is a bound instruction divided into parts so that the reader can reference specific topics easily. A manual is divided into the front, inside front, inside back, and back cover, and the body that contains the tasks and information.

Manufacturing company: The place where raw materials or parts of machines are worked and assembled into machines. Each step of the manufacturing process increases the value of the raw materials that are used to make the machine.

Marketplace: The place where the machine is sold; the location where the product is traded for something of equal value.

Mass consumer: The phrase refers to the potential buyers of a product or machine that is made in the mass production process. Mass consumer markets are usually synonymous with the heterogenous audience.

Mass production: A technique that uses automation, material handling, and assembly methods to produce machines in stages. Many people are involved with the fabrication and assembly of one machine. This is opposed to the shop, where one person makes and assembles the machine. Mass production relies on interchangeability of the machine parts.

Mechanical system: A connection of parts that mechanically transmit power to a mechanical point of work.

Metric sytsem: A system of measurement based on units of ten. It is an alternative to the standard system in the United States, which is based on units of 12. The meter, gram, and liter are the basic measurements of the metric system.

Mirror-image writing: A tendency by the technical writer to assume that the audience knows as much about the machine as does the writer. Mirror-image writing often results in essential information being deleted from the instruction.

Opaquing: The process of eliminating unwanted images on a photograph or photographic negative by adding film or other blocking material.

Open vocabulary: A vocabulary that has no conscious or unconscious limitations as to how words are chosen or what words are used.

Operator: The person who controls the machine. The operator must collect and interpret the results of work done by the machine and make adjustments to the machine systems accordingly. The operator is used interchangeably with the term "consumer" in this text.

Outline: A basic sketch of ideas that gives preliminary structure to an instruction.

Pneumatic system: Air flow that is converted to work by similar parts in an unbroken system.

Point (measurement): A standard measurement of type and format dimensions of an instruction. There are 72 points in an inch.

Point of use: The place where information is necessary in order to become part of the task. A point-of-use safety message is located at the place in the instruction where it will be most useful.

Point of work: The place where work is done. A mechanical system may have a point of work at the footprint or tread of a tire.

Postsale document: The technical instruction is intended to be used after the sale of the machine is made and the operator uses it.

Power source: A system that converts fuel into power.

Power system: A system that turns power into flow, controls the flow through resistance, and transforms it into work.

Prepress makeready: A printing company must prepare the instruction for the printing process. The text must be typeset, the negatives created, and the printing plates made. These are all steps in the makeready process.

Presale document: An advertisement or fact sheet that is used to help persuade the customer to purchase a machine.

Pressdowns: These come in many shapes and forms and can be used to point out or add to the information on artwork. Pressdown arrows, numbers, and letters are the most common types.

Prime mover: A prime mover is the place where fuel is converted into power. Animals, the sun, humans, and machines can be prime movers.

Proof: A stage of the printing process where the printing company shows the technical publisher exactly what is going to be printed. The proof is the last chance for the publisher to make changes before the instruction is printed.

Prototype machine: A machine fabricated and assembled to test the manufacturing process and the machine before mass production takes place.

Quote: A set of print specifications for an instruction

that is sent to the prospective printing companies for a print job. Quotes must allow the companies to make a fair bid for a job. Quotes are used to find the most efficient and less costly printing company for a particular job.

Reciprocating engine: Fuel is transformed into heat, which creates an up-and-down movement of pistons. Piston rods connected to a driveshaft convert this up-and-down movement into torque, or circular motion.

Restricted vocabulary: An artificially limited vocabulary with a set amount of basic words that must be used in all communication.

Retouch techniques: Photographs that are not shot properly can be edged up, opaqued, or air-brushed to eliminate the image problems.

Role-play: A form of desk-checking where someone who is not familiar with the machine or who represents the lowest common denominator operates the machine for the first time by using the technical operation instruction.

Rotary engine: Fuel is converted into heat, which drives a cylinder mounted directly to the driveshaft. Both cylinder and driveshaft are in circular or rotary motion.

Rough outline: A rough outline shows both words and artwork together. It is used as a review before the final manuscript is created.

Shop floor: The area in a manufacturing company where the machine is fabricated and assembled.

Spatial logic: The correct sequence of movement by the operator on and around a machine. Spatial logic creates the correct position of the operator so that he or she can safely and efficiently perform the task at hand.

Split-brain concept: The brain is divided into two separate hemispheres that tend to dominate certain types of thought. The right brain, which controls the left side of the body, is usually spatially oriented. The left side, which controls the right side of the body, is usually verbally oriented. The important point is that both pictures and words are necessary to reach the total thought processes of consumers.

Storyboard: The first attempt to bring together the words with the pictures that have been created for a task. A storyboard eliminates spontaneous creation of the instruction by giving the technical writer a visual list of artwork required to correspond to the words.

Style sheet: A set of grammar rules and conventions that will be followed during every writing assignment.

Survival language: A basic shared language that is developed by children in order to obtain their immediate needs. It is an unlayered language that has direct relationships with the material world.

Technical writer: The professional writer who creates technical operation instructions for a manufacturing company.

Technology: Machines used to make other machines and the manufacturing techniques employed are all part of technology. The remaining part is the machine or product that is made. Technology builds upon past technology.

Triune-brain concept: The concept that the human brain developed in three layers. Human beings use these three abilities to interpret and react to outside stimulus.

Typeface: The physical appearance of a letter of type.

Verbal logic: The listing of a set of verbal or written tasks in the correct sequence so that the job can be done safely and efficiently.

Vignette: A visual block of information where words and a picture mutually support each other. Vignettes are as old as written communication.

Vocabulary: The list of words that an individual develops in layers from the exposure to training, education, and social stimulus.

Warranty: A written agreement that the seller and consumer enter into when a machine is sold. The seller generally sets the time limit and parameters that will be followed in regards to machine dependability. The seller tells the consumer exactly what will be fixed if the part breaks, and how long the warranty contract will last.

Writing style: The individual way that the writer selects and assembles information. Every technical writer may have a slightly different style and yet express the same ideas and tasks.

ACKNOWLEDGMENTS FOR THE UNCREDITED ART IN THE TEXT

Art that appears without a credit in the text is supplied by the following sources and graphic art companies.

Courtesy Deere & Company Historical Files:
Figures: 4–1, 4–6, 4–12, 4–13, 4–14, 4–27, 4–28, 4–31, 4–32, 4–33, 4–37, 4–38, 4–39, 4–40, 4–44, 4–53, 5–21, 5–22, 6–5, 6–6, 6–7, 6–16, 6–25, 6–28, 6–31, 7–18, 9–1.

Courtesy Vicomm:
Figures: 4–15, 4–16, 4–17, 4–18, 4–19, 4–20, 4–21, 4–22, 4–23, 4–24, 4–25, 4–26, 4–29, 4–30, 4–34, 4–35, 4–45, 4–46, 4–47, 4–48, 4–49, 4–50, 4–51, 4–52, 4–54, 4–55, 5–1, 5–15, 5–26, 6–35, 10–12, 10–13, 10–19, 10–22, 10–23.

Courtesy Davis Technical Art Services, Inc.:
Figures: 3–3, 3–4, 3–5, 4–36, 5–16, 5–23, 5–24, 5–25, 6–9, 6–18, 6–22, 6–26, 6–29, 6–30, 6–32, 6–33, 6–34, 7–2, 7–3, 7–5, 7–6, 7–11, 7–14, 7–23, 7–24, 7–28.

Courtesy CDS/Wordworks:
Figures: 2–13, 2–14, 2–15, 3–1, 5–10, 5–11, 5–12, 5–13, 5–14, 5–17, 5–18, 5–19, 5–20, 8–7.

Bibliography

BUSCH, TED. *Fundamentals of Dimensional Metrology.* Albany, N.Y.: Delmar Publishers, 1966.

BUZAN, TONY. *Use Both Sides of Your Brain.* New York: E.P. Dutton, 1976.

COMMERCE CLEARING HOUSE. *Products Liability Reports,* No. 546. Chicago: June 8, 1984.

The Consumer Product Safety Act: Text, Analysis, Legislative History. Washington, D.C.: The Bureau of National Affairs, 1973.

CRAIG, JAMES. *Production for the Graphics Designer.* New York: Watson-Guptill Publications, 1974.

EDWARDS, BETTY. *Drawing on the Right Side of the Brain: A Course in Enhancing Creativity and Artistic Confidence.* Los Angeles: J.P. Tarcher, Inc., 1979.

——. "Fundamentals of Service" Series. Moline, Ill.: Deere & Company: *Engines,* 1986; *Hydraulics,* 1987; *Electrical Systems,* 1984; *Power Trains,* 1984.

——. "Fundamentals of Service Compact Equipment." Moline, Ill.: Deere & Company: *Engines,* 1983; *Electrical Systems,* 1982; *Hydraulics,* 1983; *Power Trains,* 1983.

——. *Machinery Maintenance.* Moline, Ill.: Deere & Company, 1987.

GRAVES, ROBERT, and ALAN HODGE. *The Reader Over Your Shoulder: A Handbook for Writers of English Prose.* New York: Random House, 1979.

HOUNSHELL, DAVID. *From the American System to Mass Production 1800–1932: The Development of Manufacturing Technology in the United States.* Baltimore: The Johns Hopkins University Press, 1984.

Information Please Almanac. Boston: Houghton Mifflin, 1986.

JOHNSTON, DONALD F. *Copyright Handbook.* New York: R.R. Bowker Company, 1982.

KLEMM, FREDRICH. *A History of Western Technology.* Cambridge, Mass.: The MIT Press, 1964.

LACEY, ROBERT. *Ford: The Men and the Machine.* Boston: Little, Brown, 1986.

MACLEAN, PAUL D. *The Triune Concept of the Brain and Behaviour.* Toronto: University of Toronto Press, 1973.

MCCRUM, ROBERT, WILLIAM CRAN, and ROBERT MACNEIL. *The Story of English.* New York: Viking, 1986.

NOEL, DIX W., and JERRY J. PHILLIPS. *Products Liability Cases and Materials.* St. Paul, Minn.: West Publishing Company, 1982.

NORMAN, COLIN. *The God That Limps: Science and Technology in the Eighties.* New York: W.W. Norton & Company, 1981.

O'BRIEN, MARK. *High-Tech Jobs for Non-tech Grads.* Englewood Cliffs, N.J.: Prentice Hall, 1986.

Pocket Pal: A Graphic Arts Production Handbook. New York: International Paper Company, 1981.

SCHWARTZ, VICTOR E. *Comparative Negligence.* New York: The Allen Smith Company, 1986.

SEBRANEK, PATRICK, and VERNE MEYER. *Basic English Revisited: A Student Guide.* Burlington, Wisc.: Basic English Revisited, 1985.

STEVENSON, GEORGE A. *Graphics Art Encyclopedia.* New York: McGraw-Hill, 1979.

STRUNK, WILLIAM JR., and E. B. WHITE. *The Elements of Style.* New York: Macmillan Publishing Company, 1979.

TURNER, ROLAND, and STEVEN GOULDEN (Eds.). *Great Engineers and Pioneers in Technology.* New York: St. Martin's Press, 1981.

VELEY, VICTOR F., and JOHN J. DULIN. *Modern Electronics: A First Course.* Englewood Cliffs, N.J.: Prentice Hall, 1983.

WILCOCK, KEITH D. *The Corporate Tribe.* New York: Warner Books (paperback), 1984.

WITHERELL, CHARLES E. *How to Avoid Products Liability Lawsuits and Damages: Practical Guidelines for Engineers and Manufacturers.* Park Ridge, N.J.: Noyes Publications, 1985.

ZDENEK, MARILEE. *The Right-Brain Experience.* New York: McGraw-Hill, 1983.

FILMS AVAILABLE FOR RENTAL

"The Triune Brain." Indiana University Audio-Visual Center, Bloomington, Ind.

"Left Brain, Right Brain, Part 1: Left Brain." "Part 2: Right Brain." Indiana University Audio-Visual Center, Bloomington, Ind.

ADDRESS OF THE SOCIETY FOR TECHNICAL COMMUNICATION

Society for Technical Communication, Inc.
815 15th Street, N.W., Suite 516
Washington, D.C. 20005

Index